2

Risk assessments of *Salmonella* in eggs and broiler chickens

WORLD HEALTH ORGANIZATION

FOOD AND AGRICULTURE ORGANIZATION OF THE UNITED NATIONS

2002

WHO Library Cataloguing-in-Publication Data

Risk assessments of Salmonella in eggs and broiler chickens.

(Microbiological risk assessment series No. 2)
1. Salmonella – pathogenicity 2.Salmonella enteritidis – pathogenicity
3. Eggs – microbiology 4. Chickens – microbiology 5. Risk assessment – methods 6. Risk management – methods 7.Guidelines I. World Health Organization
II. Food and Agriculture Organization of the United Nations III. Series

ISBN 92 9 156229 3 (WHO)   (LC/NLM classification: QW 138.5.S2)
ISBN 92 5 104872 X (FAO)
ISSN 1726-5274

CONTENTS

Acknowledgements

The Food and Agriculture Organization of the United Nations and the World Health Organization would like to express their appreciation to all those who contributed to the preparation of this document through the provision of their time and expertise, data and other relevant information and by reviewing the document and providing comments. Special appreciation is extended to the risk assessment drafting group for the time and effort that they freely dedicated to the elaboration of this risk assessment.

The designations employed and the presentation of the material in this publication do not imply the expression of any opinion whatsoever on the part of the World health Organization nor of the Food and Agriculture Organization of the United Nations concerning the legal status of any country, territory, city or area or of its authorities, or concerning the delimitation of its frontiers or boundaries.

The designations "developed" and "developing" economies are intended for statistical convenience and do not necessarily express a judgement about the stage reached by a particular country, country territory or area in the development process.

The views expressed herein are those of the authors and do not necessarily represent those of the World Health Organization nor of the Food and Agriculture Organization of the United Nations nor of their affiliated organization(s).

The World Health Organization and the Food and Agriculture Organization of the United Nations do not warrant that the information contained in this publication is complete and correct and shall not be liable for any damages incurred as a result of its use.

Responsible technical units:

Food Safety Department
World Health Organization

Food Quality and Standards Service
Food and Nutrition Division
Food and Agriculture Organization of the United Nations

RISK ASSESSMENT DRAFTING GROUP

(in alphabetical order)

Wayne ANDERSON
Food Safety Authority of Ireland,
Ireland

Eric EBEL
Office of Public Health and Science, Food Safety and Inspection Service,
United States Department of Agriculture, United States of America

Aamir M. FAZIL
Health Protection Branch, Health Canada,
Canada.

Fumiko KASUGA
National Institute of Infectious Diseases, Department of Biomedical Food Research,
Japan.

Louise KELLY
Department of Risk Research, Veterinary Laboratories Agency,
United Kingdom

Anna LAMMERDING
Microbial Food Safety, Risk Assessment, Health Protection Branch, Health Canada,
Canada

Roberta MORALES
Health and Human Resource Economics, Centre for Economic Research,
Research Triangle Institute, United States of America

Wayne SCHLOSSER
Office of Public Health and Science, Food Safety and Inspection Service,
United States Department of Agriculture, United States of America

Emma SNARY
Department of Risk Research, Veterinary Laboratories Agency,
United Kingdom

Andrea VICARI
FAHRM, College of Veterinary Medicine, North Carolina State University,
United States of America

Shigeki YAMAMOTO
National Institute of Infectious Diseases, Department of Biomedical Food Research,
Japan.

REVIEWERS

The risk assessment was reviewed on several occasions both during and after its elaboration, including through expert consultations and selected peer reviews, and by members of the public in response to a call for public comments.

PARTICIPANTS IN EXPERT CONSULTATIONS

Amma Anandavally, Export Quality Control Laboratory, Cochin, India.

Robert Buchanan, Center for Food Safety and Applied Nutrition, United States Food and Drug Administration, United States of America.

Olivier Cerf, Ecole Nationale Vétérinaire d'Alfort (ENVA), France.

Jean-Yves D'Aoust, Health Protection Branch, Health Canada, Canada.

Paw Dalgaard, Department of Seafood Research, Danish Institute for Fisheries Research, Ministry of Food Agriculture and Fisheries, Denmark.

Michael Doyle, Center for Food Safety, University of Georgia, United States of America.

Emilio Esteban, Food Safety Initiative Activity, Centres for Disease Control and Prevention, United States of America.

Lone Gram, Department of Seafood Research, Danish Institute for Fisheries Research, Ministry of Food Agriculture and Fisheries, Denmark.

Steve Hathaway, MAF Regulatory Authority (Meat and Seafood), New Zealand.

Matthias Hartung, National Reference Laboratory on the Epidemiology of Zoonoses, Joint FAO/WHO Collaborating Centre, Federal Institute for Health Protection of Consumers and Veterinary Medicine, Germany.

Inocencio Higuera Ciapara, Research Centre for Food and Development (CIAD), Mexico.

John Andrew Hudson, The Institute of Environmental Science and Research Ltd, New Zealand.

David Jordan, New South Wales Agriculture, Wollongbar Agricultural Institute, Australia.

Julia A. Kiehlbauch, Microbiology Consultant, United States of America.

Günter Klein, Division of Food Hygiene, Joint FAO/WHO Collaborating Centre, Federal Institute for Health Protection of Consumers and Veterinary Medicine, Germany.

Susumu Kumagai, Graduate School of Agriculture and Life Sciences, The University of Tokyo, Japan.

Roland Lindqvist, National Food Administration, Sweden.

Xiumei Liu, Department of Microbiology and Natural Toxins, Institute of Nutrition and Food Hygiene, Chinese Academy of Preventative Medicine, Ministry of Health, China.

Carol Maczka, Office of Public Health and Science, Food Safety and Inspection Service, United States Department of Agriculture, United States of America.

Patience Mensah, Noguchi Memorial Institute for Medical Research, University of Ghana, Ghana.

George Nasinyama, Department of Epidemiology and Food Safety, Faculty of Veterinary Medicine, Makerere University, Uganda.

Gregory Paoli, Decisionalysis Risk Consultants Inc., Canada.

Irma N.G. Rivera, Departamento de Microbiologia, Instituto de Ciências Biomédicas, Universidade de São Paulo, Brazil.

Son Radu, Department of Biotechnology, Faculty of Food Science and Biotechnology, University Putra Malaysia, Malaysia.

Tom Ross, School of Agricultural Science, University of Tasmania, Australia.

Dulce Maria Tocchetto Schuch, Laboratorio Regional de Apoio Animal - Lara / RS, Agriculture Ministry, Brazil.

Eystein Skjerve, Department of Pharmacology, Microbiology and Food Hygiene, The Norwegian School of Veterinary Science, Norway.

Ewen C.D. Todd, The National Food Safety and Toxicology Center, Michigan State University, United States of America.

Robert Bruce Tompkin, ConAgra Inc., United States of America.

Suzanne Van Gerwen, Microbiology and Preservation Unit, Unilever Research, The Netherlands.

Michiel Van Schothorst, Wageningen University, The Netherlands.

Kaye Wachsmuth, Office of Public Health and Science, Food Safety and Inspection Service, United States Department of Agriculture, United States of America.

Helene Wahlström, National Veterinary Institute, Sweden.

Richard C. Whiting, Center for Food Safety and Applied Nutrition, Food and Drug Administration, United States of America.

Charles Yoe, Department of Economics, College of Notre Dame of Maryland, United States of America.

REVIEWERS IN RESPONSE TO CALL FOR PUBLIC COMMENT

The United States Food and Drug Administration.

The Industry Council for Development.

Professor O.O. Komolafe, Department of Microbiology, College of Medicine, Malawi.

PEER REVIEWERS

Fred Angulo, Centers for Disease Control and Protection, United States of America.

Dane Bernard, Keystone Foods, United States of America.

Tine Hald, Danish Veterinary Laboratory, Denmark.

Thomas Humphrey, University of Bristol, United Kingdom.

Hilde Kruse, National Veterinary Institute, Norway.

Geoffrey Mead, Food Hygiene consultant,United Kingdom.

Robert T. Mitchell, Public Health Laboratory Service, United Kingdom.

Maarten Nauta, RIVM, The Netherlands.

Terry A. Roberts, Consultant, Food Hygiene and Safety, United Kingdom.

John Sofos, Colorado State University, United States of America.

Katharina Stark, Swiss Federal Veterinary Office, Switzerland.

Isabel Walls, International Life Sciences Institute, United States of America.

THE JOINT FAO/WHO SECRETARIAT ON RISK ASSESSMENT OF MICROBIOLOGICAL HAZARDS IN FOODS

Lahsen Ababouch, FAO

Peter Karim Ben Embarek, WHO

Sarah Cahill, FAO

Maria de Lourdes Costarrica, FAO

Françoise Fontannaz, WHO

Allan Hogue, WHO

Jean-Louis Jouve, (from June 2001) FAO

Hector Lupin, FAO

Jeronimas Maskeliunas, FAO

Jocelyne Rocourt, WHO

Jørgen Schlundt, WHO

Hajime Toyofuku, WHO

EDITOR

Thorgeir Lawrence, Technical editor, Iceland.

ABBREVIATIONS USED IN THE TEXT

AIDS	Acquired Immunodeficiency Syndrome
ANOVA	Analysis of variance
CAC	FAO/WHO Codex Alimentarius Commission
CCFH	Codex Committee on Food Hygiene
CDC	United States Center for Disease Control and Prevention
CFIA	Canadian Food Inspection Agency
CFU	Colony-forming unit
EGR	Exponential growth rate
EU	European Union
FDA	Food and Drug Administration [of the United States of America]
FSIS	Food Safety and Inspection Service [USDA]
HIV	Human Immunodeficiency Virus
HLA-B27	Human Leukocyte Antigen B27
IgG	Immunoglobulin G
IgM	Immunoglobulin M
IUNA	Irish Universities Nutrition Alliance
MPN	Most probable number
MRA	Microbiological risk assessment
SE	*Salmonella enterica* serotype Enteritidis (*S.* Enteritidis)
US SE RA	USDA-FSIS *Salmonella* Enteritidis Risk Assessment
USDA	United States Department of Agriculture
YMT	Yolk membrane breakdown time

FOREWORD

The Members of the Food and Agriculture Organization of the United Nations (FAO) and of the World Health Organization (WHO) have expressed concern regarding the level of safety of food both at the national and the international levels. Increasing foodborne disease incidence over the last decades seems, in many countries, to be related to an increase in disease caused by microorganisms in food. This concern has been voiced in meetings of the Governing Bodies of both Organizations and in the Codex Alimentarius Commission. It is not easy to decide whether the suggested increase is real or an artefact of changes in other areas, such as improved disease surveillance or better detection methods for microorganisms in foods. However, the important issue is whether new tools or revised and improved actions can contribute to our ability to lower the disease burden and provide safer food. Fortunately new tools, which can facilitate actions, seem to be on their way.

Over the past decade Risk Analysis, a process consisting of risk assessment, risk management and risk communication, has emerged as a structured model for improving our food control systems with the objectives of producing safer food, reducing the numbers of foodborne illnesses and facilitating domestic and international trade in food. Furthermore we are moving towards a more holistic approach to food safety where the entire food chain needs to be considered in efforts to produce safer food.

As with any model, tools are needed for the implementation of the risk analysis paradigm. Risk assessment is the science based component of risk analysis. Science today provides us with indepth information on life in the world we live in. It has allowed us to accumulate a wealth of knowledge on microscopic organisms, their growth, survival and death, even their genetic make-up. It has given us an understanding of food production, processing and preservation and the link between the microscopic and the macroscopic world and how we can benefit from as well as suffer from these microorganisms. Risk assessment provides us with a framework for organising all this data and information and to better understand the interaction between microorganisms, foods and human illness. It provides us with the ability to estimate the risk to human health from specific microorganisms in foods and gives us a tool with which we can compare and evaluate different scenarios as well as identify what type of data is necessary for estimating and optimising mitigating interventions.

Microbiological risk assessment can be considered as a tool that can be used in the management of the risks posed by food-borne pathogens and in the elaboration of standards for food in international trade. However, undertaking a microbiological risk assessment (MRA), particularly quantitative MRA, is recognized as a resource-intensive task requiring a multidisciplinary approach. Yet, food-borne illness is among the most widespread public health problems creating social and economic burdens as well as human suffering, making it a concern that all countries need to address. As risk assessment can also be used to justify the introduction of more stringent standards for imported foods, a knowledge of MRA is important for trade purposes, and there is a need to provide countries with the tools for understanding and, if possible, undertaking MRA. This need, combined with that of the

Codex Alimentarius for risk based scientific advice led FAO and WHO to undertake a programme of activities on MRA at the international level.

The Food Quality and Standards Service, FAO and the Food Safety Department, WHO are the lead units responsible for this initiative. The two groups have worked together to develop the area of MRA at the international level for application at both the national and international levels. This work has been greatly facilitated by the contribution of people from around the world with expertise in microbiology, mathematical modelling, epidemiology and food technology to name but a few.

This Microbiological Risk Assessment series provides a range of data and information to those who need to understand or undertake MRA. It comprises risk assessments of particular pathogen-commodity combinations, interpretative summaries of the risk assessments, guidelines for undertaking and using risk assessment and reports addressing other pertinent aspects of MRA.

We hope that this series will provide a greater insight into MRA, how it is undertaken and how it can be used. We strongly believe that this is an area that should be developed in the international sphere, and have already from the present work clear indications that an international approach and early agreement in this area will strengthen the future potential of use of this tool in all parts of the world as well as in international standard setting. We would welcome comments and feedback on any of the documents within this series so that we can endeavour to provide Member States, Codex Alimentarius and other users of this material with the information they need to use risk based tools with the ultimate objective of ensuring that safe food is available for all consumers.

<table>
<tr><td>Jean-Louis Jouve</td><td>Jørgen Schlundt</td></tr>
<tr><td>Food Quality and Standards Service</td><td>Food Safety Department</td></tr>
<tr><td>FAO</td><td>WHO</td></tr>
</table>

EXECUTIVE SUMMARY

BACKGROUND

FAO and WHO undertook a risk assessment of *Salmonella* in eggs and broiler chickens in response to requests for expert advice on this issue from their member countries and from the Codex Alimentarius Commission. Guidance on this issue is needed, as salmonellosis is a leading cause of foodborne illness in many countries, with eggs and poultry being important vehicles of transmission.

The risk assessment had several objectives.

1. To develop a resource document of all currently available information relevant to risk assessment of *Salmonella* in eggs and broiler chickens and also to identify the current gaps in the data that need to be filled in order to more completely address this issue.

2. To develop an example risk assessment framework and model for worldwide application.

3. To use this risk assessment work to consider the efficacy of some risk management interventions for addressing the problems associated with *Salmonella* in eggs and broiler chickens.

This document could be used as a resource document that includes currently available information relevant to risk assessment of *Salmonella* in eggs and broiler chickens. Although a cost–benefit analysis of potential mitigations would assist risk managers in determining which mitigations to implement, it was not within the scope of this work and is not considered here.

In order to develop the model, the risk assessment was divided into two risk assessments with a shared hazard identification and hazard characterization. These two risk assessments included the four steps of risk assessment: hazard identification, hazard characterization, exposure assessment, and risk characterization.

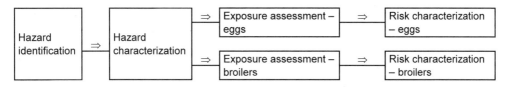

One hazard identification and one hazard characterization, including a dose-response model, and two exposure assessment models – one for *S.* Enteritidis in eggs and one for *Salmonella* in broiler chickens – were elaborated. For *S.* Enteritidis in eggs, the risk characterization estimates the probability of human illness due to *S.* Enteritidis following the ingestion of a single food serving of internally contaminated shell eggs, consumed as either whole eggs, egg meals or as ingredients in more complex food (e.g. cake). This work addressed selected aspects of egg production on farms; further processing of eggs into egg products; retail and consumer egg handling; and meal preparation practices. For *Salmonella* in broiler chickens, the risk characterization estimates the probability of illness in a year due to the ingestion of *Salmonella* on fresh whole broiler chicken carcasses with the skin intact, and which are cooked in the domestic kitchen for immediate consumption. This work commenced at the conclusion of slaughterhouse processing and considers in-home handling and cooking practices. The effects of pre-slaughter interventions and the slaughter process are not currently included in this model.

The inputs for this risk assessment were obtained from a variety of sources. Information was compiled from published literature, national reports and from unpublished data submitted to FAO/WHO by various interested parties.

The main outputs from the risk assessment are summarized below. It should also be noted that, in the course of the work, efforts were made to identify features that have an impact on the acceptability of findings and the appropriateness of extrapolating findings to scenarios not explicitly investigated in the risk assessments, and these are identified in the risk assessment document.

HAZARD IDENTIFICATION

During the past two decades, *Salmonella* Enteritidis has emerged as a leading cause of human infections in many countries, with hen eggs being a principal source of the pathogen. This has been attributed to this serovar's unusual ability to colonize ovarian tissue of hens and be present within the contents of intact shell eggs. Broiler chicken is the main type of chicken consumed as poultry in many countries. Large percentages are colonized by salmonellae during grow-out and the skin and meat of carcasses are frequently contaminated by the pathogen during slaughter and processing. Considering the major role eggs and poultry have as vehicles of human cases of salmonellosis, an assessment of different factors affecting the prevalence, growth and transmission of *Salmonella* in eggs and on broiler chicken carcasses and the related risk of human illness would be useful to risk managers in identifying the intervention strategies that would have the greatest impact on reducing human infections.

HAZARD CHARACTERIZATION

The hazard characterization provides a description of the public health outcomes, pathogen characteristics, host characteristics, and food-related factors that may affect the survival of *Salmonella* through the stomach. It also presents a review of information on relevant dose-response models describing the mathematical relationship between an ingested dose of *Salmonella* and the probability of human illness. An extensive review of available outbreak data was also conducted. From these data, a new dose-response model was derived using a re-sampling approach, and this was used in both risk characterizations in preference to existing models that are defined within this component of the risk assessment. Finally, an attempt was made to discern whether separate dose-response curves could be justified for different human sub-populations defined on the basis of age and "susceptibility", and whether a dose-response for *S*. Enteritidis was distinguishable from a dose-responses for other *Salmonella.*

Three existing dose-response models for *Salmonella* were identified:

1. Fazil, 1996, using the Beta-Poisson model (Haas, 1983) fitted to the naive human data from *Salmonella* feeding trials (McCullough and Eisele, 1951a, b, c).

2. United States *Salmonella* Enteritidis Risk Assessment (US SE RA) (USDA-FSIS, 1998), based on the use of human feeding trial data for a surrogate pathogen (*Shigella dysenteriae*) with illness as the measured endpoint to describe the dose-response relationship.

3. *Salmonella* Enteritidis Risk Assessment conducted by Health Canada (2000, but unpublished) based on a Weibull-Gamma dose-response relationship. The model uses data from many different pathogen-feeding trials and combines the information with key *Salmonella* outbreak data, using a Bayesian relationship.

These dose-response models for *S*. Enteritidis and *Salmonella* were found to inadequately characterize the dose-response relationship observed in the outbreak data. A new dose-response model was developed in the course of this work. It was derived from outbreak data and was considered to be the most appropriate estimate for the probability of illness upon ingestion of a dose of *Salmonella*. The model was based on observed real world data, and as such was not subject to some of the flaws inherent in using purely experimental data. Nevertheless, the current outbreak data also have uncertainties associated with them and some of the outbreak data points required assumptions to be made. The outbreak data are also from a limited number of developed countries and may not be applicable to other regions.

From the outbreak data used to examine the dose-response relationship, it could not be concluded that *S*. Enteritidis has a different likelihood from other serovars of producing illness. In addition, comparing the attack rates of *Salmonella* for children less than five years of age, against those for the rest of the population in the outbreak database, did not reveal an overall trend of increased risk for this subpopulation. Although some indication for a difference in attack rates for the two populations had been noted in two of the outbreaks examined, the database of outbreak information might lack the potential to reveal the existence of any true differences. Severity of illness as a function of patient age, *Salmonella* serovar or pathogen dose were not evaluated, although severity could potentially be influenced by these factors and by pathogenicity. However, the current database of information was insufficient to derive a quantitative estimate for these factors.

EXPOSURE ASSESSMENT AND RISK CHARACTERIZATION OF *SALMONELLA* ENTERITIDIS IN EGGS

The exposure assessment section for *S*. Enteritidis in eggs compares and contrasts previously completed models. It describes the general framework of these models, the data used, and the analysis completed for modelling analysis. Generally, these models comprise a production module, a module for the processing and distribution of shell eggs, a module for the processing of egg products, and a module for preparation and consumption. The production module predicts the likelihood of a *S*. Enteritidis-contaminated egg occurring. This depends on the flock prevalence, within-flock prevalence, and the frequency that infected hens lay contaminated eggs. The flock prevalence (i.e. the likelihood of a flock containing one or more infected hens) further depends on factors that serve to introduce *S*. Enteritidis into flocks (e.g. replacement pullets, environmental carryover from previously infected flocks, food contamination, etc.). The shell egg processing and distribution, and preparation and consumption modules predict the likelihood of human exposures to various doses of *S*. Enteritidis from contaminated eggs. The dose consumed in an egg-containing meal depends on the amount of *S*. Enteritidis growth between the time the egg was laid and when it was prepared, as well as how the egg was prepared and cooked. Growth of *S*. Enteritidis in contaminated eggs is a function of storage time and temperature. The output of the exposure assessment, in general, feeds into the hazard characterization to produce the risk characterization output. This output is the probability of human illness per serving of an egg-containing meal.

The exposure assessment included consideration of yolk-contaminated eggs and growth of *S*. Enteritidis in eggs prior to processing for egg products. These issues have not been previously addressed by exposure assessments of *S*. Enteritidis in eggs. Yolk-contaminated eggs might allow more rapid growth of *S*. Enteritidis inside such eggs compared with eggs that are not yolk-contaminated.

This risk characterization of *S*. Enteritidis in eggs was intentionally developed so as not to be representative of any specific country or region. However, some model inputs are based on evidence or assumptions derived from specific national situations. Caution is therefore required when extrapolating from this model to other countries.

Key findings

The risk of human illness from *S*. Enteritidis in eggs varies according to the different input assumptions in the model. The risk of illness per serving increases as flock prevalence increases. However, uncertainty regarding the predicted risk also increases as flock prevalence increases. **Reducing flock prevalence results in a directly proportional reduction in human health risk. For example, reducing flock prevalence from 50% to 25% results in a halving of the mean probability of illness per serving. Reducing prevalence within infected flocks also results in a directly proportional reduction in human health risk. For example, risk of illness per serving generated from eggs produced by a flock with 1% within-flock prevalence is one-tenth that of a flock with 10% within-flock prevalence.**

Adjusting both egg storage time and temperature profiles for eggs from production to consumption was associated with large effects on the predicted risk of human illness. **The risk of human illness per serving appears to be insensitive to the number of *Salmonella* Enteritidis in contaminated eggs across the range considered at the time of lay. For example, whether it is assumed that all contaminated eggs had an initial number of 10 or 100 *S*. Enteritidis organisms, the predicted risk of illness per serving was similar.** This may be because the effect of *S*. Enteritidis growth is greater than the initial contamination level in eggs.

As an example of how the efficacy of interventions aimed at reducing flock prevalence may be assessed the risk assessment examined the effect of a "test and divert" programme. Two protocols were assumed, with either one (at the beginning of egg production) or three (beginning of egg production, four months later & just before flock depopulation) tests administered to the entire population of egg production flocks and their effectiveness was estimated over a four-year period. Testing three times per year for four years reduced the risk of human illness from shell eggs by more than 90% (i.e. >1 log). Testing once a year for four years reduced risk by over 70%.

Other potential interventions evaluated included vaccination and refrigeration. To evaluate the effectiveness of vaccination against *S*. Enteritidis a single test, or two tests four months apart, with 90 faecal samples per test, was considered. The vaccine was assumed to be capable of reducing the frequency of contaminated eggs by approximately 75%. The effects of time and temperature restrictions were evaluated assuming a flock prevalence of 25%. Restricting shelf-life to less than 14 days reduced the predicted risk of illness per serving by a negligible amount (~1%). However, keeping retail storage temperature at no more than 7.7°C reduced risk of illness per serving by about 60%. Were shelf-life to be reduced to 7 days, risk per serving would also be reduced by about 60%.

Limitation

The available data on which this risk assessment was based was limited. For example, evidence regarding enumeration of the organism within eggs was based on only 63 *S*. Enteritidis-contaminated eggs, and in part on estimates of the concentration of the organism in contaminated eggs. It is difficult to represent uncertainty and variability with such limited data. Apparently, there is a lot of uncertainty and it is difficult to quantify. In addition, statistical or model uncertainty was not fully explored.

Much uncertainty attends the effectiveness of various management interventions for controlling *S*. Enteritidis. The magnitudes of uncertainty regarding test sensitivity, effectiveness of cleaning and disinfecting, and vaccination efficacy have not been measured. Some data were available to describe these inputs, but the data may not be relevant to all regions or countries where such interventions might be applied.

Statistical or model uncertainty was not fully explored in this risk characterization. For example, alternative distributions to the lognormal for within-flock prevalence were not considered. In addition, the predictive microbiology used in this model was dependent on very limited data pertaining to *S*. Enteritidis growth inside eggs. Alternative functional specifications for *S*. Enteritidis growth equations were not pursued in this analysis.

EXPOSURE ASSESSMENT AND RISK CHARACTERIZATION OF *SALMONELLA* IN BROILER CHICKENS

The risk assessment model is defined in terms of a number of parameters that describe the processes of broiler chicken carcass distribution and storage, preparation, cooking and consumption. Some of these parameters can be considered general in that they can be used to describe the situation in many countries. At the same time, some parameters are country specific, such as the prevalence of carcasses contaminated with *Salmonella* at the completion of processing. Predictions of risk for a particular country are best obtained from data relevant to that country.

The exposure assessment of *Salmonella* in broiler chickens mimics the movement of *Salmonella*-contaminated chickens through the food chain, commencing at the point of completion of the slaughter process. For each iteration of the model, a chicken carcass was randomly allocated an infection status and those carcasses identified as contaminated were randomly assigned a number of *Salmonella* organisms. From this point until consumption, changes in the size of the *Salmonella* population on each contaminated chicken were modelled using equations for growth and death. The growth of *Salmonella* was predicted using random inputs for storage time at retail stores, transport time, storage time in homes, and the temperatures the carcass was exposed to during each of these periods. Death of *Salmonella* during cooking was predicted using random inputs describing the probability that a carcass was not adequately cooked, the proportion of *Salmonella* organisms attached to areas of the carcass that were protected from heat, the temperature of exposure of protected bacteria, and the time for which such exposure occurs. The number of *Salmonella* consumed were then derived using a random input defining the weight of chicken meat consumed per serving and the numbers of *Salmonella* cells in meat as defined from the various growth and death processes. Finally, in the risk characterization, the probability of illness was derived by combining the number of organisms ingested (from the exposure assessment) with information on the dose-response relationship (hazard characterization).

Key findings

The *Salmonella* in broiler chickens risk assessment does not consider all parts of the production-to-consumption continuum, and this limits the range of control options that can be assessed. This is primarily due to the lack of representative data to analyse how much change in either the prevalence or level of *Salmonella* in poultry could be attributable to any specific treatment or action. However, the establishment of a baseline model provided a means to compare the effects on risk when prevalence and cell numbers were changed. The model parameters can be modified to evaluate the efficacy of risk mitigation strategies that target those parameters. For example, the parameter describing prevalence of *Salmonella*-contaminated broiler chickens exiting processing can be modified to evaluate the effectiveness of a processing measure such as chlorination of the chilling water to reduce the prevalence of *Salmonella*-contaminated carcasses.

Reduction in the prevalence of *Salmonella*-contaminated chicken was associated with a reduction in the risk of illness. A one-to-one relationship was estimated, with a percentage change in prevalence, assuming everything else remains constant, reducing the expected risk by a similar percentage. **For instance, a 50% reduction in the prevalence of contaminated poultry (20% to 10%) produced a 50% reduction in the expected risk of** illness per serving. **Similarly, a large reduction in prevalence (20% to 0.05%) would produce a 99.75% reduction in the expected risk of illness.** If management strategies are implemented that affect the level of contamination, i.e. the numbers of *Salmonella* on chickens, the relationship to risk of illness is estimated to be greater than a one-to-one relationship. **A shift in the distribution of *Salmonella* cell numbers on broiler chickens exiting the chill tank at the end of processing, such that the mean number of cells is reduced by 40% on the non-log scale, reduces the expected risk of illness per serving by approximately 65%.**

A small reduction in the frequency of undercooking and the magnitude of the undercooking event results in a marked reduction of the expected risk of illness per serving. The important caveat here is that altering cooking practices does not address the risk of illness through the cross-contamination pathway. The strategy of changing the consumer's cooking practices needs to be tempered by the fact that cross-contamination may in fact be the predominant source of risk of illness, and the nature of cross-contamination in the home is still a highly uncertain phenomenon.

Limitations and caveats

It was not possible to provide a perfect representation of growth of *Salmonella* in raw poultry and seasonal variations in ambient temperature were not accounted for. The model adopted also assumed that ambient temperature had no impact on the rate of change for storage temperatures used for predicting growth, and this is intuitively inappropriate in some circumstances. Similarly, limitations were present in the way the model predicts the death of *Salmonella* in broiler chicken carcasses during the cooking process.

At several steps, reliance was placed on expert opinion to estimate the value of model inputs. While often easily accessible and sometimes sufficiently accurate, occasionally, expert opinion might reduce transparency and introduce an unacceptable bias that may not be detected by the risk assessors.

Surveillance data from some countries often show a marked seasonality in the number of notifications of human salmonellosis, with peak incidence occurring in the warmer months and the current model cannot account for or explain this important phenomenon.

A lack of detailed understanding of all aspects of cross-contamination in the home hampered the ability of the risk assessment to address this process. While the uncertainty associated with several parameters in the consumption portion of the risk assessment was accounted for, a full analysis of statistical and model uncertainty was not done. Thus, the influence of uncertainty in the cross-contamination pathway was not explored.

CONCLUSIONS

This *Salmonella* risk assessment provides information that should be useful in determining the impact intervention strategies may have on reducing cases of salmonellosis from contaminated eggs and poultry. In the risk assessment of *Salmonella* in broiler chickens, for example, it was determined that there is a relationship between changing the prevalence of *Salmonella* on the broiler chickens and reducing the risk of illness per serving. In the risk assessment of *S.* Enteritidis in eggs, reducing the prevalence of *S.* Enteritidis in poultry flocks was directly proportional to the reduction in risk to human health. The model can also be used to estimate the change in risk of human illness from changing storage times or temperature of eggs. However, comparison of effects of intervention measures, i.e. sensitivity analysis, cannot be done because this risk assessment is not conducted for a specific region or country, or for global settings. Data was collected from different countries for different input parameters. If those data were changed reflecting a specific national situation, the impact of a measure would also be changed. Therefore, caution would be needed in interpreting the results of this risk assessment in Codex activities.

REFERENCES CITED IN THE EXECUTIVE SUMMARY

Fazil, A.M. 1996. A quantitative risk assessment model for salmonella. Drexel University, Philadelphia PA. [Dissertation].

Haas, C.N. 1983. Estimation of risk due to low doses of microorganisms: a comparison of alternative methodologies. *American Journal of Epidemiology*, **118**: 573–582.

Health Canada. [2000]. Risk assessment model for *Salmonella* Enteritidis. Unpublished Health Canada Document.

McCullough, N.B., & Eisele, C.W. 1951a. Experimental human salmonellosis. I. Pathogenicity of strains of *Salmonella* Meleagridis and *Salmonella anatum* obtained from spray-dried whole egg. *Journal of Infectious Diseases*, **88**: 278–289.

-- 1951b. Experimental human salmonellosis. II. Immunity studies following experimental illness with *Salmonella* Meleagridis and *Salmonella anatum*. *Journal of Immunology*, **66**: 595–608.

-- 1951c. Experimental human salmonellosis. III. Pathogenicity of strains of *Salmonella* Newport, *Salmonella derby*, and *Salmonella* Bareilly obtained from spray dried whole egg. *Journal of Infectious Diseases*, **89**: 209–213.

USDA-FSIS. 1998. *Salmonella* Enteritidis Risk Assessment. Shell Eggs and Egg Products. Final Report. Prepared for FSIS by the *Salmonella* Enteritidis Risk Assessment Team. 268 pp. Available on Internet as PDF document at: www.fsis.usda.gov/ophs/risk/contents.htm.

1. INTRODUCTION

1.1 RISK ASSESSMENT

Risk assessment, along with risk management and risk communication, is one of the components of risk analysis, which can be defined as an overall strategy for addressing risk. The importance of an overlap between these three elements (risk assessment, risk management and risk communication) is well recognized, but some functional separation is also necessary. In relation to risk assessment, such separation ensures that issues are addressed in a transparent manner with a scientific basis.

The FAO/WHO Codex Alimentarius Commission (CAC) defines risk assessment as a scientifically based process consisting of four steps:

1. *Hazard identification*, which is the identification of the biological agent that may be present in a particular food or group of foods and capable of causing adverse health effects.

2. *Hazard characterization*, which is the qualitative or quantitative, or both, evaluation of the nature of the adverse health effects associated with the biological agents that may be present in food, and in such cases a dose-response assessment should be performed if the data are obtainable.

3. *Exposure assessment*, which is the qualitative or quantitative, or both, evaluation of the likely intake of the biological agent through food, as well as through exposure from other sources, if relevant.

4. *Risk characterization*, namely the qualitative or quantitative, or both, estimation, including attendant uncertainties, of the probability of occurrence and severity of known or potential adverse health effects in a given population based on hazard identification, exposure assessment and exposure assessment.

These steps are illustrated schematically in Figure 1.1. The risk assessment process is a means of providing an estimate of the probability and severity of illness attributable to a particular pathogen-commodity combination. The four-step process enables this to be carried out in a systematic manner, but the level of detail in which each step is addressed will depend on the scope of the risk assessment. This should be defined clearly by the risk manager through ongoing dialogue with the risk assessor.

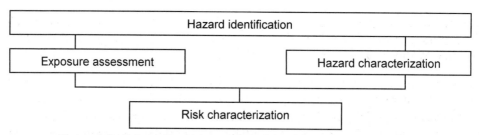

Figure 1.1. Schematic representation of the risk assessment process.

Undertaking a risk assessment is recognized as a resource-intensive task requiring a multidisciplinary approach. While MRA is becoming an important tool for assessing the risks to human health from foodborne pathogens, and can be used in the elaboration of standards for food in international trade, it is not within the capacity of many, perhaps even most, countries to carry out a complete quantitative MRA. Yet foodborne illness is among the most widespread public health problems and creates social and economic burdens as well as human suffering, making it a concern that all countries need to address, and risk assessment is a tool that can be used in the management of the risks posed by foodborne pathogens. At the same time, risk assessment can also be used to justify the introduction of more stringent standards for imported foods. A knowledge of MRA is therefore also important for trade purposes, and there is a need to provide countries with the tools for understanding and, if possible, carrying out MRA.

1.2 BACKGROUND TO THE FAO/WHO MICROBIOLOGICAL RISK ASSESSMENT WORK

Risk analysis has evolved over the last decade within CAC. Since the Uruguay Round Trade Agreement on the Application of Sanitary and Phytosanitary Measures (SPS) entered into force in 1995, the importance of risk analysis has increased. Risk analysis is now considered to be an integral part of the decision-making process of Codex. CAC has adopted definitions of risk analysis terminology related to food safety, and statements of principle relating to the role of food safety risk assessment. Furthermore, in 1999, it adopted the *Principles and Guidelines for the Conduct of Microbiological Risk Assessment* (CAC, 1999a), which were developed by the Codex Committee on Food Hygiene (CCFH).

In addition to these developments in risk assessment, the 22nd Session of CAC requested FAO and WHO to convene an international advisory body on the microbiological aspects of food safety in order to address MRA in particular (CAC, 1997). In response to this, and as follow-up to previous activities in the area of risk analysis, FAO and WHO convened an expert consultation in March 1999 to examine in an international forum the issue of MRA. The main outcome of this expert consultation was an outline strategy and mechanism for addressing MRA at the international level (WHO, 1999). Subsequently, at its 32nd Session, in November 1999, CCFH recognized that there are significant public health problems related to microbiological hazards in foods (CAC, 1999b). It identified 21 pathogen-commodity combinations of concern, and prioritized these according to criteria such as the significance of the public health problem, the extent of the problem in relation to geographical distribution and international trade, and the availability of data and other information with which to conduct a risk assessment. CCFH suggested that FAO and WHO convene ad hoc expert consultations to provide advice on MRA, and also recommended that these consultations be conducted according to the format outlined at the 1999 expert consultation (WHO, 1999).

The need by member countries for advice on risk assessment in order to reduce the risk to consumers of becoming ill from food, and to meet their obligations imposed under World Trade Organization (WTO) agreements, combined with the requests of CAC and CCFH for scientific advice on MRA, led FAO and WHO to undertake a programme of activities to address at international level the issue of MRA. The aim of the joint programme is to provide a transparent review of scientific opinion on the state of the art of MRA, and to develop the means to achieve sound quantitative risk assessments of specific pathogen-commodity combinations. It also aims to create awareness of the risk assessment process, provide information and advice, and develop tools that could be used by countries in undertaking MRA.

This programme of MRA activities aims to serve two user groups: CAC; and FAO and WHO member countries. CAC requires sound scientific advice as a basis for the development of standards, guidelines and related texts for the management of risks posed by microbiological hazards in foods. Member countries, in contrast, need adaptable risk assessment tools to use in conducting their own assessments and, if possible, some modules directly applicable to their national situation.

Taking these needs into account, FAO and WHO initiated work on three of the pathogen-commodity combinations identified as priority issues: *Salmonella* Enteritidis in eggs; *Salmonella* in broiler chickens; and *Listeria monocytogenes* in ready-to-eat foods. In order to facilitate communication with risk managers, a two-year process of work was introduced (Figures 1.2a, b). Problems such as the lack of a clear-cut risk management question at the outset, and limitations in the usefulness of a global risk estimate, were recognized and addressed to the extent possible in the course of the work.

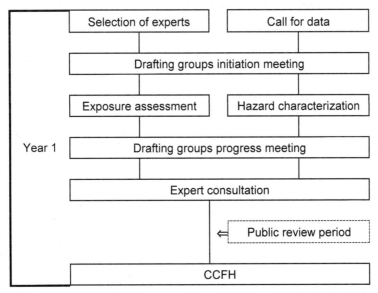

Figure 1.2a. Year 1 of the FAO/WHO process for undertaking microbiological risk assessment.

The risk assessments of *Salmonella* in eggs and broiler chickens began in January 2000. Risk assessment work on *L. monocytogenes* in ready-to-eat foods was undertaken concurrently. In the first year of the process, drafting groups were established to examine available information and prepare technical documentation on the hazard identification, exposure assessment and hazard characterization components of the risk assessment. These documents were then reviewed and evaluated by a joint expert consultation in July 2000 (FAO, 2000). That expert consultation made recommendations for the improvement of the preliminary documents, identified knowledge gaps and information requirements needed to complete the risk assessment work, and developed a list of issues to be brought to the attention of CCFH. The report of that consultation was presented to the 33rd Session of CCFH in order to inform risk managers regarding the progress of the risk assessment and to seek more precise guidance on the needs of risk managers. A number of specific risk management questions were identified by the Committee (CAC, 2000) and these issues were subsequently addressed in the completion of the risk assessment. The documentation was also made available for public comment as a means of reviewing the preliminary work.

The second year of the process focused on the completion of the risk assessment by undertaking the risk characterization step. Again, the risk characterization document that was developed was critically reviewed by an expert consultation convened in April-May 2001 in Rome (FAO, 2001). The report of this consultation, which included preliminary answers to the questions posed by CCFH, was presented to the 34th Session of the Committee.

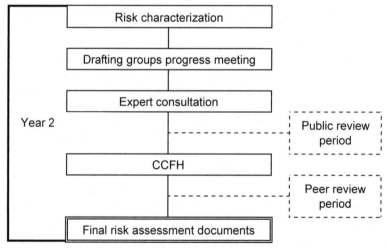

Figure 1.2b. Year 2 of the FAO/WHO process for undertaking MRA.

1.3 SCOPE OF THE RISK ASSESSMENT

The risk assessment initially set out to understand how the incidence of human salmonellosis is influenced by various factors, from the agricultural phase of chicken meat and egg production, through marketing, processing, distribution, retail storage, consumer storage and meal preparation, to final consumption. Such models are appealing because they permit the study of the broadest range of intervention strategies. However, as the work progressed it became evident that the quantity and quality of information available from all sources was not sufficient to allow the construction of a full and expansive model. Thus the final scope of the *Salmonella* risk assessment, and the components of the food production and consumption continuum that were considered, became:

1. ***Salmonella enterica* serotype Enteritidis in eggs.** This risk characterization estimates the probability of human illness due to *Salmonella* Enteritidis (SE) following the ingestion of a single food serving of internally contaminated shell eggs, either consumed as whole eggs, egg meals or as ingredients in more complex food (e.g. cake). This work addressed selected aspects of egg production on farms, further processing of eggs into egg products, retail and consumer egg handling, and meal preparation practices. Risk reductions for specific intervention strategies were also estimated.

2. ***Salmonella enterica* (multiple serotypes) in broiler chickens.** This risk characterization estimates the probability of acute gastroenteritis per person per serving and per year, due to the ingestion of *Salmonella enterica* on fresh whole broiler chicken carcasses with the skin intact and which are cooked in the domestic kitchen for immediate consumption. This work commences at the conclusion of slaughterhouse processing, and considers in-home handling and cooking practices, including cross-contamination events. The effects of pre-slaughter interventions and the slaughter process are not currently included in this model. However, for any intervention strategy, whether at farm or during processing, that reduces the prevalence or numbers, or both, of *Salmonella* on poultry or carcasses by a measurable quantity, the amount of risk reduction can be calculated from the risk model, and examples are provided.

Risk estimates for *S.* Enteritidis in eggs and *S. enterica* in broiler chickens used a common dose-response model. Within the hazard-characterization step, the objectives were to produce one or more curves describing the probability that an individual would become ill versus the dose of *Salmonella* ingested within food.

Human-host-adapted, predominantly invasive *Salmonella* serotypes (e.g. *S.* Typhi, *S.* Paratyphi) were not considered in developing the dose-response model. As noted, the outcome of interest was defined as acute gastroenteritis. Hence, disease outcomes that may occur beyond the diagnosis of gastroenteritis were not included in the risk estimations, but are described in hazard identification. Similarly, severity of disease outcomes attributable to multiresistant strains of *Salmonella* were not estimated, nor for the more highly invasive *Salmonella* serotypes that are not commonly associated with poultry, i.e. *S.* Dublin and *S.* Cholerasuis. Cost-benefit analysis of risk reduction interventions was not included in the risk management charge to the risk assessors, and thus is beyond the scope of this work.

In practice, one would start with the model and then look for information. In fact, during the course of drafting meetings, the team considered the model at an earlier stage. However, considering the purpose and the task of international risk assessment, the team needed to look at data at the same time. In this document, data used with modelling structures is first explained, followed by the mathematical modelling of inputs.

The writing format differs between the Exposure assessments and the Risk characterizations of *S.* Enteritidis in eggs and *S. enterica* in broiler chickens. This difference is derived from different approaches taken for these commodities. In the egg exposure assessment, previously reported risk assessments are critically reviewed and compared. Therefore one could look back into the original reports to see the details of input parameters to the model. In contrast, the broiler exposure assessment is written to describe a desirable structure of exposure assessment for this pathogen-commodity combination, without any referable assessment. The models and their parameters are described in Section 6.4.

1.4 REFERENCES CITED IN CHAPTER 1

CAC [Codex Alimentarius Commission]. 1997. Report of the 22nd Session of the Joint FAO/WHO Codex Alimentarius Commission. Geneva, 23-28 June 1997.

CAC. 1999a. Principles and Guidelines for the Conduct of Microbiological Risk Assessment. CAC/GL-30.

CAC. 1999b. Report of the 32nd Session of the Codex Committee on Food Hygiene. Washington, DC, 29 November-4 December 1999.

CAC. 2000. Report of the 33rd Session of the Codex Committee on Food Hygiene. Washington, DC, 23-28 October 2000.

FAO [Food and Agriculture Organization of the United Nations]. 2000. [Report of the] Joint FAO/WHO Expert Consultation on Risk Assessment of Microbiological Hazards in Foods. FAO headquarters, Rome, 17-21 July 2000. *FAO Food and Nutrition Paper,* 71.

FAO. 2001. Joint FAO/WHO Expert Consultation on Risk Assessment of Microbiological Hazards in Foods. Risk characterization of *Salmonella* spp. in eggs and broiler chickens and *Listeria monocytogenes* in ready-to-eat foods. *FAO Food and Nutrition Paper*, 72.

WHO [World Health Organization]. 1999. Risk assessment of microbiological hazards in foods. Report of a Joint FAO/WHO Expert Consultation. Geneva, 15-19 March 1999.

2. HAZARD IDENTIFICATION

2.1 SUMMARY

Over 2500 *Salmonella enterica* serotypes are recognized, and all are regarded as capable of producing disease in humans. Worldwide, salmonellosis is a leading cause of enteric infectious disease attributable to foods. Illnesses caused by the majority of Salmonella serotypes range from mild to severe gastroenteritis, and in some patients, bacteraemia, septicaemia and a variety of associated longer-term conditions. A wide range of foods has been implicated in foodborne illness due to *Salmonella enterica*. However, foods of animal origin, especially poultry and poultry products, including eggs, have been consistently implicated in sporadic cases and outbreaks of human salmonellosis.

2.2 *SALMONELLA* IN FOODS AND ASSOCIATION WITH ILLNESS

Salmonellosis is one of the most frequently reported foodborne diseases worldwide.

Each year, approximately 40 000 *Salmonella* infections are culture-confirmed, serotyped, and reported to the United States Centers for Disease Control and Prevention (CDCs). Of total salmonellosis cases, an estimated 96% are caused by foods (Mead et al., 1999).

International data summarized by Thorns (2000) provides estimated incidences of salmonellosis per 100 000 people for the year 1997: 14 in the USA, 38 in Australia, and 73 cases per 100 000 in Japan. In the Europe Union, the estimates range from 16 cases per 100 000 (The Netherlands) to 120 cases per 100 000 in parts of Germany.

A review conducted in southern Latin America on foodborne outbreaks due to bacteria between the years 1995 and 1998 indicated that *Salmonella* were responsible for most (36.8%) of the reported cases in the region (Franco et al., in press). Salmonellosis infections were 55.1% of the reported foodborne disease cases reported from 1993 to 1996 in Korea (Bajk and Roh, 1998). "Salmonella species" was the causative agent reported most often in outbreaks in the European region, being responsible for 77.1% of the outbreaks recorded in which the etiologic agent was determined (WHO, 2001).

The genus *Salmonella* is considered to be a single species named *Salmonella enterica*. Serotyping differentiates the strains, and these are referred to by name as, for example, *Salmonella enterica* serotype Typhimurium, or as *Salmonella* Typhimurium (Hohmann, 2001). Salmonellae are gram-negative, motile (with a few exceptions), facultatively anaerobic bacteria (D'Aoust, 1997). Salmonellae grow between 8°C and 45°C, and at a pH of 4 to 8. With the exception of a limited number of human-host-adapted serotypes (also referred to as the typhoidal salmonellae), the members of the genus *Salmonella* are regarded as zoonotic or potentially zoonotic (Acha and Szyfres, 2001).

Non-typhoidal *Salmonella enterica* typically cause a self-limiting episode of gastroenteritis, characterized by diarrhoea, fever, abdominal cramps, and dehydration. The most cases are mild, and are generally not reported to public health agencies. However, more severe outcomes may result from the infection, depending on host factors and *Salmonella* serotype. Severe disease may occur in healthy individuals, but is most often seen in individuals who are immunocompromised, the very young, or the elderly. In addition, a small percent of cases in healthy individuals are complicated by chronic reactive arthritis.

In the United States of America alone, it has been estimated that 1.4 million cases, 16 430 hospitalizations and 582 deaths are caused by salmonellosis annually.

Costs of foodborne salmonellosis have been calculated for the United States of America population, and are estimated to be as high as US$ 2 329 million annually (in 1998 US dollars) for medical care and lost productivity (Frenzen *et al.*, 1999).

A wide range of foods has been implicated in foodborne illness attributable to *Salmonella enterica*. Foods of animal origin, especially poultry, poultry products and raw eggs, are often implicated in sporadic cases and outbreaks of human salmonellosis (Bryan and Doyle, 1995; Humphrey, 2000). Recent years have seen increases in salmonellosis associated with contaminated fruits and vegetables. Other sources of exposure include water, handling of farm animals and pets, and human person-to-person when hand-mouth contact occurs without proper washing of hands.

Poultry is widely acknowledged to be a reservoir for *Salmonella* infections in humans due to the ability of *Salmonella* to proliferate in the gastrointestinal tract of chicken (Poppe, 2000) and subsequently survive on commercially processed broiler carcasses and edible giblets.

The evolution of the *Salmonella enterica* serotype Enteritidis (*S.* Enteritidis) pandemic beginning in the 1980s led to increased foodborne illnesses associated with poultry in many countries, specifically outbreaks and single cases associated with eggs and egg products (Levy et al., 1996; Rodrigue, Tauxe and Rowe, 1990; Thorns, 2000). Chicken, turkey and eggs were responsible for, respectively, 8.6%, 4.7% and 4.3% of 465 foodborne outbreaks caused by bacterial pathogens for which a vehicle was identified and that were reported to CDCs during the years 1988–1992 (Bean et al., 1997). *Salmonella* caused 12 of 18 outbreaks attributed to chicken, 6 of 12 turkey-associated occurrences, and 19 of 19 egg-related outbreaks. *S.* Enteritidis was responsible for the largest number of foodborne outbreaks, cases and deaths reported in the United States of America (Bean et al., 1997).

In southern Latin America, eggs and mayonnaise were the most common food products associated with outbreaks, but poultry meat was an equally important vehicle (Franco et al., in press). Of the reported foodborne outbreaks in Europe caused by an identified agent, more than one-third were confirmed to be caused by *S.* Enteritidis (WHO, 2001). Foods associated with *S.* Enteritidis outbreaks include egg and egg products (68.2%), cake and ice creams (8%), and poultry and poultry products (3%). Other vehicles include meat and meat products (4%), mixed foods (4%), fish and shellfish (2%), and milk and milk products (3%). In *S.* Typhimurium outbreaks, eggs and egg products (39%), meat and meat products (33%, frequently pork), and poultry and poultry products (10%) were reported as the vehicles of infection. A large number of other *Salmonella* serotypes were also involved in outbreaks in Europe, but specific serotypes were not reported.

2.3 PUBLIC HEALTH OUTCOMES

Over 2500 *Salmonella* serotypes, also referred to as serovars, are known to cause illness in humans. As with all enteric pathogens, outcomes of exposure to *Salmonella* can range from no effects, to colonization of the gastrointestinal tract without any symptoms of illness, to colonization with the typical symptoms of acute gastroenteritis, and – less commonly – to invasive disease characterized by bacteraemia, sequelae, and, rarely, death.

In cases of acute gastroenteritis, the incubation period is generally 12–72 hours, commonly 12–36 hour. Illness lasts 2–7 days, and is characterized by the symptoms noted in the foregoing. Patients usually recover uneventfully within a week without antibiotic treatment. In some cases, severe diarrhoea requires medical interventions such as intravenous fluid rehydration. In cases where the pathogen enters the bloodstream, i.e. septicaemia or bacteraemia, symptoms include high fever, malaise, pain in the thorax and abdomen, chills and anorexia. In some patients, long-term effects or sequelae may occur, and a variety have been identified, including arthritis, osteoarthritis, appendicitis, endocarditis, pericarditis, meningitis, peritonitis and urinary tract infections (Bell, 2002). Typhoid, or enteric fever, caused by only a small number of specific serotypes, is discussed later in this section.

Severe illness resulting from salmonellosis is further exacerbated by the emergence of strains of *Salmonella enterica* that are multiple antibiotic resistant. The effects of underlying illnesses often complicate evaluation of the added clinical impact of resistant *Salmonella* However, in a study referring to the United States of America and the years 1989–90, after accounting for prior antimicrobial exposure and underlying illness, patients with resistant *Salmonella* were more likely to be hospitalized, and for a longer period of time (Lee et al., 1994).

Antibiotic therapy is not routinely recommended for the treatment of mild to moderate presumed or confirmed salmonella gastroenteritis in healthy individuals (Hohmann, 2001). Antimicrobial therapy should be initiated for those who are severely ill and for patients with risk factors for extra-intestinal spread of infection, after appropriate blood and faecal cultures are obtained. An intermittent period of faecal shedding may follow the acute illness, lasting from days to years. Buchwald and Blaser (1984) reviewed 32 reports and showed that the median duration of shedding following acute disease was 5 weeks, with less than 1% of patients becoming chronic carriers. Children may shed up to 10^6 to 10^7 salmonellae per gram faeces during convalescence (Cruickshank and Humphrey, 1987).

From United States of America data, it is estimated that, in general, 93% of individuals with symptoms of salmonellosis recover fully without a physician visit, 5% see a physician and recover fully, 1.1–1.5% of patients require hospitalization, and 0.04–0.1% of patients will die (Buzby et al., 1996; Mead et al., 1999).

However, both sporadic cases and outbreaks demonstrate that the health impacts in specific episodes of gastroenteritis can be particularly severe. Mattila et al. (1998) described a 1994 outbreak of *S.* Bovismorbificans in southern Finland from sprouted alfalfa seeds. Out of 191 respondents, 117 (61%) of the cases required a physician's visit due to intestinal or extra-intestinal symptoms, and 21 (11%) individuals were hospitalized with a median hospital stay of 9 days. The authors state that most hospitalized patients were over 65 years of age. Of the subjects, 94 (49%) received antimicrobials (primarily fluoroquinolones) with a majority (78 out of 94 cases, or 83%) requiring antimicrobial treatment because of diarrhoea, fever, or a salmonella-positive urine sample. Duration of antimicrobial therapy (known for 70 patients) was 2 weeks or more in 44%, 10–12 days in 34% and 1 week or less in 21% of patients. The reason for the severity of the health outcomes in this outbreak was not determined, but it may have been associated with the numbers of salmonella that were consumed. Kanakoudi-Tsakalidou et al. (1998) conducted a prospective study of *S.* Enteritidis infection in nine children. Diarrhoea lasted 3–7 days, accompanied by fever in all cases. Four of the nine patients required hospitalization because of severe dehydration or bloody stools.

Inman et al. (1988) reported on a large outbreak in September 1984 of *S.* Typhimurium PT 22 in a group of police officers given a prepackaged box lunch. There were 473 individuals that fitted the case definition for salmonellosis, and they were mailed a questionnaire enquiring about symptoms associated with the gastroenteritis, with a 72% respondent rate. Out of 340 responders, 196 individuals experienced extra-enteric symptoms, including headaches (182 or 53.5%), joint pain (106 or 31.2%), redness or soreness in the eyes (37 or 10.9%), soreness in the mouth (15 or 4.4%) and skin rash (10 or 2.9%).

Mattila et al. (1998) identified a total of 210 cases with stool samples positive for *S.* Bovismorbificans for questionnaire follow-up regarding symptoms. Of the 191 (91%) respondents, 66 (35%) had articular symptoms, 52 (27%) experienced headaches, 8 (4%) had eye symptoms, and 7 (4%) had cutaneous symptoms, including one child who experienced erythema nodosum (a dermatological disorder characterized by the formation of tender, red nodules, usually located on the front of the legs). Cortazar et al. (1985) have likewise noted the association of erythema nodosum with *Salmonella* gastroenteritis.

Salmonella has been implicated as a triggering organism for reactive arthritis (ReA) and Reiter's syndrome, in otherwise healthy individuals. Reactive arthritis is characterized by the development of synovitis (joint swelling and tenderness) within a few weeks after the occurrence of gastroenteritic symptoms. Maki-Ikola and Granfors (1992) summarized the clinical, epidemiological and laboratory data on *Salmonella*-triggered ReA. A review of extra-articular manifestations reported in 55 journal publications showed that these included urethritis, conjunctivitis, entesopathy, myalgia, weight loss exceeding 5 kg, dactylitis, erythema nodosum, oral ulcers, myocarditis, acute anterior uveitis, iritis, cholecystitis, keratitis, pharyngitis and pneumonia. Reiter's syndrome is defined as the occurrence of arthritis with one or more extra-articular symptoms typical of the disease, such as conjunctivitis, iritis, urethritis and balanitis. The prognosis for ReA is usually favourable, with symptoms lasting for <1 year in most persons, although 5–18% may have symptoms that last more than 1 year and 15–48% may experience multiple episodes of arthritis.

Generally, 1–2% of a population infected by triggering organisms will develop ReA or Reiter's syndrome (Keat, 1983; Smith, Palumbo and Walls, 1993). Maki-Ikola and Granfors (1992) reviewed several published outbreaks, totalling 5525 patients with salmonellosis, and estimated an incidence of reactive arthritis of 1.2–7.3% (mean: 3.5%).

Several researchers (Aho, Leirisalo-Repo and Repo, 1985; Archer, 1985; Calin, 1988) assert that HLA-B27-positive individuals are at higher risk for developing ReA, Reiter's syndrome and ankylosing spondylitis after an enteric infection with triggering organisms. It is estimated that approximately 20% of HLA-B27-positive individuals who become ill with salmonellosis develop these chronic sequelae. However, a lack of correlation between ReA and HLA-B27 was been observed after *S.* Typhimurium and *S.* Heidelberg/*S.* Hadar outbreaks in Canada (Inman et al., 1988; Thomson et al., 1992; Thompson et al., 1995).

Ike et al. (1986) reported an incidence of ReA on physical examination in 2.3% of patients with *Salmonella*-positive stools following the 1985 Chicago milk outbreak of *S.* Typhimurium gastroenteritis. Reiter's syndrome occurred approximately 10-fold less often than ReA. In a follow-up study of the Chicago patients, Ike, Arnold and Eisenberg (1987) found that 20 out of 29 reported persistent symptoms of ReA after one year, and symptoms had actually worsened in six cases.

In September 1984, a Canadian outbreak of *S.* Typhimurium PT 22 occurred in 473 out of 1608 police officers given a prepackaged box lunch (Inman et al., 1988). A cohort of 137 out of 196 individuals experiencing extra-enteric manifestations agreed to participate in a follow-up. Questionnaires were mailed out to their physicians and were returned for 116 (85%) volunteers, further describing the acute phase of the illness, with 19 reported by the physician to have experienced joint pain. Inman et al. (1988) noted a positive correlation between duration of gastrointestinal symptoms and duration of joint symptoms. In 13 patients, symptoms were restricted to ReA, while Reiter's syndrome was present in 6 patients (Inman et al., 1988).

An outbreak in Sweden in 1990 involved 113 medical scientists attending a radiology symposium, who were exposed to food contaminated with *S.* Enteritidis (Locht, Kihlstrom and Lindstrom, 1993), with 108 (96%) developing symptoms of salmonellosis and 17 (15%) of the 108 also developing ReA. Of the individuals developing ReA, 9 (53%) were men and 8 (47%) were women, with a mean age of 48.5 years (range: 34–60 years old; Locht, Kihlstrom and Lindstrom, 1993; Smith, 1994).

In another Canadian outbreak (Thomson et al., 1992), 79 women and 4 men in attendance at a luncheon were exposed to *S.* Heidelberg and *S.* Hadar from eating contaminated potato salad, and 73 subsequently developed salmonellosis. In addition to *S.* Thompson, *S.* Hadar and *S.* Heidelberg were isolated from the stools of 21 patients. Six of the 73 ill individuals developed ReA (Thomson et al., 1992; Smith, 1994). Ages of individuals who developed ReA were not significantly different from those cases that did not develop ReA (Thomson et al., 1992).

A 1994 outbreak in Finland caused by sprouted alfalfa seeds contaminated with *S.* Bovismorbificans was recently reported by Mattila et al. (1998). Questionnaires were sent to all 210 subjects with positive stool cultures. Median age in the 191 (91%) respondents was 32 years (range: 1–90), with 80% being older than 16 years of age; 130 (68%) were female. A total of 66 (35%) subjects reported articular symptoms, and 51 of the cases reporting articular symptoms were examined and 13 were contacted by telephone. A total of 12% (22 out of 191) fulfilled the criteria for ReA: 19 adults and 3 children. The incidence of

ReA was not significantly different between children (8%) and adults (12%) (Mattila et al., 1998).

Kanakoudi-Tsakalidou et al. (1998) followed 9 cases of juvenile ReA prospectively, concluding that the disease in children is generally mild, transient and self-limiting. Five out of 9 patients carried the HLA-B27 antigen and experienced a prolonged course for arthritis (mean duration 9.5 months).

The duration of ReA illness was evaluated in several studies:

- Radiology symposium in Sweden (Locht, Kihlstrom and Lindstrom, 1993). A 6-month follow-up assessment on 13 of the 17 individuals who developed ReA showed 5 patients having complete resolution of symptoms, but arthritis persisting in 8 patients 6 months after the outbreak.

- Canadian outbreak among policemen (Inman et al., 1988). A 12-month follow-up assessment was conducted on 15 patients out of 19 patients experiencing arthritis. Symptoms resolved in 8 out of 15 patients within 12 months, while symptoms persisted in 7 patients 12 months after the outbreak.

- Canadian outbreak at a women's luncheon (Thomson et al., 1992). Duration of illness in the 6 individuals who developed ReA ranged from 4 to 24 weeks in 4 individuals to greater than 6 months in the other two.

- Sprouted seed outbreak in southern Finland (Mattila et al., 1998). The median onset of joint symptoms was 8.5 days (range: 3–30) after the first symptoms of diarrhoea. Joint symptoms lasted less than 2 months in 11 (50%) subjects, 2–4 months in 7 (32%) and more than 4 months in 4 (18%) individuals.

- Prospective study of nine children (Kanakoudi-Tsdkalidou et al., 1998). Juvenile ReA has been reported to have a milder course, with duration varying from 1 to 12 months. In contrast to adult ReA, it seldom recurs or becomes chronic.

2.4 HOST-ADAPTED *SALMONELLA*

Most, if not all *Salmonella*, are capable of causing systemic disease and can be isolated from extra-intestinal sites. For the majority of serovars, this manifestation of disease occurs infrequently and mainly in patients who are immunocompromised, in infants or the elderly. However, a small number of serovars are known to be primarily or exclusively limited in host range (host-adapted; Selander et al., 1990) and primarily cause more severe forms of disease, including in immunocompetent patients. The most important human-adapted serovar is *S.* Typhi, the agent of typhoid fever; others include *S.* Paratyphi A, *S.* Paratyphi C, and *S.* Sendai, which present a typhoid-like enteric fever (Selander et al., 1990). The incubation period for these diseases is 7–28 days after exposure, with an average of 14 days. Symptoms include: high fever, malaise, nausea, abdominal pain, anorexia, delirium, constipation in early stages, and, in later stages, approximately one-third of patients develop diarrhoea (Bell, 2002). Convalescence may take up to 8 weeks.

Genetically, these differ from the majority of *Salmonella* serovars that typically cause gastroenteritis, and have distinctly different virulence attributes (Bäumler, Tsolis and

Heffron, 2000). In the United States of America, 70% of the estimated 824 cases of typhoid fever per year have been associated with foreign travel (Mead et al., 1999). The principal source of the human-adapted serovars is human faecal contamination of water or prepared foods. Other host adapted strains of human importance include *S.* Dublin (cattle-adapted), and *S.* Cholerasuis (pig-adapted) both of which are markedly more frequently isolated from blood or other extragastrointestinal sites in humans than other typically foodborne serovars (McDonough et al., 1999; Olsen et al., 2001; Sockett, 1993). In some parts of the world, humans are a secondary host for *S.* Cholerasuis, producing severe enteric fever and high mortality (Selander et al., 1990).

2.5 DEFINING THE SCOPE OF THE RISK ASSESSMENTS

The disease outcome of concern in the risk management request for risk assessment was acute gastroenteritis associated with *Salmonella enterica* in poultry. Human-host-adapted, predominantly invasive *Salmonella* serotypes (e.g. *S.* Typhi, *S.* Paratyphi) were not considered in developing the dose-response model. Disease outcomes that may occur beyond the diagnosis of gastroenteritis were not included in the risk estimations. Similarly, severity of disease outcomes attributable to antibiotic resistant strains of *Salmonella* were not estimated, nor for the more highly invasive *Salmonella* serotypes that are not commonly associated with poultry, i.e. *S.* Dublin and *S.* Cholerasuis.

2.6 REFERENCES CITED IN CHAPTER 2

Acha, P.N., & Szyfres, B. 2001. Salmonellosis. pp. 223–246, *in: Zoonoses and communicable diseases common to man and animals.* Pan American Health Organization, Washington, DC. Scientific and Technical Publication No. 580.

Aho, K., Leirisalo-Repo, M., & Repo, H. 1985. Reactive arthritis. *Clinical Rheum. Dis.*, **11**: 25–40.

Archer, D.L. 1985. Enteric microorganisms in rheumatoid diseases: causative agents and possible mechanisms. *Journal of Food Protection*, **48**: 538–545.

Archer, D.L., & Young, F.E. 1988. Contemporary Issues: Diseases with a food vector. *Clinical Microbiology Reviews*, **1**: 377–398.

Bajk, G.J., & Roh, W.S. 1998. Estimates of cases and social economic costs of foodborne salmonellosis in Korea. *Journal of Food Hygiene and Safety*, **13**: 299–304.

Bäumler, A.J., Tsolis, R.W, & Heffron, F. 2000. Virulence mechanisms of *Salmonella* and their genetic basis.

Bean, N.H., Goulding, J.S., Daniels, M.T., & Angulo, F.J. 1997. Surveillance for foodborne disease outbreaks – United States, 1988–1992. *Journal of Food Protection*, **60**: 1265–1286.

Bell, C. 2002. Salmonella. pp. 307–335, *in:* C. de W. Blackburn and P.J. McClure (eds). *Foodborne pathogens: Hazards, risk analysis and control.* Boca Raton, FL: Woodhead Publishing and CRC Press.

Bryan, F.L., & Doyle, M.P. 1995. Health risks and consequences of *Salmonella* and *Campylobacter jejuni* in raw poultry. *Journal of Food Protection*, **58**: 326–344.

Buchwald, D.S., Blaser, M.J. 1984. A review of human salmonellosis: II. Duration of excretion following infection with nontyphi Salmonella. *Reviews of Infectious Diseases*, **6**: 345–356.

Buzby, J.C., Roberts, T., Lin, J.C.-T., & MacDonald, J.M. 1996. Bacterial Foodborne Disease, Medical Costs and Productivity Losses. *USDA-ERS Report,* No. 741.

Calin, A. 1988. Ankylosing spondylitis. *Clinics in Rheumatic Diseases*, **11**: 41–60.

Cortazar, A.C., Rodriguez, A.P., Baranda, M.M., & Errasti, C.A. 1985. Salmonella gastroenteritis and erythema nodosum. *Canadian Medical Association Journal*, **133**: 120.

Cruickshank, J.G., & Humphrey, T.J. 1987. The carrier foodhandler and non-typhoid salmonellosis. *Epidemiology and Infection*, **98**: 223–230.

D'Aoust, J.Y. 1997. *Salmonella* Species. *In:* M.P. Doyle, L.R. Beuchat and T.J. Montville (eds). *Food microbiology: Fundamentals and frontiers*. Washington, DC: American Society for Microbiology Press.

Franco, B.D.G., Landgraf, M., Destro, M.T., & Gelli, D. (in press). Foodborne diseases in Southern South America. *In:* M.D. Miliotis and J. Bier (eds). *International handbook on foodborne pathogens*. New York, NY: Marcel Dekker.

Frenzen, P.D., Riggs, T.L., Buzby, J.C., Breuer, T., Roberts, T., Voetsch, D., Reddy, S., & FoodNet Working Group. 1999. Salmonella cost estimate updated using FoodNet data. *Food Review*, **22**: 10–15.

Hohmann, E.L. 2001. Nontyphoidal salmonellosis. *Clinical Infectious Diseases,* **32**: 263–269.

Humphrey, T. 2000. Public health aspects of *Salmonella* infection. pp. 245–262, *in:* C. Wray and A. Wray (eds). *Salmonella in Domestic Animals*. Wallingford, UK: CABI Publishing.

Ike, R., Arnold, W., Simon, C., Eisenberg, G., Batt, M., & White, G. 1986. Reactive arthritis syndrome (RAS) following an epidemic of Salmonella gastroenteritis (SG). *Clinical Research*, **34**: A618–A618.

Ike, R.W., Arnold, W.J., & Eisenberg, G.M. 1987. p. S245, in 1st Annual Meeting of the American Rheumatolgy Association, 1987. Cited in Archer and Young, 1988.

Inman, R.D., Johnston, M.E., Hodge, M., Falk, J., & Helewa, A. 1988. Postdysenteric reactive arthritis. A clinical and immunogenetic study following an outbreak of salmonellosis. *Arthritis and Rheumatism*, **31**: 1377–1383.

Kanakoudi-Tsakalidou, F., Pardalos, G., Pratsidou-Gertsi, P., Kansouzidou-Kanakoudi, A., & Tsangaropoulou-Stinga, H. 1998. Persistent or severe course of reactive arthritis following *Salmonella enteritidis* infection: A Prospective study of 9 cases. *Scandinavian Journal of Rheumatology*, **27**: 431–434.

Keat, A. 1983. Reiter's syndrome and reactive arthritis in perspective. *New England Journal of Medicine*, **309**: 1606–1615.

Lee, L.A., Puhr, N.D., Maloney, E.K., Bean, N.H., & Tauxe, R.V. 1994. Increase in antimicrobial-resistant Salmonella infections in the United States, 1989-1990. *Journal of Infectious Diseases*, **170**: 128–134.

Levy, M., Fletcher, M., Moody, M., and 27 others. (1996) Outbreaks of *Salmonella* serotype enteritidis infection associated with consumption of raw shell eggs – United States, 1994–1995. *Morbidity and Mortality Weekly Report*, **45**: 737–742.

Locht, H., Kihlstrom, E., & Lindstrom, F.D. 1993. Reactive arthritis after Salmonella among medical doctors: Study of an outbreak. *Journal of Rheumatology*, **20**: 845–848.

Maki-Ikola, O., & Granfors, K. 1992. Salmonella-triggered reactive arthritis. *Scandinavian Journal of Rheumatology*, **21**: 265–270.

Mattila, L., Leirisalo-Repo, M., Pelkonen, P., Koskimies, S., Granfors, K., & Siitonen, A. 1998. Reactive arthritis following an outbreak of *Salmonella bovismorbificans* infection. *Journal of Infection*, **36**: 289–295.

McDonough, P.L., Fogelman, D., Shin, S.J., Brunner, M.A., & Lein, D.H. 1999. *Salmonella enterica* serotype Dublin infection: an emerging infectious disease for the Northeastern United States. *Journal of Clinical Microbiology*, **37**: 2418–2427.

Mead, P.S., Slutsker, L., Dietz, V., McCraig, L.F., Bresee, J.S., Shapiro, C., Griffin, P.M., & Tauxe, R.V. 1999. Food-related illness and death in the United States. *Emerging Infectious Diseases*, **5**: 607–625.

Olsen, S.J., Bishop, R., Brenner, F.W., Roels, T.H., Bean, N., Tauxe, R.V., & Slutsker, L. 2001. The changing epidemiology of *Salmonella*: Trends in serotypes isolated from humans in the United States, 1987–1997. *The Journal of Infectious Diseases,* **183**: 753–761.

Poppe, C. 2000. *Salmonella* infections in the domestic fowl. pp. 107–132, *in:* C. Wray and A. Wray (eds). *Salmonella in Domestic Animals*. New York, NY: CAB International.

Rodrigue, D.C., Tauxe, R.V., & Rowe, B. 1990. International increase in *Salmonella enteritidis*: a new pandemic? *Epidemiology and Infection*, **105**: 21–27.

Selander, R.K., Beltran, P., Smith, N.H., Helmuth, R., Rubin, F.A., Kopecko, D.J., Ferris, K., Tall, B.D., Cravioto, A., & Musser, J.M. 1990. Evolutionary genetic relationships of clones of *Salmonella* serovars that cause typhoid and other enteric fevers. *Infection and Immunity,* **58**: 2262–2275.

Smith, J.L. 1994. Arthritis and foodborne bacteria. *Journal of Food Protection*, **57**: 935–941.

Smith, J.L., Palumbo, S.A., & Walls, I. 1993. Relationship between foodborne bacterial pathogens and the reactive arthritides. *Journal of Food Safety*, **13**: 209–236.

Sockett, P.N. 1993. The economic and social impact of human salmonellosis in England and Wales: A study of the costs and epidemiology of illness and the benefits of prevention. University of London. [Thesis]

Thomson, G.T.D., Chiu, B., Derubeis, D., Falk, J., & Inman, R.D. 1992. Immunoepidemiology of post-Salmonella reactive arthritis in a cohort of women. *Clinical Immunology and Immunopathology*, **64**: 227–232.

Thompson, G.T.D., DeRubeis, D.A., Hodge, M.A., Rajanaygam, C., & Inman, R.D. 1995. Post-salmonella reactive arthritis: late clinical sequelae in a point source cohort. *American Journal of Medicine*, **98**: 13–21.

Thorns, C.J. 2000. Bacterial food-borne zoonoses. *Revenue scientifique et technique Office international des epizooties*, **19**(1): 226–239.

WHO. 2001. World Health Organization Surveillance Programme for Control of Foodborne Infections and Intoxications in Europe. Seventh Report, 1993-1998. Edited by K. Schmidt and C. Tirado. Published by the Federal Institute for Health Protection of Consumers and Veterinary Medicine (BgVV), Berlin, Germany. (FAO/WHO Collaborating Centre for Research and Training in Food Hygiene and Zoonoses) (see pages 415, 422–423.

3. HAZARD CHARACTERIZATION OF *SALMONELLA*

3.1 SUMMARY

This section reviews the basic characteristics of the organism, human host factors, and composition factors of the food matrix that influence the outcome of exposure to non-typhoidal *Salmonella enterica*. Human volunteer feeding trial data for various *Salmonella* serotypes and dose-response models that have been developed based on those studies are reviewed. Limitations in the results from human feeding trials are discussed. Additional data were collected from salmonellosis outbreak reports that provided detailed information on parameters such as the numbers of the pathogen in the contaminated food, approximate amount of food eaten, numbers of people who consumed the food, numbers of people exposed who developed the clinical symptoms of acute gastroenteritis, age information, and, in some cases, prior health information.

The existing dose-response models were compared with the outbreak data as a validation step. These models failed to adequately represent the observed outbreak data. Consequently, a new dose-response model was developed, based on the outbreak data, and was used with exposure assessment information for eggs and broiler chickens to derive the risk estimates. In addition, an analysis of the outbreak data was done to attempt to derive quantitative estimates for the effect of host age and *Salmonella* serotype on the probability of acute gastroenteritis. No differentiation could be made on the basis of the dose-response outbreak data available at this time. The dose-response relationship derived from the outbreak data measured the host response in terms of acute gastroenteritis. Follow-up patient information on progression of the primary illness to more severe consequences was not detailed in the outbreak reports; in addition, the severity of illness – i.e. severity characterized by hospitalization, bacteraemia, reactive arthritis, other symptoms or death – is often complicated by factors that are difficult to quantify, and hence the corresponding risk estimates were not calculated.

3.2 ORGANISM, HOST AND MATRIX CHARACTERISTICS

3.2.1 Characteristics of the organism

In order for infection with a non-typhoid *Salmonella* to occur, the organism must survive a rather hostile environment. It must adapt to differences in growth conditions between the outside environment and the host, and within highly variable microenvironments within the host. The invasive journey towards illness in the host must negotiate distinct temperature differences, osmolarity, oxidation-reduction potentials, environmental iron concentrations, pH and organic and inorganic nutrient environments (Slauch, Taylor and Maloy, 1997). An infective *Salmonella* must then survive peristalsis, the epithelial surface and the host immune response.

Non-typhoid salmonellae possessing certain adaptive characteristics are more likely to produce foodborne disease. First, they must be acid tolerant to survive the pH of the stomach. They must also be able to attach themselves to and invade the intestinal epithelia and Peyer's patches (D'Aoust, 1997). Bacterial virulence factors include those that promote adhesion to host cells in the intestines: specific fimbriae, chromosome-coded bacterial surface adhesins, haemagglutinins, and epithelial cell induction of bacterial polypeptides that can promote colonization and adhesion.

Resistance of *Salmonella* to lytic action of complement varies with the length of the O side chains of lipopolysaccharide (LPS) molecules (D'Aoust, 1991). Smooth varieties are more resistant than rough types. The O side chains of the lipopolysaccharide molecules have also been shown to affect invasiveness and enterotoxin production (Murray, 1986).

Siderophores, which chelate iron, are necessary for the accumulation of sufficient environmental iron to allow growth of *Salmonella*. Siderophores include hydroxamate, phenolate and catechol types. Porins are hydrophobic bacterial cell proteins that enhance the virulence of *Salmonella* by repression of macrophage and polymorphonuclear-dependent phagocytosis. *Salmonella* porins may, however, have a limited importance in pathogenicity. Chromosomal determinants include specific virulence genes whose potential for action is tightly controlled by regulatory genes. Gene expression is determined by the environment and invasion occurs by the two-component regulatory system PhoPQ, which enables survival of *Salmonella* within the hostile environment of phagocytes (Slauch, Taylor and Maloy, 1997).

Virulence plasmids in the range of 50–100 kilobases been associated with the ability to spread after colonization, invasion of the intestine, ability to grow in the spleen, and a general suppression of the host immune response (Slauch, Taylor and Maloy, 1997). The presence of virulence plasmids in *Salmonella* is limited. Chiu, Lin and Ou (1999) studied virulence plasmids in 436 clinical human samples in Taiwan: 287 isolates were from faeces, 122 from blood and the remaining were isolated from other sites. Of the non-faecal isolates, 66% contained a virulence plasmid, compared with 40% of the faecal isolates. All the isolates (n=50) of the three highly invasive serotypes – S. Enteritidis, *S.* Dublin and *S.* Choleraesuis contained virulence plasmids. Virulence plasmids have also been confirmed in *S.* Typhimurium, *S.* Gallinarum-pullorum and *S.* Abortusovis, but are notably absent in *S.* Typhi, which is host-adapted and highly infectious.

Other factors that affect the ability of the organism to cause disease include the presence of cytotoxins and diarrhoeagenic enterotoxins. The enterotoxin is released into the lumen of the intestine and results in the loss of intestinal fluids (D'Aoust, 1991).

Antimicrobial resistance can have two effects on the outcome of exposure: there can be an accompanying change in the virulence of the organism, or there can be a poorer response to treatment because of the empirical choice of an antimicrobial to which the organism is resistant (Travers and Barza, 2002). An increase in virulence could result from linkage of resistance factors to other virulence genes, such as those for adherence, invasion and toxin production. A study by the United States Centers for Disease Control and Prevention (CDCs) (Lee et al., 1994) revealed that subjects with infections caused by antimicrobial-resistant *Salmonella* were significantly more likely to be hospitalized than those with antimicrobial-susceptible infections (35% vs 27%, $P = 0.006$) and this difference persisted even after correction for underlying illness. Patients infected with resistant strains also tended to be ill longer (median: 10 vs 8 days) and hospitalized longer (median: 5 vs 4 days). Most subjects were treated with an agent to which the organism was susceptible, and therefore the difference in hospitalization rates probably reflected increased virulence of the infecting organism rather than inappropriate choice of treatment. Thus, the data suggest that antimicrobial-resistant strains are somewhat more virulent than susceptible strains, in that they cause more prolonged or more severe illness than do antimicrobial-susceptible strains (Travers and Barza, 2002).

Two potentially confounding factors in the study were the host susceptibility in terms of age, and potential differences in virulence between serotypes. Neither factor was controlled for in the study (Travers and Barza, 2002). Black race and less than one year of age appeared to be host characteristics associated with a resistant infection, although differences in the distribution of infecting serovars among ethnic and age groups contributed to the occurrence of such effects. Varying food preferences or methods of food preparation might have been at the basis of different serovar distribution. The same consideration may explain the results of an earlier study, which associated infection with *S.* Heidelberg, penicillin intake, Hispanic origin, more than 60 years of age and antacid use to infection with a multi-resistant *Salmonella* (Riley et al., 1984). The conclusion of this study – that multi-resistant organisms are more dependent on host characteristics than sensitive organisms to cause disease – should be qualified accordingly.

3.2.2 Host characteristics

Literature tends to be biased towards reporting statistically significant and positive results. This review can only reflect such a bias, and the focus is evidently on host factors for which a statistically significant association to salmonella gastroenteritis and related complications has been reported. Where clear indication of a non-significant finding is made in the original study, such a finding is also reported. In addition, since not all studies considered the same factors, the significance of one factor in a given study may merely depend on the presence or absence of other ones. For instance, while a Swiss study considered travel abroad an important source of resistant *Salmonella* (Schmid et al., 1996), such an association was not seen in a United States of America study (Lee et al., 1994). Such apparent inconsistencies may have various explanations, but their discussion is beyond the scope of this review.

Host factors that can affect the outcome of exposure to the pathogen by ingestion, and which are considered in this review, are the following:

Demographic and socio-economic factors	Age
	Gender
	Race and ethnicity
	Nutritional status
	Social, economic and environmental factors
	Foreign travel
Genetic factors	HLA-B27 gene
Health factors	Immune status
	Previous exposure
	Concurrent infections
	Underlying diseases
	Concurrent medications
	Pregnancy

Demographic and socioeconomic factors

The following factors are considered in this section: age; gender; race and ethnicity; nutritional status; socioeconomic and environmental factors; and travel abroad.

Age

A common observation is that the age of patients with *Salmonella* infections is distributed according to a bimodal distribution with peaks in children and elderly. In a Belgian hospital-based study covering isolates for a 20-year period (1973–1992), *S.* Typhimurium and *S.* Enteritidis were mainly isolated in children under 5 years of age (Le Bacq, Louwagie and Verhagen, 1994). The age distribution was, however, less accentuated for *S.* Enteritidis than for *S.* Typhimurium. Both serovars were more likely to lead to bacteraemia in middle and older age groups than in those younger than 5 years of age (Le Bacq, Louwagie and Verhagen, 1994), confirming a previous observation made in the United States of America (Blaser and Feldman, 1981). Another study reports on *Salmonella* isolates from a Hong Kong hospital for the period 1982–1993 (Wong et al., 1994). Among both intestinal and extra-intestinal isolates, *S.* Typhimurium, *S.* Derby and *S.* Saintpaul predominated in infants. In patients older than 1 year of age, *S.* Derby and *S.* Typhimurium were the most common intestinal isolates, while *S.* Typhi, *S.* Typhimurium and *S.* Enteritidis were the most common extra-intestinal isolates. In a British population-based study, highest age-specific isolation rates for *S.* Enteritidis were observed in children aged under 2 years, and *S.* Typhimurium in those under 1 year (Banatvala et al., 1999).

In children in their first year, the peak incidence is generally observed in the second and third months (Ryder et al., 1976; Davis, 1981). The study from Hong Kong showed, however, a peak at 12 months of age (Wong et al., 1994). In a study of Peruvian children, the IgG and IgM titres against *Salmonella* serogroups AO, BO and DO were higher at 12 months of age than at 2 or 3 months of age, which was interpreted as an indication of acquired immunity (Nguyen et al., 1998). In the United States of America, infants under the age of 1 year have the highest reported incidence rate of salmonellosis, with the highest rate in infants 2 months of age, and an abrupt decrease after infancy (Olsen et al., 2001). Most cases are relatively mild. However, as with the immunocompromised and the elderly, children also face a relatively higher rate of severe outcomes, including death, than other demographic categories. Olsen et al. (2001) note a 4–13-fold higher rate of invasive disease in young children than other age groups. Buzby (2001) noted that most children who contract salmonellosis are believed to have been infected from contaminated food, as outbreaks in childcare facilities are rare. However, a matched case-control study among children in France found that cases were more likely to report a case of diarrhoea in the household 3–10 days before onset of illness, particularly in the age group less than 1 year old, indicating a role of person-to-person transmission of salmonellosis in infants (Delarocque-Astagneau et al., 1998).

It is noted that age associations may be influenced by other factors. In the very young, this includes increased susceptibility upon first exposure, but also that medical care is quickly sought for infants and incidents reported, and they are also more likely to be tested than adults with foodborne illness. Similarly, the very elderly with diarrhoea may also be expected to be more frequently cultured than other age groups (Banatvala et al., 1999). As mentioned earlier, differences in the distribution of infecting serovars among age groups was considered the reason for an apparent increased risk of resistant *Salmonella* infection in infants (Lee et al., 1994). When exposed to the same contaminated food in an outbreak, with the assumption that the individuals involved were exposed to a similar dose, no significant age-related difference was observed between those who became ill and those who remained healthy (range: 1–61 years old; median, 30; 12 children under the age of 15 years, 4 of whom became ill) in an outbreak investigated by Rejnmark et al. (1997). Similarly, no age-related association with hospitalization was noted in that investigation. Cowden and Noah (1989) postulated that the popularity of eggs and egg dishes in the diets of weaned and older children poses a serious problem. This suggests an increased rate of exposure to *S.* Enteritidis. Moreover, age association may reflect behavioural characteristics. For instance, eating snow, sand, or soil – a behaviour more likely in children – was found to be associated with infection by *S.* Typhimurium O:4-12 (Kapperud, Stenwig and Lassen, 1998). Handling pets, including reptiles, and farm animals, followed by hand-to-contact without washing increases exposure opportunities.

Gender

In terms of number of isolates, several studies indicate that men seem to be generally more affected than are women. A male-to-female ratio of 1.1 has been reported on various occasions (Blaser and Feldman, 1981; Le Bacq, Louwagie and Verhagen, 1994; Wong et al., 1994). However, in other studies, the isolation rate for women exceeded that for men between the ages of 20 and 74 years, although boys 15 years or under had a slightly higher age-specific isolation rate than girls (Olsen et al., 2001). The significance of such a findings does not appear to have been addressed. Several factors may play an important role, such as proportion of the two genders, as well as different age distributions for males and females within a country or hospital catchment area. In the evaluation of a single study, it should be pointed out that the occurrence of other factors, e.g. pregnancy or use of antacids, relates to one gender more often or exclusively, and gender may thus have the effect of a confounder. Furthermore, differences in food handling practices and hygiene during food preparation, and amount of food consumed, may also be contributors to any apparent gender differences.

Race and ethnicity

The potential role of race and ethnicity has seldom been considered. As mentioned above, an association with black race and Hispanic origin was reported for resistant *Salmonella* infections (Lee et al., 1994; Riley et al., 1984). In the former case, the association was explained by differences in the distribution of infecting serovars among ethnic groups, which in turn depended on varying food preferences or methods of food preparation.

Nutritional status

An association between altered nutritional status and acute gastroenteritis has been shown in AIDS patients (Tacconelli et al., 1998). Apart from this report, no direct reference to the role of nutritional status was found in the literature.

Social, economic and environmental factors

Isolation rates of several *Salmonella* serovars among groups of different socioeconomic extraction have been compared on the basis of the Townsend score, an index for deprivation

(Banatvala et al., 1999). While isolation rates for *S.* Typhimurium were not related to the Townsend score, the highest isolation rates of *S.* Enteritidis were observed in more prosperous areas. It was advanced that populations living in such areas more frequently ingested vehicles harbouring *S.* Enteritidis.

Sanitation deficiencies have been associated with high rates of enteric disease but direct reference to the potential role of *Salmonella* is scarce. In the 1950s, lack of sanitation, poor housing, limited water supply and poor personal hygiene were associated with high *Shigella* rates in Guatemala (Beck, Muñoz and Scrimshaw, 1957). A similar observation was made in the United States of America where, in areas of inadequate sanitary facilities, poor housing and low income, *Shigella* infections were the major causes of diarrhoeal disease. In particular, there were nearly twice as many cases of diarrhoea among persons living in dwellings having outhouses than among those whose houses had indoor lavatories (Schliessmann et al., 1958). In certain Guatemalan villages, the habits of the people and the density of the population were found to be more important determinants than type of housing (Bruch et al., 1963). In a study conducted in Panama, six representative types of dwellings were considered as an index of social and economic influences on the prevalence of enteric pathogens among infants with diarrhoeal disease (Kourany and Vasquez, 1969). Each dwelling type differed characteristically from one another but five of the six types were considered substandard and their occupants were of low socioeconomic status. Infection rates for enteropathogenic *Escherichia coli*, *Shigella* and *Salmonella* among infants from the various groups of substandard dwellings ranged from 6.0 to 10.2%, in contrast to the zero infection rate observed in infants from the better-type housing. It is worth noting that the literature on sanitation and housing was mainly published in the 1950s and 1960s. It is possible that safety improvement in the water supply consequent to economic development has sensibly diminished the importance of those factors in several countries.

A French study on sporadic *S.* Enteritidis infections in children investigated the influence of diarrhoea in another household member in the 3 to 10 days before a child shows clinical symptoms. The strength of the association with such a factor appeared stronger for cases in infants (1 year of age or less) compared with cases in children between 1 and 5 years of age (Delarocque-Astagneau et al., 1998). On the basis of this observation, as well as other results of the study, it was postulated that *S.* Enteritidis infection in children of less than 1 year of age may arise from person-to-person contact, while children between 1 and 5 years of age contract the infection by consuming raw or undercooked egg products or chicken.

A seasonal pattern in isolations, which generally shows increased rates during hotter months, has been documented. For instance, increased isolation rates for *S.* Enteritidis, *S.* Typhimurium, *S.* Virchow and *S.* Newport were observed in summer in a British study (Banatvala *et al.*, 1999). The French study mentioned in the previous paragraph noted that the association between *S.* Enteritidis infection and prolonged storage of eggs was stronger during the summer period.

Travel abroad

Travel abroad is a risk factor for *Salmonella* gastroenteritis that has been consistently demonstrated in both North America and Europe. For California residents, Kass et al. (1992) demonstrated an association between sporadic salmonellosis and travel outside the United

States of America within 3 weeks prior to the onset of illness. Possible variations related to serovar in sporadic salmonellosis were indicated by a study concerning residents of Switzerland (Schmid et al., 1996). Having been abroad within three days prior to clinical onset of the illness was found to be associated with both *S.* Enteritidis and serovars other than Enteritidis, although to a greater extent for the latter case. Little difference was seen between the results of all *S.* Enteritidis phage types (PTs) and of *S.* Enteritidis PT4. While most patients with *S.* Enteritidis were more likely to have travelled within Europe, the majority of non-Enteritidis infections might have been imported from outside Europe. Individuals of a British region with *Salmonella* infection were more likely to have reported travel abroad in the week before the onset of illness (Banatvala et al., 1999). Frequency of overseas travel between patients with *S.* Enteritidis or *S.* Typhimurium was not different, but it was among patients with other serovars. Indication of how travel abroad may lead to salmonellosis can be found in a study referring to residents of Norway (Kapperud, Lassen and Hasseltvedt, 1998). This study suggested that about 90% of the cases from whom a travel history was available had acquired their infection abroad, but failed to show an association to either foreign travel among household members or consumption of poultry. However, consumption of poultry purchased abroad during holiday visits to neighbouring countries was the only risk factor considered by the study that remained independently associated with the disease. Only cases of *S.* Typhimurium allowed for a separate analysis that showed an association with both poultry purchased abroad and foreign travel among household members.

Genetic factors

As far as acute gastroenteritis caused by *Salmonella* is concerned, no host genetic factors have been reported. Reports concerning race and ethnicity should be considered in the light of eating habits.

The putative association of the gene Human Leukocyte Antigen B27 (HLA-B27) for patients with spondyloarthropathies, in particular reactive arthritis and Reiter's syndrome, has been described. The HLA-B27 gene has a very high prevalence among the native peoples of the circumpolar arctic and sub-arctic regions of Eurasia and North America, and in some regions of Melanesia. In contrast, it is virtually absent among the genetically unmixed native populations of South America, Australia, and among equatorial and southern African Bantus and Sans (Bushmen) (Khan, 1996). Fifty percent of Haida Indians living on the Queen Charlotte Islands of the Canadian province of British Columbia have the HLA-B27 gene, which is the highest prevalence ever observed in a population. The prevalence among Americans of African descent varies between 2 to 3%, while 8% of the Americans of European descent posses the gene (Khan, 1995).

Health factors

Immune status

The host immune status is, as in any other infectious disease, a very important factor in determining both infection and clinical illness. In general terms, its importance does not seem to have been the direct goal of any formal work and has thus to be indirectly assessed though other factors, such as age or acquired immunodeficiency. Evidence for the development of immunity against non-typhoidal *S. enterica* was recognized in human volunteer experiments (McCullough and Eisele, 1951b). When subjects who became ill on the first challenge were later re-challenged, if they became ill again the severity of the illness was usually less than that of the initial illness, despite higher challenge doses being used. This is in contrast to experiments with typhoid, where vaccines gave protection against low- but not high-challenge doses, and once clinical disease occurred, the severity was not altered by previous vaccination. Evidence that immunity is partially serotype specific is suggested by the increased incidence of salmonellosis amongst people who have travelled, and are presumably exposed to different serotypes and strains of *Salmonella* in food and water in other countries. There is a need to examine country- or region-specific population immunities in general to better understand the applicability of dose-response models to populations, countries and regions other than those where dose-response data were acquired.

Concurrent infections

Persons infected with Human Immunodeficiency Virus (HIV) tend to have recurrent enteric bacterial infections. Such infections are often virulent and associated with extraintestinal disease (Smith et al., 1988; Angulo and Swerdlow, 1995). Six risk factors for enteric salmonellosis have been identified in HIV-infected patients: increasing value on the prognostic scoring system APACHE II (Acute Physiology and Chronic Health Evaluation); altered nutritional status; previous antibiotic therapy; ingestion of undercooked poultry or eggs, or of contaminated cooked food; previous opportunistic infections; and stage C of HIV infection (Tacconelli et al., 1998).

Underlying diseases

The significance of Acquired Immunodeficiency Syndrome (AIDS) has been discussed in the previous paragraph. The risk represented by other underlying conditions was evaluated in a large nosocomial foodborne outbreak of *S.* Enteritidis that occurred in 1987 in New York (Telzak et al., 1991). Gastrointestinal and cardiovascular diseases, cancer, diabetes mellitus and alcoholism as well as use of antacids and antibiotics were the factors considered. However, diabetes was the only condition that was independently associated with infection after exposure to the contaminated meal. Although diabetic cases were more likely to develop symptomatic illness compared with non-diabetic, the difference was not statistically significant. Decreased gastric acidity and autonomic neuropathy of the small bowel (which leads to reduced intestinal motility and prolonged gastrointestinal transit time) are the two biologically plausible mechanisms for the increased risk of *S.* Enteritidis infection among diabetics. Among patients with sporadic salmonellosis in Northern California, diabetes mellitus and cardiac disease were both associated with clinical illness (Kass et al., 1992). This study contemplated 14 health conditions. Non-gastrointestinal medical conditions and, to a larger extent, a recent history of gastrointestinal disorder were associated with sporadic *S.* Typhimurium O:4-12 infection in Norway (Kapperud, Stenwig and Lassen, 1998). It was, however, noted that physicians are more likely to require a stool culture from patients with preceding illness. In a British epidemiological study, cases of *Salmonella* infection were more likely to report a long-term illness (including gastroduodenal conditions) than controls (Banatvala et al., 1999). All individuals with diabetes mellitus, malignancy or immunodeficiency were cases.

Concurrent medications

A number of investigations have examined the effects of antacids and prior or concurrent antimicrobial usage as factors influencing likelihood of contracting salmonellosis or affecting the severity of the outcome. The evidence found in the literature concerning their association with human salmonellosis is contrasting. While some studies have shown an association with antacid use (Banatvala et al., 1999), others have failed to do so (Telzak et al., 1991; Kapperud, Stenwig and Lassen, 1998). A similar situation is found for the use of antibiotics in the weeks or days preceding the infection or disease onset: some studies have demonstrated an association (Pavia et al., 1990; Kass et al., 1992; Bellido Blasco et al., 1998) but others have not (Telzak et al., 1991; Kapperud, Stenwig and Lassen, 1998; Banatvala et al., 1999). Having a resistant *Salmonella* infection has been associated with previous antibiotic use (Lee et al., 1994). A delay between antimicrobial use and onset of symptoms suggests that the effect may be due to prolonged alteration of the colonic bacterial flora, resulting in decreased resistance to colonization (Pavia et al., 1990).

Among the 11 different medical therapies considered by a North California, USA, study on sporadic clinical salmonellosis, which included antacids and antibiotics, only hormonal replacement therapy (principally conjugated estrogen) in older women was found to be associated with clinical salmonellosis (Kass et al., 1992). An association between serovars other than *S.* Enteritidis and intake of medications other than antacids was shown in Switzerland (Schmid et al., 1996). Regular use of medications was a risk factor for *S.* Typhimurium O:4-12 infection in Norway (Kapperud, Stenwig and Lassen, 1998). In the same study, use of antacids and antibiotics were not risk factors.

Pregnancy

There is a little information concerning the effect of salmonellosis specifically on pregnant women and foetuses or neonates. No studies were found to indicate that pregnant women are at an increased risk for *Salmonella*-induced enteritis. However, when a pregnant woman suffers from foodborne infection the foetus or neonate may also be affected. A recent review by Smith (2002) of *Campylobacter jejuni* infection during pregnancy summarizes the small amount of available data on the consequences of maternal *C. jejuni* enteritis or bacteraemia, or both. Outcomes may include abortion, stillbirth, premature labour, bacteraemic newborn infants, and neonates with diarrhoea or bloody diarrhoea. Similar outcomes might be expected for some cases of salmonellosis in pregnant women.

3.2.3 Factors related to the conditions of ingestion

Empirical observation, mainly from outbreak investigations, shows that foodborne salmonellosis can be related to a variety of food items. Table 3.1 lists major foodborne outbreaks of human salmonellosis and shows the wide range of foods implicated in these outbreaks (D'Aoust, 1997).

Table 3.1. Major foodborne outbreaks of human salmonellosis and the food items implicated (Adapted from D'Aoust, 1997)

Year	Country(ies)	Vehicle	Serovar
1973	Canada; USA	Chocolate	S. Eastbourne
1973	Trinidad	Milk powder	S. Derby
1974	USA	Potato salad	S. Newport
1976	Spain	Egg salad	S. Typhimurium
1976	Australia	Raw milk	S. Typhimurium PT9
1977	Sweden	Mustard dressing	S. Enteritidis PT4
1981	The Netherlands	Salad base	S. Indiana
1981	Scotland (UK)	Raw milk	S. Typhimurium PT204
1984	Canada	Cheddar cheese	S. Typhimurium PT10
1984	Canada		S. Typhimurium PT22
1984	France; England	Liver pate	S. Goldcoast
1985	USA	Pasteurized milk	S. Typhimurium
1985	Scotland (UK)	Turkey	S. Thompson, S. Infantis
1987	Republic of China	Egg drink	S. Typhimurium
1987	Norway	Chocolate	S. Typhimurium
1988	Japan	Cuttlefish	S. Champaign
1988	Japan	Cooked eggs	Salmonella (unspecified).
1988	England (UK)	Mayonnaise	S. Typhimurium DT49
1990	Sweden		S. Enteritidis
1991	Germany	Fruit soup	S. Enteritidis
1993	France	Mayonnaise	S. Enteritidis
1993	Germany	Paprika chips	S. Saintpaul, S. Javiana, S. Rubislaw
1994	USA	Ice cream	S. Enteritidis
1994	Finland; Sweden	Alfalfa sprouts	S. Bovismorbificans
1998	USA	Breakfast cereal	S. Agona
1998	England (UK)	Chopped liver	S. Enteritidis PT4
1999	USA	Orange juice	S. Muenchen

Gastric acidity is recognized as an important defence against foodborne pathogens. Pathogen, host and food factors interact in determining whether sufficient bacteria are able to withstand stomach acidity and go on to colonize the gut. Such an interaction appears extremely dynamic. Although *Salmonella* prefer to grow in neutral pH environments, they have evolved complex, inducible acid survival strategies that allow them to face the dramatic pH fluctuations encountered in nature and during pathogenesis (Bearson, Bearson and Foster, 1997). While the human stomach is normally pH 2, several host factors may cause decreased gastric acidity. Examples reported in the previous section are older age, diabetes mellitus, and use of antacid drugs. As for factors specifically related to food, it appears that a systematic treatment of this topic has not yet been carried out. Circumstantial evidence suggests that the following elements are of particular relevance: amount of food ingested; nutrient composition, including fat content of the food; buffering capacity of the food at the time of the meal; and nature of contamination. The reference to "food" rather than to "food item" emphasizes the importance of considering the whole meal.

In an *S.* Typhimurium outbreak, it was observed that persons who had eaten two or more pieces of chicken tended to have shorter incubation periods. However, both attack rate and illness severity did not appear to be a function of the amount of chicken consumed. It was concluded that the amount of food consumed provides only a crude estimate of dose because a homogenous distribution of the pathogen among the chicken pieces is unlikely (Glynn and Palmer, 1992). This also means that since infectivity is not uniformly distributed within a food, a larger meal may increase the chances of ingesting an infected portion. D'Aoust (1985) noted that in foodborne outbreaks involving fatty vehicles, relatively low doses can lead to substantial numbers of illness (chocolate: <100 cells of *S.* Eastbourne, 50 cells of *S.* Napoli; cheddar cheese: 100–500 cells of *S.* Heildelberg, 1–6 cells of *S.* Typhimurium). Microorganisms trapped in hydrophobic lipid moieties may survive the acidic conditions of the stomach and thus the fat content of contaminated foods may play a significant role in human salmonellosis. In contrast, experimental evidence in rats shows that *Salmonella* infection is not affected by milk fat (Sprong, Hulstein and van der Meer, 1999). *Salmonella* were actually protected from acid killing when inoculated onto boiled egg white – a food source high in protein and low in fat (Waterman and Small, 1998). The same study shows that the pH of the microenvironment occupied by the bacteria on the surface of a food source is critical to their survival.

The effect of substrate was studied in volunteers challenged with *Vibrio cholerae* fed in a medium with buffering capacity (Cash et al., 1974). The group of subjects that overcame the effect of a bicarbonate vehicle in less than 30 minutes (approximately half of the challenged individuals) experienced a lower attack rate than the group experiencing a prolonged buffering effect. Ingestion of low numbers of *Salmonella* between meals, i.e. on an empty stomach, was associated with an increased attack rate (Mossel and Oei, 1975). It was postulated that at such moments the pyloric barrier would initially fail. The authors also speculated that some food items, such as chocolate and ice cream, are more likely to be ingested between meals and thus lead to illness even with only a few organisms. A protective effect of alcoholic beverages was observed in an *S.* Enteritidis outbreak (Bellido Blasco et al., 1996). Besides the direct effect of ethanol on bacteria, alcohol may stimulate secretion of gastric acid. Last, but not least, an important factor in determining the survival of bacteria in the stomach may be how uniformly a food is contaminated. Although a uniform distribution is usually assumed, the very nature of bacterial growth in colonies would suggest that agglomerations of bacteria occur within the food. It can be speculated that the outer layers of bacteria would protect the inner ones, allowing some pathogen to survive the gastric passage.

3.3 HUMAN FEEDING TRIALS

Nine studies have been published of experimentally induced salmonellosis, conducted between 1936 and 1970 using a variety of serotypes and strains. Serotypes and strains used in these series of feeding trials are listed in Table 3.2.

Table 3.2. Human feeding trials that have been performed using *Salmonella*

	Serotype(s)	Strain(s)	Reference
1	*S.* Typhimurium		Hormaeche, Peluffo and Aleppo, 1936
2	*S.* Anatum		Varela and Olarte, 1942
3	*S.* Meleagridis	I, II & III	McCullough and Eisele, 1951a
	S. Anatum	I, II & III	McCullough and Eisele, 1951a
4	*S.* Newport		McCullough and Eisele, 1951c
	S. Derby		McCullough and Eisele, 1951c
	S. Bareilly		McCullough and Eisele, 1951c
5	*S.* Pullorum	I, II, III & IV	McCullough and Eisele, 1951d
6	*S.* Typhi		Sprinz et al., 1966
7	*S.* Sofia		Mackenzie and Livingstone, 1968
	S. Bovismorbificans		
8	*S.* Typhi	Quailes, Zermatt, Ty2V, 0-901	Hornick et al., 1970
9	*S.* Typhi	Quailes	Woodward, 1980

Although the list of human feeding trials for *Salmonella* in humans is more extensive than may exist for other bacterial pathogens, some of these studies were deemed to be unsuitable and were not used in further analysis to derive conclusions about the pathogenicity of *Salmonella* in general in humans. The earliest study used 5 subjects, who were all fed a dose of approximately 9-logs in water and all exposed individuals were subsequently infected (Hormaeche, Peluffo and Aleppo, 1936). In a later study (Varela and Olarte, 1942), apparently only one volunteer was used, who became ill after ingesting a dose of 10-logs in water. The study conducted by MacKenzie and Livingstone (1968) involved a nasal inoculation of approximately 25 cells in one volunteer, who subsequently became ill. These three studies were not informative due to the use of only large doses with 100% attack rates, the testing of only one dose with one subject, or the method of inoculation. Studies conducted using *S.* Typhi (Sprinz et al., 1966; Hornick et al., 1970; Woodward, 1980), were considered to be inappropriate in the current analysis, primarily because of the difference between the illnesses caused by typhoid and non-typhoid *Salmonella*. *S.* Typhi is highly invasive and causes typhoid fever, a systemic bacteraemic illness, as opposed to non-typhoid salmonellosis, characterized by gastroenteritis and marked by diarrhoea, fever and abdominal pain, with rare systemic invasion.

The most extensive human feeding trials of non-typhoid *Salmonella* were conducted in the late 1940s to early 1950s (McCullough and Eisele, 1951a, b, c & d). Six different *Salmonella* serotypes were used, with up to 3 or 4 different strains of some of the serotypes. The subjects used in the feeding trials were healthy males from a penal institution. Feeding trials using *S.* Pullorum I, II, III & IV were considered to be inappropriate for deriving estimates about the infectivity of non-typhoid *Salmonella* for humans, because, as noted by other researchers (Blaser and Newman, 1982; Coleman and Marks, 1998) this is primarily a fowl-adapted strain. It was noted that a dramatically higher dose was required to produce illness using *S.* Pullorum and the clinical picture of illness, when it did occur, was characterized by an explosive onset and fast recovery

(McCullough and Eisele, 1951d). At dosages producing illness, the organism could only be isolated from the stools for the first day or two, and not thereafter. In addition, Fazil (1996) conducted an evaluation of the feeding trial data and found that the dose-response relationship for *S.* Pullorum was significantly different from the other strains used in the feeding trials.

In order to evaluate the data derived from the human feeding trials, the experimental design used by the researchers is briefly described (McCullough and Eisele, 1951a & c).

Human volunteers

- The subjects selected for the experimental feeding trials were healthy males from a penal institution.

- According to the authors, chronic complainers and those who had frequent gastrointestinal disturbances in the past were eliminated from the trials.

- After an initial selection of volunteers, at least three weekly stool cultures were done.

- Only those individuals with no *Salmonella* or other easily confused organisms in the stools were carried further in the experiment.

- An initial serum agglutination test was done against the organism to be administered.

- Subjects that showed a moderate or high agglutination titre against a particular organism were in general not used in the experiments with that species.

Source of Salmonella *Strains*

- Strains of *Salmonella* used in the feeding trials were obtained from market samples of high-moisture spray-dried whole egg powder.

Method of feeding

- Cultures for feeding trials were subcultured on trypticase soy agar.

- After 24 hours of incubation, the resulting growth was suspended in saline and standardized turbidimetrically.

- The dose was administered in a glass of eggnog shortly following the noon meal.

- A group of men, usually consisting of six, received the same experimental feeding dose.

- Control feedings were provided by eggnog alone or by prior feeding of the test organisms at what the authors observed to be non-infective levels.

Observations after feeding (Figure 3.1)

- Following the feeding, men were interviewed and observed three times a week for a period of two weeks, and once a week thereafter.

- Additional visits were made when required by the condition of the volunteer.

- Men were questioned with regard to symptoms.

- Temperatures were recorded.

- Faecal cultures were obtained.

- When indicated, blood counts and cultures were also done.

- Blood samples for agglutination were drawn at weekly intervals for 4 weeks following feeding.

- Faecal samples were collected and cultures were done on all men 3 times a week for the first 2 weeks, after that once a week until at least three consecutive negative samples had been obtained.

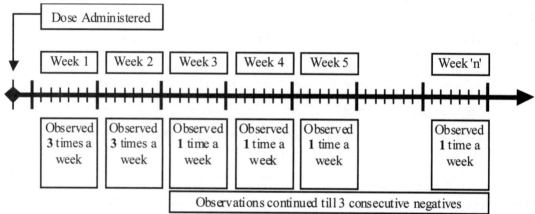

Figure 3.1. Scheme for observations during human feeding trial experiments of McCullough and Eisele (1951a, c).

Infection definition (faecal shedding)

- Infection was defined as the recovery of the administered strain from faecal samples.

Illness definition criteria

- Illness was characterized by the existence of the following two conditions:
  - ✓ documentation of symptoms; and
  - ✓ recovery of the organism from stool (infection);
- And one or more of the following:
  - ✓ diarrhoea or vomiting,
  - ✓ fever,
  - ✓ rise in specific agglutination titre, or
  - ✓ other, unspecified, signs.

The feeding trial data have been reviewed and critiqued by various researchers. Blaser and Newman (1982) reviewed the infective dose data for *Salmonella* and identified several deficiencies:

[1]   The feeding of the pathogen to the volunteers was conducted after their noon meal, when gastric acid was probably high.

[2]   It was observed that over half the volunteers who became ill had earlier been fed lower doses of the same serotype. These earlier feedings may have confounded the results by introducing a degree of immunity, thus making infection less likely, or, alternatively, the earlier feedings may have had a cumulative effect that made infection more likely.

[3]   A failure to assess the minimal infective dose.

[4]   The use of too few volunteers at low doses.

In the United States of America *Salmonella* Enteritidis Risk Assessment in eggs report (USDA-FSIS, 1998) (hereafter generally referred to as US SE RA), additional deficiencies in the feeding trial data were identified:

[1]   The use of healthy male volunteers could probably underestimate the true pathogenicity to the overall population.

[2]   The size of the groups used at each dose level was relatively small, with 18 of the 22 test doses using less than 6 people.

[3]   There were no low doses tested. The smallest dose that was tested was greater than 10^4 CFU *Salmonella* bacteria.

[4]   The lowest dose that caused an infection was also the lowest dose tested.

Additional points related to some of the critiques should also be noted. While it is true that the feeding of the dose after the noon meal when gastric acid was high could potentially reduce the estimated infectivity of the pathogen (Blaser and Newman, 1982), the dose was administered using eggnog, a high-fat-content medium. The eggnog could have conferred a level of protection against the effects of gastric acid, thus potentially negating the acid effects. It seems reasonable, however, to assume that, given the fact that the subjects used in the feeding trials were healthy males, the infectivity estimated for this population will be some factor less than for the general population, and more so for the more susceptible members of the general population. Overall, the criticisms of the feeding trial data are for the most part fair in their assessment of the potential biases in the results that may be expected.

The human feeding trial, as described earlier, measured both infection and illness. Most dose-response relationships are developed using infection (faecal shedding) as the dependent variable, primarily out of necessity due to the nature of the data. It should be noted that the use of the infection endpoint in deriving a dose-response relationship could introduce a level of conservatism into the dose-response relationship, depending on how the conditional dependence of illness – which is essentially the output of ultimate interest – following infection is treated. In the human feeding trial, it was also pointed out that approximately 40% of the volunteers that were shedding were reported to be last positive on or before the second day following administration, apparently clearing the infection two days post-administration (Coleman and Marks, 1998). These authors noted that there is some ambiguity in estimating infection based on faecal shedding for less than two days. The available data measuring illness as the endpoint is sparse, without any response being observed until a dose of approximately 6-logs. It has been noted (Blaser and Newman, 1982) that the strict criteria used by the researchers to define illness may have resulted in volunteers with mild complaints being classified as asymptomatic excretors rather than ill subjects. Although concerns have been raised as to the experimental design of the human feeding trials, it is appropriate to consider it at this juncture as still holding value in providing a basis upon which to at least start exploring the dose-response relationship.

Tables 3.3 to 3.7 present the original data from the McCullough and Eisele studies. These data are also summarized in Figure 3.2.

Table 3.3. Feeding trial data for *S.* Anatum I, II and III (McCullough and Eisele, 1951a)

Serotype	Dose	Log Dose	Positive (Inf)[1]	Total	Proportion
S. Anatum I	1.20E+04	4.08	2	5	0.40
S. Anatum I	2.40E+04	4.38	3	6	0.50
S. Anatum I	6.60E+04	4.82	4	6	0.67
S. Anatum I	9.30E+04	4.97	1	6	0.17
S. Anatum I	1.41E+05	5.15	3	6	0.50
S. Anatum I	2.56E+05	5.41	5	6	0.83
S. Anatum I	5.87E+05	5.77	4	6	0.67
S. Anatum I	8.60E+05	5.93	6	6	1.00
S. Anatum II	8.90E+04	4.95	5	6	0.83
S. Anatum II	4.48E+05	5.65	4	6	0.67
S. Anatum II	1.04E+06	6.02	6	6	1.00
S. Anatum II	3.90E+06	6.59	4	6	0.67
S. Anatum II	1.00E+07	7.00	6	6	1.00
S. Anatum II	2.39E+07	7.38	5	6	0.83
S. Anatum II	4.45E+07	7.65	6	6	1.00
S. Anatum II	6.73E+07	7.83	8	8	1.00
S. Anatum III	1.59E+05	5.20	2	6	0.33
S. Anatum III	1.26E+06	6.10	6	6	1.00
S. Anatum III	4.68E+06	6.67	6	6	1.00

NOTE: (1) Number found positive (infected).

Table 3.4. Feeding trial data for *S.* Meleagridis I, II and III (McCullough and Eisele, 1951a).

Serotype	Dose	Log Dose	Positive (Inf)[1]	Total	Proportion
S. Meleagridis I	1.20E+04	4.08	3	6	0.50
S. Meleagridis I	2.40E+04	4.38	4	6	0.67
S. Meleagridis I	5.20E+04	4.72	3	6	0.50
S. Meleagridis I	9.60E+04	4.98	3	6	0.50
S. Meleagridis I	1.55E+05	5.19	5	6	0.83
S. Meleagridis I	3.00E+05	5.48	6	6	1.00
S. Meleagridis I	7.20E+05	5.86	4	5	0.80
S. Meleagridis I	1.15E+06	6.06	6	6	1.00
S. Meleagridis I	5.50E+06	6.74	5	6	0.83
S. Meleagridis I	2.40E+07	7.38	5	5	1.00
S. Meleagridis I	5.00E+07	7.70	6	6	1.00
S. Meleagridis II	1.00E+06	6.00	6	6	1.00
S. Meleagridis II	5.50E+06	6.74	6	6	1.00
S. Meleagridis II	1.00E+07	7.00	5	6	0.83
S. Meleagridis II	2.00E+07	7.30	6	6	1.00
S. Meleagridis II	4.10E+07	7.61	6	6	1.00
S. Meleagridis III	1.58E+05	5.20	1	6	0.17
S. Meleagridis III	1.50E+06	6.18	5	6	0.83
S. Meleagridis III	7.68E+06	6.89	6	6	1.00
S. Meleagridis III	1.00E+07	7.00	5	6	0.83

NOTE: (1) Number found positive (infected).

Table 3.5. Feeding trial data for *S.* Newport (McCullough and Eisele, 1951c).

Serotype	Dose	Log Dose	Positive (Inf)[1]	Total	Proportion
S. Newport	1.52E+05	5.18	3	6	0.50
S. Newport	3.85E+05	5.59	6	8	0.75
S. Newport	1.35E+06	6.13	6	6	1.00

NOTE: (1) Number found positive (infected).

Table 3.6. Feeding trial data for *S.* Bareilly (McCullough and Eisele, 1951c).

Serotype	Dose	Log Dose	Positive (Inf)[1]	Total	Proportion
S. Bareilly	1.25E+05	5.10	5	6	0.83
S. Bareilly	6.95E+05	5.84	6	6	1.00
S. Bareilly	1.70E+06	6.23	5	6	0.83

NOTE: (1) Number found positive (infected).

Table 3.7. Feeding trial data for *S.* Derby (McCullough and Eisele, 1951c).

Serotype	Dose	Log Dose	Positive (Inf)[1]	Total	Proportion
S. Derby	1.39E+05	5.14	3	6	0.50
S. Derby	7.05E+05	5.85	4	6	0.67
S. Derby	1.66E+06	6.22	4	6	0.67
S. Derby	6.40E+06	6.81	3	6	0.50
S. Derby	1.50E+07	7.18	4	6	0.67

NOTE: (1) Number found positive (infected).

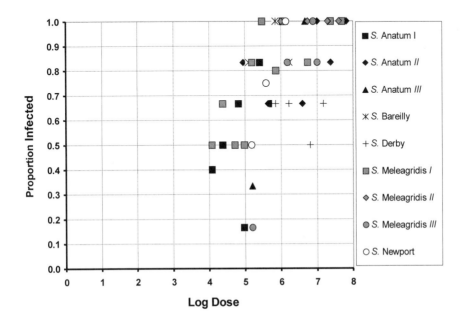

Figure 3.2. Summary of feeding trial data (McCullough and Eisele, 1951a; 1951c).

It has also been noted that, in the feeding trials, some of the volunteers were administered doses more than once. The earlier doses, which were lower and at which no response was observed, may have resulted in either a cumulative or an immunity effect. In order to attempt to remove this bias, the doses and subjects at which repeat feedings were conducted were edited out and the data re-evaluated. The edited data for naive subjects only are presented in Tables 3.8 to 3.12, and summarized in Figure 3.3.

Table 3.8. Feeding trial data for *S.* Anatum I, II and III for naive subjects.

Serotype	Dose	Log Dose	Positive (Inf)[1]	Total	Proportion
S. Anatum I	1.20E+04	4.08	2	5	0.40
S. Anatum I	6.60E+04	4.82	4	6	0.67
S. Anatum I	5.87E+05	5.77	4	6	0.67
S. Anatum I	8.60E+05	5.93	4	4	1.00
S. Anatum II	8.90E+04	4.95	3	4	0.75
S. Anatum II	4.48E+05	5.65	4	6	0.67
S. Anatum II	2.39E+07	7.38	3	3	1.00
S. Anatum II	4.45E+07	7.65	3	3	1.00
S. Anatum III	1.59E+05	5.20	1	3	0.33
S. Anatum III	1.26E+06	6.10	6	6	1.00
S. Anatum III	4.68E+06	6.67	3	3	1.00

NOTE: (1) Number found positive (infected).

Table 3.9. Feeding trial data for *S.* Meleagridis I, II and III for naive subjects.

Serotype	Dose	Log Dose	Positive (Inf)[1]	Total	Proportion
S. Meleagridis I	1.20E+04	4.08	3	6	0.50
S. Meleagridis I	2.40E+04	4.38	4	6	0.67
S. Meleagridis I	5.20E+04	4.72	3	6	0.50
S. Meleagridis I	1.15E+06	6.06	6	6	1.00
S. Meleagridis I	5.50E+06	6.74	5	6	0.83
S. Meleagridis I	2.40E+07	7.38	4	4	1.00
S. Meleagridis II	1.00E+06	6.00	6	6	1.00
S. Meleagridis II	5.50E+06	6.74	6	6	1.00
S. Meleagridis II	2.00E+07	7.30	3	3	1.00
S. Meleagridis III	1.58E+05	5.20	1	3	0.33
S. Meleagridis III	1.50E+06	6.18	5	6	0.83
S. Meleagridis III	7.68E+06	6.89	4	4	1.00

NOTE: (1) Number found positive (infected).

Table 3.10. Feeding trial data for *S.* Newport for naive subjects.

Serotype	Dose	Log Dose	Positive (Inf)[1]	Total	Proportion
S. Newport	1.52E+05	5.18	3	6	0.50
S. Newport	3.85E+05	5.59	4	4	1.00
S. Newport	1.35E+06	6.13	3	3	1.00

NOTE: (1) Number found positive (infected).

Table 3.11. Feeding trial data for *S.* Bareilly for naive subjects.

Serotype	Dose	Log Dose	Positive (Inf)[1]	Total	Proportion
S. Bareilly	1.25E+05	5.10	5	6	0.83
S. Bareilly	6.95E+05	5.84	3	3	1.00
S. Bareilly	1.70E+06	6.23	3	3	1.00

NOTE: (1) Number found positive (infected).

Table 3.12. Feeding trial data for *S.* Derby for naive subjects.

Serotype	Dose	Log Dose	Positive (Inf)[1]	Total	Proportion
S. Derby	1.39E+05	5.14	3	6	0.50
S. Derby	7.05E+05	5.85	2	3	0.67
S. Derby	1.66E+06	6.22	3	4	0.75
S. Derby*	6.40E+06	6.81	1	3	0.33

NOTE: (1) Number found positive (infected).

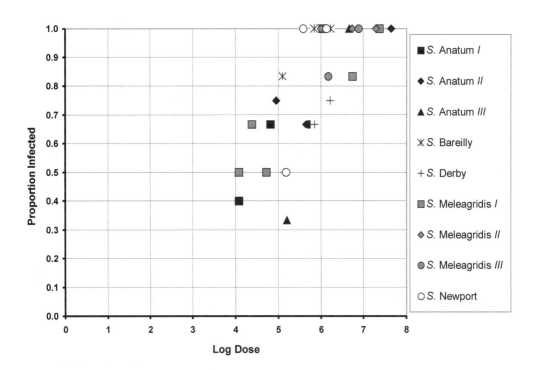

Figure 3.3. Summary of feeding trial data for naive subjects.

3.4 DOSE-RESPONSE ASSESSMENT

This section presents the quantitative information that is available for *Salmonella* infectivity or illness, from which dose-response relationships can be estimated. It is not possible to provide all the details necessary to give a complete coverage of the theory behind the dose-response relationships in this document. However, a comprehensive treatment of dose-response models and assumptions related to the mathematical derivation of the various equations is given in the FAO/WHO Hazard Characterization Guidelines document (currently in preparation)

3.4.1 Dose-response models for *Salmonella*

Several approaches and models to characterize the dose-response relationship for *Salmonella* have been presented in the literature or in official reports and documents. This report discusses three different approaches for modelling *Salmonella.* The first model is the beta-Poisson model fitted to the human feeding trial data for *Salmonella* (Fazil, 1996). The second model was proposed in the US SE RA and was based on the use of a surrogate pathogen to describe the dose-response relationship. The third model, introduced in the Health Canada *Salmonella* Enteritidis risk assessment, used a Weibull dose-response relationship updated to reflect outbreak information using Bayesian techniques. In addition to these models, the current analysis also explores the effect of fitting the beta-Poisson model to the human feeding trial data for naive subjects only.

Dose-response model fitted to non-typhi Salmonella *human feeding trial data*

The human feeding trial data have been analysed using the beta-Poisson, lognormal (log-probit) and exponential dose-response functional forms (Fazil, 1996). Three doses in the data set were identified as "outliers" (i.e. *S.* Anatum I: 9.3E+5; *S.* Meleagridis III: 1.58E+5; *S.* Derby: 6.4E+6) and were subsequently removed from the analysis. The analysis concluded that both the lognormal and beta-Poisson functional forms fit the majority of the data. However, based upon theoretical considerations (threshold vs non-threshold, where threshold models assume that there is some finite minimum dose below which no response can occur, while non-threshold models assume that the minimum possible dose that can cause a response is one cell, even though the probability may be very low for one cell to successfully survive all the host defences), the beta-Poisson model was proposed as the model to describe the dose-response relationship for *Salmonella.* In addition, it was reported that all the serotypes could be adequately described using a single beta-Poisson dose-response curve. The parameters of the beta-Poisson dose-response model for non-typhi *Salmonella* in general were reported as alpha = 0.3126, and beta = 2885. The uncertainty in the parameters was estimated using a bootstrap approach, which generated sets of parameters that satisfied the model fitting conditions. The potential for a greater probability of illness for susceptible and normal populations was not addressed in the analysis.

$$P_{ill} = 1 - \left(1 + \frac{Dose}{\beta}\right)^{-\alpha}$$

Model Used:	**Beta-Poisson**
Parameters:	Alpha = 0.3126
	Beta = 2885
Comment:	Uncertainty in the parameters estimated using a bootstrap approach, which generated a set of alpha and beta parameters that could be randomly sampled in order to incorporate uncertainty.

Dose-response model fitted to non-typhi **Salmonella** *naive human feeding trial data*

The model parameters reported by Fazil (1996) did not consider the effect that multiple feedings may have on the dose-response relationship. As a result, for this present review, the data using only naive subjects (Tables 3.8 to 3.12 and Figure 3.3) were re-fitted to the beta-Poisson model and the parameters for this model were estimated. The data were fitted using maximum likelihood techniques, as described by various authors (Haas, 1983; Haas et al., 1993; Regli et al., 1991; Teunis et al., 1996). The parameters of the beta-Poisson dose-response model fitted to the data for naive subjects was estimated to be alpha = 0.4047 and beta = 5587. The uncertainty in the parameters was estimated using the bootstrap approach.

$$P_{ill} = 1 - \left(1 + \frac{Dose}{\beta}\right)^{-\alpha}$$

Model Used:	**Beta-Poisson**
Parameters:	Alpha = 0.4047
	Beta = 5587
Comment:	Uncertainty in the parameters estimated using a bootstrap approach, which generated a set of alpha and beta parameters that could be randomly sampled in order to incorporate uncertainty.

The beta-Poisson dose-response curves generated using the original dose-response data and the data edited to reflect only naive subjects are shown in Figure 3.4. Also shown in the figure are the feeding trial data to illustrate the fit to the data.

As shown in Figure 3.4, both models fit the feeding trial data well and the difference between the curves using the original data and the data that reflects only naive subjects is small. Interestingly, the curve fitted to the naive data tends to estimate a greater probability of infection at doses above approximately 10^4 than does the curve fitted to the original data, perhaps reflecting a tendency in the data for a slightly greater susceptibility for naive subjects. Within the lower dose regions, the two curves are very similar, and the dose translating to a probability of infection for 50% of the population is virtually identical for the two curves (2.36E4 vs 2.54E4 for the original and naive models). The low dose extrapolation for the two dose-response curves was also very similar. As a result of the similarities between the models and the concerns that have been raised about potential immunity or cumulative effects, the beta-Poisson model fitted to the data of naive subjects is used in the remainder of this analysis as the representation of the human feeding trial data fitted to the beta-Poisson model.

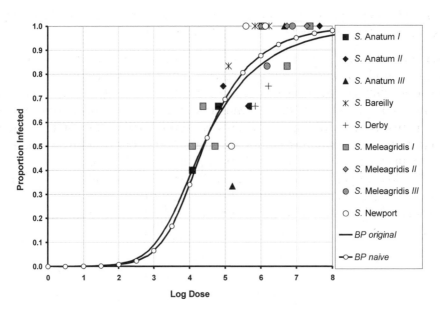

Figure 3.4. Comparison between dose-response model fitted to original feeding trial data and feeding trial data for naive subjects.

USDA-FSIS Salmonella *Enteritidis Risk Assessment*

The hazard characterization in the US SE RA evaluates the public health impacts of exposure to *S.* Enteritidis through shell eggs and egg products in terms of numbers of illnesses and specific public health outcomes on an annual basis. Considerations in quantifying the dose-response relationship included the selection of an appropriate functional form, extrapolation of fitted curves to low-dose ranges, and the use of surrogate organisms in the absence of feeding trial data specific for *S.* Enteritidis.

In the initial quantification of a dose-response relationship for *S.* Enteritidis, a beta-Poisson dose-response curve was fitted to the pooled data from all *Salmonella* feeding trials (McCullough and Eisele, 1951a, c) and the fitted model compared with epidemiological information from available *S.* Enteritidis outbreak data. The model validation on the epidemiological data showed that outbreaks associated with *S.* Enteritidis exhibited a higher attack rate than would be estimated using the pooled human feeding trial data for *Salmonella.* Furthermore, an analysis of variance (ANOVA) on the *Salmonella* human feeding trial data for dose and serotype effects revealed two distinct, statistically significant dose-response patterns (representative of doses $>10^3$ organisms) among the *Salmonella* serotypes in the human feeding studies data (Morales, Jaykus and Cowen, 1995; Jaykus, Morales and Cowen, 1997).

The inability of several dose-response models, fitted to the *Salmonella* data, to predict the high attack rates associated with low doses, such as the 1994 *S.* Enteritidis outbreak from ice cream (Hennessy et al., 1996) was likewise previously noted by Morales, Jaykus and Cowen (1995). In order to capture the region of concern (i.e. the low-dose range with corresponding high attack rates evident in the outbreak investigation data), human feeding study data utilizing a low-dose organism was selected for subsequent dose-response modelling as a surrogate for *S.* Enteritidis. The absence of human feeding study data for *S.* Enteritidis prompted the selection of *Shigella dysenteriae* (Levine and DuPont, 1973) as a proxy for modelling "low-dose" *Salmonella* serotypes (attack rates >0 with doses = 10^3 organisms).

Epidemiological evidence from outbreak investigations was once again used to conduct a model validation check on the two dose-response models generated (beta-Poisson curves fitted to human feeding trial data for pooled *Salmonella* species and to the low-dose proxy *Shigella dysenteriae*). A review of the epidemiological outbreak investigations showed that many of the reported doses resulting in illnesses were several orders of magnitude lower than the doses reported in the *Salmonella* feeding trials. Further, the doses which caused outbreaks were likewise several orders of magnitude lower than the doses which were predicted by the dose-response models constructed from the *Salmonella* feeding trial data. Model validation to the available outbreak investigation data subsequently served as the basis for selection of a dose-response relationship (Figure 3.5). The outbreak investigation data used for dose-response model validation are detailed in Table 3.13.

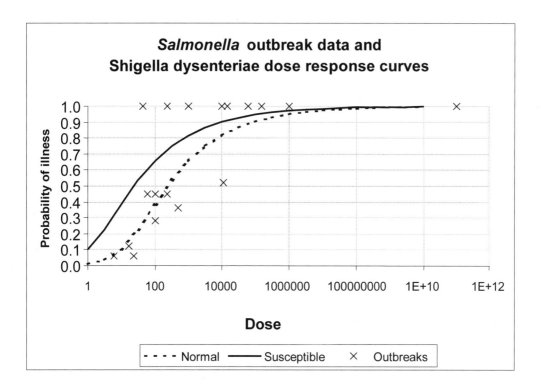

Figure 3.5. USDA comparison of available *Salmonella* outbreak investigation data and beta-Poisson dose-response curves for *Shigella dysenteriae* estimated for normal and susceptible subpopulations (USDA-FSIS, 1998).

Table 3.13. *Salmonella* outbreak investigation data used in the US SE RA to compare with *Shigella* dose-response curves (USDA-FSIS, 1998).

Serovar	Dose	Log dose	Number ill	Attack rate	Reference
Typhimurium	1.7E+01	1.23	16 000	12%	Boring, Martin and Elliott, 1971
Schwarzengrund	4.4^E+01	1.64	1	100%	Lipson, 1976
Newport	6.0^E+01	1.78	48	45%	Fontaine et al., 1978
Eastbourne	1.0^E+02	2.00	95	45%	D'Aoust et al., 1975
Heidelberg	1.0E+02	2.00	339	28%	Fontaine et al., 1980
Heidelberg	2.0^E+02	2.30	1	100%	George, 1976
Newport	2.34^E+02	2.36	46	45%	Fontaine et al., 1978
Heidelberg	5.0E+02	2.70	339	36%	Fontaine et al., 1980
Typhimurium	1.1^E+04	4.04	1 790	52%	Armstrong et al., 1970
Cubana	1.5^E+04	4.18	28	100%	Lang et al., 1967
Cubana	6.0^E+04	4.78	28	100%	Lang et al., 1967
Zanzibar	1.5^E+05	5.18	6	100%	Reitler, Yarom and Seligmann, 1960
Infantis	1.0^E+06	6.00	5	100%	Angelotti et al., 1961
Zanzibar	1.0^E+11	11.00	8	100%	Reitler et al., 1960
Enteritidis	6.0E+00	0.77	>1 000	6%	Hennessy et al., 1996
Enteritidis	2.4^E+01	1.38	>1 000	6%	Vought and Tatini, 1998
Enteritidis	1.0^E+03	3.00	39	100%	Levy et al., 1996
Enteritidis	1.0E+04	4.00	39	100%	Levy et al., 1996

The dose-response relationship subsequently used was a beta-Poisson model fitted to the human feeding trial data for *Shigella dysenteriae* M131, with parameters alpha = 0.2767 and beta = 21.159 (www.fsis.usda.gov/OPHS/risk/semodel.htm, July 2000). Uncertainty was introduced into the beta parameter by characterizing it as a normal distribution truncated at zero, with a maximum of 60 and a mean and standard deviation of 21.159 and 20 respectively for the proportion of the population assumed to be in good health (normal subpopulation). In addition, the beta parameter of the *S. dysenteriae* beta-Poisson model was reduced by a factor of 10, thus shifting the curve to the left to estimate a higher probability of illness for susceptible individuals (susceptible subpopulation). Uncertainty in the beta parameter for the susceptible subpopulation was therefore introduced using a normal distribution with a mean and variance of 2.116 and 2.0 respectively, and a minimum of 0 and maximum of 6.

$$P_{ill} = 1 - \left(1 + \frac{Dose}{\beta}\right)^{-\alpha}$$

Model Used:	**Beta-Poisson**
Parameters: Normal	Alpha = 0.2767
	Beta = Normal (μ:21.159, σ:20, min:0, max:60)
Susceptible	Alpha = 0.2767
	Beta = Normal (μ:2.116, σ:2, min:0, max:6)
Comment:	Human feeding trial data for *Shigella dysenteriae* used as a surrogate. Susceptible population characterized by reducing beta parameter by a factor of 10. Simulation of public health outcomes for normal and susceptible subpopulations incorporates the uncertainty represented in the beta parameters.

Health Canada Salmonella *Enteritidis dose-response relationship*

The Health Canada *Salmonella* Enteritidis risk assessment used a re-parameterized Weibull dose-response model. Bayesian methods were employed as a means to provide a consistent framework for combining information from various sources including feeding and epidemiological studies (Health Canada, 2000, but unpublished). The Canadian *Salmonella* Enteritidis risk assessment had not been published at the time of preparing this report, but a brief description of the procedure used is provided, and the model generated is compared with the other alternatives.

The Canadian model begins with the Weibull dose-response model:

$$P = 1 - \exp\left(-\theta \times d^b\right)$$

where d is the dose.

The model was re-parameterized as summarized below (Health canada, 2000, but unpublished), and this is the equation that is referred to in the remainder of this section.

$$P = 1 - \exp\left(-\exp\left\{b\left[\ln(d) - \kappa\right]\right\}\right)$$

$$\beta = \ln(b)$$

$$\kappa = \frac{-\ln(\theta)}{b}$$

The parameter b in the model was characterized by performing a meta-analysis of all the bacterial feeding trial data. This analysis determined that the log transformed value of b, termed β ($\beta = \ln[b]$) could be well described using a normal distribution with mean of -1.22 and a standard deviation of 0.025. This characterization of β for all bacterial pathogens represents between-study variability, which is used as a reference input (Health Canada, 2000, but unpublished). Epidemiological data – specifically information generated from the Schwanns ice cream outbreak (Hennessy et al., 1996; Vought and Tatini, 1998) – was incorporated into the model by adjusting the parameter θ.

In order to adjust the parameter θ, the following equation in terms of epidemiological information was used (Health Canada, 2000, but unpublished):

$$\theta = \frac{-\ln(1-P)}{X^b}$$

where P represents the attack rate reported in an epidemiological outbreak and X represents the dose estimated to have caused the outbreak.

Within the model, the dose ingested was defined stochastically so as to reflect the uncertainty associated with the data. A single value for the attack rate P was used, and this was estimated to be 6% (Hennessy et al., 1996). The dose was estimated based on the concentration reported and the amount of ice cream consumed. The concentration (in CFU/g) was characterized using a lognormal distribution with a mean of 0.15 and a standard deviation of 0.1. The amount of ice cream consumed was estimated using a PERT distribution with a minimum of 60, a mode of 130, and a maximum of 260.

A separate dose-response relationship was generated for the susceptible population, which was based on epidemiological information. Specifically, information from a waterborne outbreak of *S.* Typhimurium in Riverside, California (Boring, Martin and Elliott, 1971), which reported on age-specific attack rates, was used to shift the value of θ according to the following equation (Health Canada, 2000, but unpublished):

$$\theta_{susceptible} = \theta_{normal} \left(\frac{\ln\left[1 - beta_{sus}\{a_s, b_s\}\right]}{\ln\left[1 - beta_{norm}\{a_n, b_n\}\right]} \right)$$

where the parameters (a and b) for the beta distributions are estimated from the reported epidemiological data on the total number of individuals exposed and the number that became ill. The subscripts s and n refer to the data for susceptible and normal populations respectively.

$$\theta_{susceptible} = \theta_{normal} \left(\frac{\ln\left[1 - beta_{sus}\{a_s, b_s\}\right]}{\ln\left[1 - beta_{norm}\{a_n, b_n\}\right]} \right)$$

Model Used:	**Re-parameterized Weibull**
Parameters:	Beta = Normal (μ: -1.22, σ:0.025)
	Concentration = Lognormal (μ:0.15, σ:0.1)
	Amount consumed = PERT (min:60, mode:130, max:260)
	Attack Rate = 6.6%. a_s = 231; b_s = 987; a_n = 749; b_n = 5 966

Several parameters in the dose-response models described incorporated uncertainty into their characterization. In order to display the dose-response curves in the following sections, the uncertainty in those parameters has been simulated and the specified moments displayed.

The following abbreviations are introduced and will be used when referring to the dose-response curves: Can-norm = Canadian normal population dose-response; Can-susc = Canadian susceptible population dose-response; US-norm = the United States of America normal population dose-response; US-susc = the United States of America susceptible population dose-response; and Naive-BP = beta-Poisson dose-response curve fitted to naive subject human feeding trial data. These are shown in Figures 3.6 to 3.8.

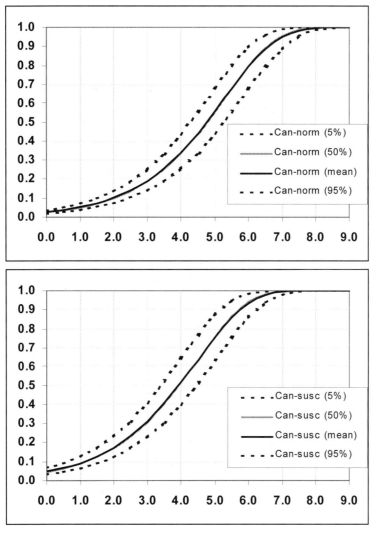

Figure 3.6. Dose-response curves for normal (Can-norm – upper panel) and susceptible (Can-susc – lower panel) populations, as estimated in Canadian *Salmonella* Enteritidis risk assessment.

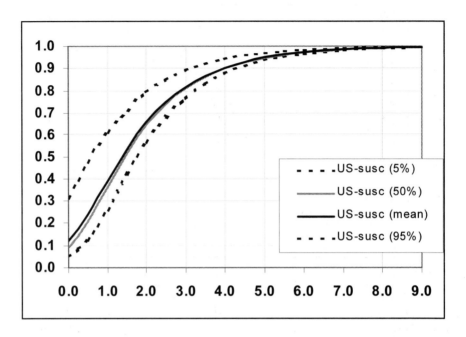

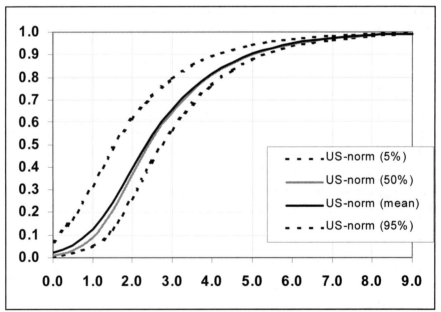

Figure 3.7. Dose-response curves for normal (US-norm – upper panel) and susceptible (US-susc – lower panel) populations, as estimated in the US SE RA.

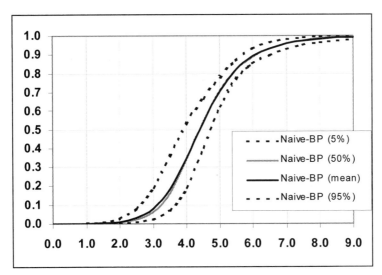

Figure 3.8. Beta-Poisson dose-response curve fitted to naive subject non-typhi *Salmonella* human feeding trial data (Naive-BP).

The five dose-response curves are plotted together in Figure 3.9 to assist in the comparison of the curves. Since the 50th percentile and the mean are very similar in all five dose-response curves (Figures 3.6 to 3.8), only the mean values for the curves are plotted. The 95th percentile and 5th percentile boundaries for the curves are omitted from this figure for visual reasons.

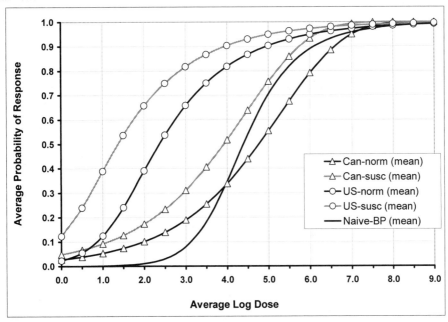

Figure 3.9. Comparison between five dose-response curves: Can-norm, Can-susc, US-norm, US-susc and Naive-BP.

There is some overlap between the Can-norm and the Naive-BP dose-response curves. However, the Naive-BP curve estimates a higher probability of response than the Can-norm for individuals exposed to a dose greater than approximately 10^4 cells. At an average dose of less than approximately 10^4 cells, the Can-norm dose-response curve estimates a greater probability of response than does the Naive-BP. In fact, at an average dose of 2 log (100 cells) the Can-norm dose-response curve estimates a probability of response of approximately 10% compared with approximately 1% for the Naive-BP dose-response curve. The adjustment of the Canadian dose-response curve to reflect epidemiological information, specifically the 6% response rate at a dose of approximately 1 log (Hennessy et al., 1996; Vought and Tatini, 1998) is evident in the behaviour of the curve in that lower-dose region.

The US-norm and US-susc dose-response curves, which are based on using *Shigella* as a surrogate pathogen, estimate a higher probability of illness at a given dose than the other dose-response curves across almost the entire dose range, except the lowest (≤ 10 organisms). At the 2 log (100 cells) average dose level, the normal population using the United States of America dose-response curve would be estimated to have approximately a 40% average probability of response and the susceptible population would be estimated to have approximately a 65% probability of response. This can be compared with 10% and 18% for normal and susceptible populations using the Canadian dose-response curves.

The dose-response curves thus have a significant degree of deviation from each other. Selecting a dose-response curve from this information would have to be based on several considerations that include: the level of conservatism that one wishes to employ; the theoretical acceptability of using a surrogate pathogen; the biological plausibility of various functional forms for modelling dose-response relationships; the biological endpoint or public health outcome of interest; or the acceptability of the human feeding trial data in capturing the overall response for a population.

In order to gain additional insight into the pathogenesis of *Salmonella*, the available data from epidemiological information were explored.

3.5 EPIDEMIOLOGICAL INFORMATION

Epidemiological data can provide valuable insight into the pathogenicity of microorganisms as it applies to the general population. In a sense, outbreaks represent realistic feeding trials with the exposed population often representing a broad segment of society. The doses are essentially real-world levels, and the medium carrying the pathogen represents a range of characteristics (protective, fatty, long residence time, etc.). Ideally, an epidemiological investigation should attempt to collect as much quantitative information as possible in order to lend itself to better characterizing the dose-response relationship for microbial pathogens. In order to refine the dose-response relationship so that it has greater applicability to the general population, various information is required in an epidemiological investigation: the dose, the population exposed, and the number of people exhibiting a response (illness, fever, etc.).

The dose that is suspected to have caused illness in a specific outbreak is often the most difficult measure in an investigation. The lack of dose estimates can be attributed to either the inability to obtain samples of contaminated food or the lack of emphasis being placed on the value of such information. Often, contaminated food is tested and only the presence or absence of the suspected pathogen is reported. This information is often viewed as sufficient to incriminate the food, but it does little to further knowledge of the dose-response relationship.

The attack rate represents the response in a dose-response relationship. In order to estimate the attack rate, an accurate estimate is required of not only the population that was exposed to the contaminated food, but also the number of individuals that became ill. In addition, it is valuable to know the characteristics of the exposed and affected population, in order to account for potential susceptibility issues.

3.5.1 Summary of epidemiological and outbreak information

The following sections present and summarize outbreaks found in the literature that included quantitative information from which the dose and attack rate could be estimated. It is important to note that although these outbreaks include quantitative data, some assumptions had to be made, depending on the nature of the information. In the interest of transparency, the following sections present the information from the original epidemiological reports in as much detail as possible, and, where appropriate, the assumptions that are used are clearly indicated.

In addition, reports that were currently unpublished at the time of drafting this report were received from Japan (Ministry of Health and Welfare, 1999). Although these reports have not been published, and the details of the methods used in the investigations have not been stated (other than through personal communication), they represent a valuable source of information on the real-world dose-response relationship and expand our database of *Salmonella* pathogenicity considerably. The data in these reports are generated as part of the epidemiological investigations that take place in Japan following an outbreak of foodborne illness. In accordance with a Japanese notification released on March 1997, large-scale cooking facilities that prepare more than 750 meals per day or more than 300 dishes of a single menu at a time are advised to save food for future possible analysis in the event of an outbreak. Thus, 50-g portions of each raw food ingredient and each cooked dish are saved for more than 2 weeks at a temperature below -20°C. Although this notification is not mandatory, it is also applicable to smaller-scale kitchens with social responsibility, such as those in schools, day care centres and other child-welfare and social-welfare facilities. Some of the local governments in Japan also have local regulations that require food saving, but the duration and the storage temperature requirements vary.

In the evaluation of the outbreak data, whenever sufficient information was available, susceptible and normal populations were separated out of the database to aid in further analysis. Children aged 5 years or younger were considered to be a susceptible population. The criteria or assumptions used to identify potentially susceptible populations are noted in the individual outbreak summaries.

In addition, the uncertainty associated with each of the outbreak parameters are also summarized and defined at the end of each outbreak description. The published reports were used as a basis upon which to derive a reasonable characterization of the uncertainty. However, it should be recognized that since only rarely is sufficient information given upon which to derive a range of uncertainty for the parameters, the uncertainty ranges used are only a crude estimate. In addition, in several reports there is no information whatsoever to use as a basis for uncertainty estimates; in these cases a consistent default assumption was used. To capture the dose uncertainty, 25% over- and under-estimates for the reported concentration and amount consumed were used.

——— § ———

Case Number: 1
Reference: Boring, Martin & Elliott, 1971
Serovar: *S.* Typhimurium
Setting: Citywide municipal water
Medium: Water

Concentration		Amount Ingested		Dose	Comments
Value	Units	Value	Units		
17	#/litre	0.75	litre	1.28E+01	Concentration found in tap water using composite sample
1000	#/litre	0.75	litre	7.50E+02	Order of magnitude for concentration found in tap water based on single sample collected independently

Exposed	Response	Attack Rate	Comments
8 788	1 035	11.78%	Reported average attack rate for all individuals
7 572	805	10.63%	Attack rate reported for individuals >5 years old (assumed "normal" population)
1 216	230	18.91%	Attack rate reported for individuals <5 years old (assumed "susceptible" population)

Comments

Composite water samples were collected late in the epidemic (9 days after initial case) and water in the composite samples had been stored for 1 to 4 days at room temperature prior to culturing. Since varying amounts of water, from a few millilitres to as much as 500 ml, were pooled from several sample bottles, it is possible that numbers in some samples were greatly diluted by negative samples. The pooled sample consisted of water from 74 different samples, and only 5 of the 74 samples were actually positive. The concentration of 1000/litre was an order of magnitude estimate following a single isolation made independently from a 1-ml sample (suggesting an order of magnitude of 1000 organisms/litre). The concentration in the water was therefore assumed to range between the two reported concentration estimates (between 50/litre and 500/litre was the range for concentrations), with water consumption of 0.75 litres which results in a dose range of between 37 and 375 cells.

A house-to-house survey was conducted that comprised 8788 people, with 1035 reporting gastroenteric illness. The report also identified attack rates according to age, which was used in the current analysis as an estimate of the potential attack rate for susceptible and normal populations. Children under 5 years (assumed potentially susceptible) were reported to have an 18.9% attack rate, compared with approximately 11% for the rest of the population. The uncertainty in the average attack rate was calculated allowing for 5% under- or over-reporting. Given 1035 people reporting gastroenteric illness, only 983 may have actually been sick, with the other 5% claiming to be sick; alternatively, 1087 people may have actually been sick, with the additional people not reporting sickness. It was assumed that the contamination in the water supply was randomly distributed throughout, such that all 8788 people that reported having drunk water were exposed. It should also be noted that the attack rates listed in this table assume exposure to the pathogen only once during the outbreak.

Outbreak parameter uncertainty

Dose Uniform Distribution		Exposed population	Positive Pert Distribution		
Min	Max	Value	Min	ML	Max
37	375	7 572	765	805	845
37	375	1 216	219	230	242
37	375	8 788	983	1 035	1 087

—— § ——

Case Number: 2
Reference: Fontaine et al., 1980
Serovar: *S.* Heidelberg
Setting: Restaurant
Medium: Cheddar cheese

Concentration		Amount Ingested		Dose	Comments
Value	Units	Value	Units		
0.36	#/100 g	28	g	0.10	Concentration reported by Food Research Institute, Wisconsin, USA
1.8	#/100 g	28	g	0.50	Concentration reported by CDC, Atlanta, USA
1.08	#/100 g	28	g	0.30	Average of two reported concentrations
108	#/100 g	28	g	30.24	Average concentration adjusted for a 99% die-off prior to culturing
1 080	#/100 g	28	g	302.40	Average concentration adjusted for a 99.9% die-off prior to culturing

Exposed	Response	Attack Rate	Comments
205	68	33.17%	Attack rate based on exposed employees in incriminated restaurants, consumers at incriminated restaurants, and employees at restaurants that received contaminated cheese lot shipments and at which employee cases existed

Comments

Samples analysed by CDC, Atlanta, were reported to have an MPN of 1.8 organisms/100 g, while the Food Research Institute in Wisconsin reported an MPN of 0.36 organisms/100 g. According to the restaurant, the serving size was approximately 28 g of cheese per meal.

The potentially very low infectious dose for this outbreak was noted by the researchers, and the potential for the occurrence of up to a 99.9% die-off prior to culturing was acknowledged. The concentration in the food at consumption was assumed to range between 108 and 1080 cells per 100 g (99% to 99.9% die-off prior to culture). The dose ingested, based on the nominal amount consumed, was estimated to range between 30 and 300 cells.

The attack rate in this outbreak was reported to range from 28% to 36%. The exposed population was estimated to be 205, consisting of employees in incriminated restaurants, consumers at incriminated restaurants, and employees at restaurants that received contaminated cheese lot shipments and where employee cases existed. The number of positives (57 to 74) was back calculated from the reported attack rate range and the exposed population.

Outbreak parameter uncertainty

Dose		Exposed population	Positive		
Uniform distribution			Pert distribution		
Min	Max	Value	Min	ML	Max
30	300	205	57	68	74

— § —

Case Number: 3
Reference: Lang et al., 1967
Serovar: *S.* Cubana
Setting: Hospital
Medium: Carmine dye capsules

Concentration		Amount Ingested		Dose	Comments
Value	Units	Value	Units		
30 000	#/capsule	0.5	capsule	15 000	Lower dose estimate based on some patients being given ½ a capsule
30 000	#/capsule	2.0	capsule	60 000	Upper dose estimate based on some patients being given up to 2 capsules

Exposed	Response	Attack Rate	Comments
?	21	?	Recognized cases during outbreak
?	12	?	Confirmed cases as a result of dye capsule ingestion

Comments

This outbreak involved a susceptible population that consisted of debilitated and aged people, infants and persons with altered gastrointestinal function. Carmine dye capsules are used as a faecal dye marker for such things as the collection of timed stool specimens, gastrointestinal transit time and the demonstration of gastrointestinal fistulas. The number of capsules given to patients ranged from 0.5 to 2; as a result, the dose ingested was assumed to range from 15 000 to 60 000.

There were a total of 21 recognized cases during this outbreak, but 4 were reported to have been infected prior to admission and 5 cases were suspected to have been secondary transmission. Therefore there were 12 confirmed cases directly as a result of carmine dye capsule ingestion. Unfortunately, for attack rate estimation, the total number of exposed individuals was not determined, although the authors of the report note that there were some people who received carmine but were not infected. It was thus inferred that the attack rate was some value less than 100%. As an upper and lower bound, it was assumed that 14 to 20 individuals received dye capsules.

Outbreak parameter uncertainty

Dose		Exposed population			Positive
Uniform distribution		Pert distribution			
Min	Max	Min	ML	Max	Value
15 000	60 000	14	17	20	12

—— § ——

Case Number: 4
Reference: Angelotti et al., 1961
Serovar: *S.* Infantis
Setting: Home
Medium: Ham

Concentration		Amount Ingested		Dose	Comments
Value	Units	Value	Units		
23 000	#/g	50	g	1 150 000	Lower dose estimate based on lower weight of slice and only one slice consumed
23 000	#/g	200	g	4 600 000	Upper dose estimate based on higher weight of slice and up to 2 slices consumed.

Exposed	Response	Attack Rate	Comments
8	8	100%	

Comments

This outbreak occurred in a family consisting of adults and at least two children of grade school age. Smoked ham purchased from a supermarket was taken home and refrigerated for approximately 5 hours. Eight people in the family ate either raw or fried slices of ham, and all 8 experienced acute diarrhoea with gastroenteritis symptoms within 8 to 24 hours. An uneaten portion of ham was obtained and examined in laboratory 2 days after the outbreak occurred. Various bacteria were isolated from the raw ham: total aerobic plate count (268 000 000/g), coliform bacteria (15 000/g), *Streptococcus faecalis* (31 000 000/g), staphylococci (200 000 000/g) and *Salmonella* Infantis (23 000/g). Staphylococci were negative for coagulase production and negative for enterotoxin production. Stools from 4 of the 8 persons affected were examined 10 days after the outbreak: mother, father and two grade-school-age sons. *S. faecalis* was isolated from both parents and one son. *S. faecalis var. liquefacies* was isolated from another son, and *S.* Infantis was isolated from both parents but not the sons. The researchers noted that *S.* Infantis in the ham, stools and the long incubation period implies infection of *Salmonella* aetiology. However, a mixed infection is a possibility.

The weight of a slice of ham was estimated to range from 50 to 100 g, with 1 to 2 slices consumed. The dose was thus estimated to range from 1 150 000 to 4 600 000 cells of *S.* Infantis. The exposed and positive populations in this case were quite well established, therefore accounting for uncertainty in these parameters was unnecessary.

Outbreak parameter uncertainty

Dose		Exposed population	Positive
Uniform distribution			
Min	Max	Value	Value
1 150 000	4 600 000	8	8

—— § ——

Case Number: 5
Reference: Armstrong et al., 1970
Serovar: *S.* Typhimurium
Setting: Various parties and banquets
Medium: Imitation ice cream

Concentration		Amount ingested		Dose	Comments
Value	Units	Value	Units		
113	#/75 g	75	g	113	Reported concentration and amount consumed at limited menu venue

Exposed	Response	Attack rate	Comments
1 400	770	55%	Reported attack rate at limited menu venue

Comments

This episode involved 14 outbreaks with a total of 3450 people attending various events at which imitation ice cream (chiffonade) was identified as the vehicle of infection. The authors estimated a 52% attack rate based on a survey of persons attending seven of the events. The menus at the various events were relatively extensive, but one of the outbreaks involved a large affair with a limited menu where the authors cite that nearly all those attending had eaten all of the foods offered. Using this outbreak, the attack rate was estimated to be 55% (1400 people attending and 770 people sick).

The chiffonades were stored at -20°C for 1 month before quantitative cultures were done, and the MPN was reported to be 113 or less salmonellae per 75-g serving. The reduction in numbers that could be expected due to freezing was experimentally determined by artificial inoculation of *S.* Typhimurium into chiffonade and storing of the samples at -20°C. Artificial inoculation experiments indicated that log reductions would have occurred during the storage period, but no more than a 2-log reduction was likely to have occurred during the 1-month storage. As a result, the concentration was estimated to range between 1130 (1 log reduction) and 11 300 (2 log reduction) per serving. In this outbreak, the exposed population was reasonably well established, but the positive population was assumed to have 5% under- and over-reporting (a range of 732 to 809).

Outbreak parameter uncertainty

Dose		Exposed population	Positive		
Uniform distribution			Pert distribution		
Min	Max	Value	Min	ML	Max
1 130	11 300	1400	732	770	809

—— § ——

Case Number: 6
Reference: Fontaine et al., 1978
Serovar: *S.* Newport
Setting: Interstate (Maryland: households; Colorado: households; Florida: naval base)
Medium: Hamburger

Concentration		Amount ingested		Dose	Comments
Value	Units	Value	Units		
6	#/100 g	100	g	6	Lowest reported concentration
23	#/100 g	100	g	23	Highest reported concentration

Exposed	Response	Attack rate	Comments
?	48	?	Total number of people affected over entire geographic area of outbreak.

Comments

The concentration of *S.* Newport in ground beef was determined from MPN to be between 6 and 23 per 100 g. Accounting for freezing, the authors cite that experimental evidence would indicate a 1- to 2-log reduction due to freezing, which would place the concentration at 60–2300/100 g. However, cooking, even undercooking, is likely to produce a reduction prior to consumption. If the effects of cooking are conservatively assumed to be 1 to 2 log, then the concentration prior to consumption is again estimated to be 6–23/100 g. Assuming consumption of 100 g, the dose that was capable of causing an infection in some people can be estimated to be approximately 6–23 organisms. Unfortunately, this outbreak was geographically widespread and the authors did not report the total number of individuals exposed. The attack rate is therefore undetermined in this outbreak.

—— § ——

Case Number: 7
Reference: Fazil, 1996
Serovar: S. Newport
Setting: Naval Base
Medium: Hamburger

Concentration		Amount ingested		Dose	Comments
Value	Units	Value	Units		
4	#/100 g	100	g	4	Low reported concentration (6 CFU/100 g) with 25% allowance for uncertainty (approx. 4 CFU/100 g)
30	#/100 g	100	g	30	High reported concentration (23 CFU/100 g) with 25% allowance for uncertainty (approx. 30 CFU/100 g)

Exposed	Response	Attack rate	Comments
7254	19	0.3%	Attack rate estimated with all recruits at the base exposed
3627	19	0.5%	Attack rate assuming 50% were actually exposed
1813	19	1.0%	Attack rate assuming 25% were actually exposed
725	19	2.6%	Attack rate assuming 10% were actually exposed

Comment

The data in this outbreak is derived from the previous episode described (Case Number 6, reported by Fontaine et al., 1978). However, Fazil (1996) examined the naval outbreak in greater detail through a series of personal communications with the United States Navy, to attempt to determine an attack rate. A total of 21 cases occurred at the naval training centre: 2 were asymptomatic food handlers and 19 were trainees.

The entire complex had a population of 12 483, with the military population listed as 9904 (full time military personnel and trainees). Meals were served at several locations, and included the galley, the staff galley and the exchange cafeteria. The outbreak was reported to have occurred at the "Training Station", which is a separate area within the centre, where training is conducted. There were 7254 recruits who were fed at the galley that serviced the trainees. Therefore, depending on the assumed number of people that ate a contaminated hamburger, an attack rate can be estimated. Assuming 7254 individuals exposed (all present, which is unlikely), the attack rate is estimated to be 0.3% (19/7254). It was assumed that a more likely exposure population was 25% (1813), with an uncertainty range of between 10% (725) and 50% (3627). It was assumed that the positive population was well characterized given the nature of the location. At the naval base the trainees would have had access to convenient medical attention. It should be noted that if there was reporting bias it is more likely to be under-reporting as opposed to over-reporting.

Outbreak parameter uncertainty

Dose		Exposed population			Positive
Uniform distribution		Pert distribution			
Min	Max	Min	ML	Max	Value
4	30	725	1 813	3 627	19

—— § ——

Case Number: 8
Reference: Narain and Lofgren, 1989
Serovar: *S.* Newport
Setting: Restaurant
Medium: Pork and ham sandwiches

Concentration		Amount ingested		Dose	Comments
Value	Units	Value	Units		
4.40E+07	#/g				Concentration found in pork sandwich stored by one of the patients. No indication of how much pork or ham sandwiches individuals consumed.

Exposed	Response	Attack rate	Comments
200 ?	105	52.5%	Total number of people who became ill and were exposed at the restaurant. The 200 people listed as exposed were actually the number of people that ate at the restaurant during the period, according to the owner's recollection.

Comments

A total of 105 people were reported to have become ill during this episode, which was attributed to ham and pork sandwiches. The sandwiches were suspected to have been contaminated at the restaurant and a refrigerated portion of a pork sandwich from a patient yielded 44×10^6 *S.* Newport per gram.

The attack rate that might be inferred from information provided in this report is unknown. The restaurant reported serving approximately 200 people during the period, of whom 105 became ill. However, the information required is an estimate of the number of people that actually ate ham and pork sandwiches and were thus exposed to contaminated food. It can be assumed that not everyone ate the ham and pork sandwiches. If it is assumed that 60% of the people visiting the restaurant ate the contaminated food, then 120 people may have been exposed. At the other extreme, it could also be assumed that only 105 people were actually exposed and 105 became sick, the attack rate then being 100%.

—— § ——

Case Number: 9
Reference: Craven et al., 1975
Serovar: *S.* Eastbourne
Setting: Interstate; homes
Medium: Chocolate balls

Concentration		Amount ingested		Dose	Comments
Value	Units	Value	Units		
2.5	#/g	450	g	1 130	Reported concentration with dose estimate based on the consumption of an entire bag of chocolates (approximately 50 chocolate balls)
2.5	#/g	225	g	563	Reported concentration with dose estimate based on the consumption of half a bag (25) of chocolate balls
2.5	#/g	45	g	113	Reported concentration with dose estimate based on the consumption of approximately 5 chocolate balls

Exposed	Response	Attack rate	Comments
?	80	?	Total number of cases in geographically widespread outbreak. No information on exposed population

Comments

This outbreak involved a potentially susceptible population and involved chocolate balls. The median age of the cases in this outbreak was 3 years. The attack rate cannot be determined in this case because no information was provided in the report, and the geographically widespread nature of the outbreak makes inferences difficult. The outbreak occurred simultaneously in the United States of America and in Canada. The description of the Canadian portion of the outbreak is described in the next section (Case Number 10).

The New Jersey health department reported a mean concentration of 2.5 salmonellae per gram of chocolate from samples obtained from homes where cases occurred. A bag of the chocolate was reported to be 1 lb or approximately 450 g; therefore the maximum dose causing infection in some people was estimated to be no more than approximately 1000 cells (2.5/g × 450 g). Alternatively, the dose could be as low as 100 cells if only 40 g was consumed (2.5/g × 40 g).

—— § ——

Case Number: 10
Reference: D'Aoust et al., 1975
Serovar: *S.* Eastbourne
Setting: National; homes
Medium: Chocolate balls

Concentration		Amount Ingested		Dose	Comments
Value	Units	Value	Units		
2	#/ball	50	balls	100	Lower reported concentration and dose estimate based on consumption of entire bag
9	#/ball	50	balls	450	Upper reported concentration and dose estimate based on consumption of entire bag
2	#/ball	5	balls	10	Lower reported concentration and dose estimate based on consumption of entire bag
9	#/ball	5	balls	45	Upper reported concentration and dose estimate based on consumption of entire bag

Exposed	Response	Attack rate	Comments
?	95	?	Total number of cases in geographically widespread outbreak. No information on exposed population

Comments

This outbreak again involved a potentially susceptible population, as 46% of the cases were children aged 1 to 4 years old. There were a total of 95 reported cases. The outbreak was attributed to chocolate balls. Each ball was reported to weigh approximately 10 g, with a bag of chocolate balls containing approximately 50 balls. The contamination of the chocolate balls was estimated to be 2 to 9 salmonellae per chocolate ball. This outbreak was the Canadian part of the outbreak that also occurred simultaneously in the United States of America, and described previously (Case Number 9; Craven et al., 1975).

The dose causing illness in some of the exposed population was estimated by the authors based on the consumption of a bag of chocolate. This estimate, which might be high in view of the assumed consumption of 50 chocolate balls, would place the dose at approximately 100 to 450 cells. Depending on the assumption of the amount of chocolate that was consumed, the dose causing illness could be as low as 2 cells if only 1 ball was consumed at the lowest concentration. However, it is difficult to determine from the information given exactly how much chocolate sick individuals consumed, and what the concentration was in the chocolate that was consumed. The overall attack rate for this outbreak is also difficult to estimate, as for the previous report (Case Number 9; Craven et al., 1975), due to the geographically widespread nature of the outbreak.

—— § ——

Case Number: 11
References: Levy et al., 1996; USDA-FSIS, 1998
Serovar: *S.* Enteritidis
Setting: Hotel
Medium: Raw shell eggs (hollandaise sauce)

Concentration		Amount ingested		Dose	Comments
Value	Units	Value	Units		
1 000	#/g	10	g	10 000	Concentration reported from informal quantitation; dose estimated from consumption of 2 tablespoons of sauce
10 000	#/g	10	g	100 000	1-log higher concentration from informal quantitation results; dose estimated from consumption of 2 tablespoons of sauce

Exposed	Response	Attack rate	Comments
39	39	100.00%	Attack rate estimated from all individuals consuming hollandaise sauce becoming ill

Comment

In this outbreak, a total of 56 persons who ate at a Washington D.C. hotel had onset of diarrhoea. The Washington D.C. public health department conducted an investigation into the outbreak and identified hollandaise sauce as the likely vehicle. According to the USDA (USDA-FSIS, 1998), only 39 persons ate the hollandaise sauce, and all 39 became ill, which would imply a 100% attack rate. The attack rate in this case was assumed to be 100%, with a good characterization of the exposed and positive populations.

The actual concentration of *S.* Enteritidis causing illness in this outbreak was not reported in the publication describing the outbreak (Levy et al., 1996), but the USDA-FSIS (1998) reported the results of some testing. This informal quantitation, which was not performed to extinction, tested a sample of sauce recovered from a patron who had taken it home in a "doggy bag" and refrigerated it for 72 hours. The concentration in this sample was reported to be 10^3 per gram. It was assumed that 2 tablespoons (approximately 10 g) were consumed by the patrons of the restaurant, placing the dose at approximately 10^4 (USDA-FSIS, 1998). To allow for the uncertainty associated with the concentration estimates and the potential underestimate, an additional 1 log was allowed for in the concentration range.

Outbreak parameter uncertainty

Dose		Exposed population	Positive
Uniform distribution			
Min	Max	Value	Value
10 000	100 000	39	39

—— § ——

Case Number: 12
References: Vought and Tatini, 1998; Hennessy et al., 1996; USDA-FSIS, 1998
Serovar: *S.* Enteritidis
Setting: Interstate USA
Medium: Ice cream

Concentration		Amount ingested		Dose	Comments
Value	Units	Value	Units		
0.152	#/g	65	g	9.88	Low expected concentration using Bayesian analysis with dose calculated using smallest reported consumption amount
0.152	#/g	260	g	39.52	Low expected concentration using Bayesian analysis with dose calculated using highest reported consumption amount
0.894	#/g	65	g	58.11	High expected concentration using Bayesian analysis with dose calculated using smallest reported consumption amount
0.894	#/g	260	g	232.44	High expected concentration using Bayesian analysis with dose calculated using highest reported consumption amount

Exposed	Response	Attack rate	Comments
452	30	6.6%	Attack rate calculated based on a cross-section study for which exposure and response details were available

Comment

This was an interstate outbreak attributed to ice cream. Hennessy et al. (1996) provide details on the epidemiological characteristics of the outbreak and the concentration of *S.* Enteritidis found in samples of ice cream, using traditional MPN techniques. The effect of frozen storage was also experimentally investigated. The authors found no evidence of a decrease in numbers during storage at -20°C for 16 weeks, unlike the work of Armstrong et al. (1970), described previously (Case Number 5). A re-analysis of the quantitative MPN results was performed at a later date using alternative statistical tools to better estimate the concentration in the ice cream (Vought and Tatini, 1998). The expected concentration was reported as 0.152 MPN/g at the lower range, and 0.894 MPN/g at the upper range. In addition, a small group of people that were investigated in more detail were reported to have consumed from 65 to 260 g. The uncertainty in the dose was therefore assumed to range from 10 cells to 235 cells.

The outbreak was reported to have affected a large number of people, and from the report by Hennessy et al. (1996) there are details on a smaller cross-section of the group, with whom interviews were conducted. A total of 541 people were interviewed that had purchased the incriminated ice cream, of which 452 were reported to have consumed the product. To allow for some uncertainty in the exposed population, it was assumed that this could be 10% less than the number that reported eating the ice cream. A total of 30 individuals became ill in the population in the cross-section study. The numbers of positives were assumed to have 5% under- and over-reporting.

Outbreak parameter uncertainty

Dose		Exposed population			Positive		
Uniform distribution		Pert distribution			Pert distribution		
Min	Max	Min	ML	Max	Min	ML	Max
10	235	407	451	452	29	30	32

§

Case Number: 13
Reference: Taylor et al., 1984
Serovar: *S.* Typhimurium
Setting: Home
Medium: Ice cream

Concentration		Amount ingested		Dose	Comments
Value	Units	Value	Units		
1.00E+06	#/ml	1000	ml	1.00E+09	Dose estimated for fatality in 13-year-old boy
1.00E+06	#/ml	750	ml	7.50E+08	Dose for individuals consuming 750 ml
1.00E+06	#/ml	250	ml	2.50E+08	Dose for individuals consuming 250 ml
1.00E+06	#/ml	100	ml	1.00E+08	Dose for 2-year-old girl consuming 100 ml

Exposed	Response	Attack rate	Comments
7	7	100%	

Comments

This outbreak involved a family and one neighbour, and was attributed to home-made ice cream. The ages (years) of the exposed population were: father, 35; mother, 30; sons, 13, 9 and 8; daughters, 6 and 2; and a male neighbour, 22. Ice cream was obtained from the freezer at the farm and found to have 10^6 salmonellae/ml. One of the sons, aged 13 years, who ate the most ice cream (1000 ml) died from his illness. Various amounts of ice cream, ranging from 100 ml to 1000 ml, were reported to have been consumed by the family members. Since the actual sample of ice cream was obtained from the freezer only a few days after the event, the concentration reported was assumed to reflect that at the time of consumption. The uncertainty in the dose was modelled using the given concentration, and accounting for the different amounts consumed. In the current analysis, the child of 2 years of age was assumed to be potentially more susceptible, while the other individuals were assumed to represent a normal population. In this particular case, there was no uncertainty in the exposed and positive populations.

Outbreak parameter uncertainty

Dose		Exposed population	Positive
Uniform distribution			
Min	Max	Value	Value
1.0E+8	7.5E+8	7	7
2.5E+8	7.5E+8	6	6

——— § ———

Case Number: 14
Reference: D'Aoust, 1985; D'Aoust, Warburton & Sewell, 1985
Serovar: *S.* Typhimurium
Setting: Nationwide
Medium: Cheddar cheese

Concentration		Amount ingested		Dose	Comments
Value	Units	Value	Units		
0.36	#/100 g	100	g	0.36	Minimum concentration in samples from plant
9.3	#/100 g	100	g	9.3	Maximum concentration in samples from plant
3.5	#/100 g	100	g	3.5	Average concentration in samples from plant
1.5	#/100 g	100	g	1.5	Minimum concentration in food samples from patients
9.1	#/100 g	100	g	9.1	Maximum concentration in food samples from patients
4.2	#/100 g	100	g	4.2	Average concentration in food samples from patients

Exposed	Response	Attack rate	Comments
?	1500	?	

Comments

This outbreak involved more than 1500 people, with cheddar cheese implicated as the vehicle of infection. Cheese samples were obtained from the production plant, as well as from homes of some of the individuals that were ill. The level of contamination in the cheese from the plant was found to be between 0.36 and 9.3 salmonellae per 100 g (D'Aoust, Warburton & Sewell, 1985), while the level of contamination in cheese from individual homes was found to be between 1.5 and 9.1 salmonellae per 100 g (D'Aoust, 1985). The average concentration from cheese plant samples was estimated to be 3.5/100 g while those from homes was estimated to be 4.2/100 g. The authors noted that the number of salmonellae probably did not change substantially during storage, and the levels estimated reflect the levels at the time of consumption. It was estimated that approximately 100 g of cheese was consumed, based on the level of consumption reported for six individuals, ranging from 20 g to 170 g.

The attack rate in this case is again difficult to estimate due to a lack of information on the exposed population and the inability to make reasonable assumptions given the information and the widespread distribution of the outbreak.

—— § ——

Case Number: 15
Reference: George, 1976
Serovar: *S.* Schwarzengrund
Setting: Hospital
Medium: Pancreatin

Concentration		Amount ingested		Dose	Comments
Value	Units	Value	Units		
1000	#/g	0.2	g	200	Reported concentration and dose estimated from consumption of 200 mg by single susceptible individual

Exposed	Response	Attack rate	Comments
1	1	100%	

Comments

This case involved a susceptible individual (1-year-old child) who developed diarrhoea when treated with pancreatic extract (pancreatin is an extract from the pancreas of mammals, and used to assist in the digestion of food) that was contaminated with *S.* Schwarzengrund. The pancreatic extract was found to contain 1000 salmonellae per gram, and the child became ill following ingestion of 200 mg. It should be noted that this case involves only one individual and the 100% attack rate quoted for this dose could skew the true attack rate, which could be less for a group of individuals receiving this dose. For example, it could be possible that this one individual might be the only one that got sick if 20 similarly susceptible individuals were given the same dose. In that hypothetical situation, the attack rate would be estimated to be only 5%.

—— § ——

Case Number: 16
Reference: Lipson, 1976
Serovar: *S.* Schwarzengrund
Setting: Hospital
Medium: Pancreatin

Concentration		Amount ingested		Dose	Comments
Value	Units	Value	Units		
8	#/g	5.6	g	44.8	Reported concentration and dose estimate based on last 24 hours of feedings, comprising 4 × 1.4-g amounts.

Exposed	Response	Attack rate	Comments
1	1	100%	

Comments

This case involved a single susceptible individual (9-month-old child with cystic fibrosis), who was fed pancreatin contaminated with *S.* Schwarzengrund. The pancreatin was found to be contaminated at a level of 8 salmonellae per gram. The child was given approximately 700 mg with each 6-hourly feed for the first 10 days, increasing to approximately 1.4 g in the 36 hours before the onset of symptoms. The authors note that the child had therefore ingested less than 22 organisms per day initially and less than 44 organisms per day in the last 36 hours. If the dose is not cumulative over 24 hours, then the infective dose would be approximately 44 organisms (24 hours, fed every 6 hours, which translates to 4 feedings; each feeding is 1.4 g, which translates to 5.6 g. 5.6 g × 8/g = approximately 44 cells). The points raised about one individual exposed and the attack rate estimates in the previous case (Number 15; George, 1976) also apply in this case.

——— § ———

Case Number: 17
Reference: Greenwood and Hooper, 1983
Serovar: *S.* Napoli
Setting: Nationwide
Medium: Chocolate bars

Concentration		Amount ingested		Dose	Comments
Value	Units	Value	Units		
16	#/10 g	64	g	102	Average reported concentration and consumption amount by one individual that became ill
58.5	#/10 g	64	g	374	Highest average concentration reported in a packet of 6 bars
240	#/10 g	64	g	1540	Highest concentration reported in an individual bar

Exposed	Response	Attack rate	Comments
1	1	100%	Widespread outbreak geographically, with a large potentially exposed and sick population, but details only available on one individual

Comments

This was a nationwide outbreak attributed to chocolate bars (16 g each) contaminated with *S.* Napoli. Although the overall attack rate in the population exposed cannot be determined, details were given on three individuals: a mother and two sons. All three ate two bars on the first day, and one son ate two more bars on the second day. The son that ate chocolate bars on both days became ill. He may have received a larger dose, or, alternatively, not all the bars were contaminated and the ill child ingested a single contaminated bar. We can only state that the attack rate for the one child that ate four chocolate bars was 100%.

A box of chocolates, which consisted of 8 packets with 6 bars in each packet, was obtained from a retailer from whom two patients had purchased chocolate. This box of chocolates was analysed and 42 of the 48 bars examined were positive, with the average concentration for the positive bars reported to be 16 organisms per 10 g. The highest concentration for one bar was 240 organisms per 10 g, and the lowest was 3 organisms per 10 g. It was also observed that the level of contamination per packet was not consistent. Packets consisting of 6 bars that were all positive also tended to have a higher contamination level. Of the 8 packets examined, the packet with the highest average concentration was 58.5 organisms per 10 g.

Since information is only known about one case, these data were not considered for further analysis.

—— § ——

Case Number: 18
Reference: Ministry of Health and Welfare [Japan], 1999
Serovar: *S.* Enteritidis (PT4)
Setting: Restaurant
Medium: Roasted beef

Concentration		Amount ingested		Dose	Comments
Value	Units	Value	Units		
2000	#/g	120	g	240 000	Reported concentration and consumption

Exposed	Response	Attack rate	Comments
5	3	60%	Reported exposed and positive numbers

Comments

In order to incorporate uncertainty in the dose, the concentration and amount consumed were assumed to have a potential range of 25% of the one reported. The lower and upper bounds for the dose were estimated to be 135 000 (1500 CFU/g × 90 g) to 375 000 (2500 CFU/g × 150 g). Since the size of the exposed population was reasonably small, it can be assumed that the uncertainty associated with the exposed and positive populations is minimal.

Outbreak parameter uncertainty

Dose		Exposed population	Positive
Uniform distribution			
Min	Max	Value	Value
135 000	375 000	5	3

—— § ——

Case Number: 19
Reference: Ministry of Health and Welfare [Japan], 1999
Serovar: *S.* Enteritidis
Setting: Caterer
Medium: Grated yam diluted with soup

Concentration		Amount ingested		Dose	Comments
Value	Units	Value	Units		
32 000	#/g	60	g	1 920 000	Reported concentration and consumption

Exposed	Response	Attack rate	Comments
123	113	91.87%	Reported exposed and positive numbers

Comments

In order to incorporate uncertainty in the dose, the concentration and amount consumed were assumed to have a potential range of 25% of the one reported. The lower and upper bounds for the dose were estimated to be 1 080 000 (24 000 CFU/g × 45 g) to 3 000 000 (40 000 CFU/g × 15 g). The exposed and positive populations in this case were potentially uncertain. Since the degree of uncertainty is unknown, it was assumed that the reported exposed population could not have been exceeded; however, there could have been 10% fewer people actually exposed. The number of positives reported was assumed to represent the most likely number, but a 5% under- and over-reporting were allowed for.

Outbreak parameter uncertainty

Dose		Exposed population			Positive		
Uniform distribution		Pert distribution			Pert distribution		
Min	Max	Min	ML	Max	Min	ML	Max
1 080 000	3 000 000	111	122	123	107	113	119

—— § ——

Case Number: 20
Reference: Ministry of Health and Welfare [Japan], 1999
Serovar: *S.* Enteritidis (PT22)
Setting: School lunch
Medium: Beef and bean sprouts

Concentration		Amount ingested		Dose	Comments
Value	Units	Value	Units		
40	#/g	22	g	880	Reported concentration and consumption

Exposed	Response	Attack rate	Comments
10 552	967	9.16%	Reported number of potentially exposed population
5 276	967	18.33%	Attack rate with 1/2 of the population exposed
3 517	967	27.50%	Attack rate with 1/3 of the population exposed
2 638	967	36.66%	Attack rate with 1/4 of the population exposed

Comments

The number of potentially exposed elementary school students (6 to12 years old) was very large, since a central cooking facility served 15 schools. Patients were found from almost all the schools, but there was an indication that most of the exposures occurred at 5 schools. It is highly unlikely that all 10 775 people were exposed to contaminated food. As a result, it was assumed that only a proportion, ranging from 1/2 to 1/4 of the total potentially exposed population, were actually exposed. There could also be uncertainty in the number of positives, but given the size of the denominator (exposed population) and the size of the numerator (positives), incorporating a 5% allowance for under- and over-reporting has minimal effect on the attack rate uncertainty range.

In order to incorporate uncertainty in the dose, the concentration and amount consumed were assumed to have a potential range of 25% of the one reported. The lower and upper bounds for the dose were estimated to be 495 (30 CFU/g × 16.5 g) and 1375 (50 CFU/g × 27.5 g), respectively.

Outbreak parameter uncertainty

Dose		Exposed population			Positive
Uniform distribution		Pert distribution			
Min	Max	Min	ML	Max	Value
495	1375	2638	3517	5276	967

—— § ——

Case Number: 21
Reference: Ministry of Health and Welfare [Japan], 1999
Serovar: *S.* Enteritidis
Setting: Home
Medium: Egg

Concentration		Amount ingested		Dose	Comments
Value	Units	Value	Units		
<0.03	#/g	60	g	<1.8	

Exposed	Response	Attack rate	Comments
5	3	[60.00%]	

—— § ——

Case Number: 22
Reference: Ministry of Health and Welfare [Japan], 1999
Serovar: *S.* Enteritidis
Setting: Hotel
Medium: Scallop roasted with egg yolk (product 1);
                 Shrimp roll in bread (product 2);
                 Hamburg steak (product 3)

Concentration		Amount Ingested		Dose	Comments
Value	Units	Value	Units		
47 000	#/g	40	g	1 880 000	Concentration and consumption amount reported for product 1

Exposed	Response	Attack rate	Comments
115	63	54.78%	

Comments

In order to incorporate uncertainty in the dose, the concentration and amount consumed were assumed to have a potential range of 25% of the values reported. The lower and upper bounds for the dose were estimated to be 1 057 500 (35 250 CFU/g × 30 g) and 2 937 500 (58 750 CFU/g × 50 g). The exposed and positive populations in this case were also potentially uncertain. Since the degree of uncertainty is unknown, it was assumed that the reported exposed population could not have been exceeded, and also that there could have been 10% fewer people actually exposed. The number of positives reported was assumed to represent the most likely number, but 5% under- and over-reporting was allowed for.

Outbreak parameter uncertainty

Dose		Exposed population			Positive		
Uniform distribution		Pert distribution			Pert distribution		
Min	Max	Min	ML	Max	Min	ML	Max
1 057 500	2 937 500	104	114	115	60	63	66

—— § ——

Case Number: 23
Reference: Ministry of Health and Welfare [Japan], 1999
Serovar: *S.* Enteritidis
Setting: Confectionery
Medium: Cake

Concentration		Amount ingested		Dose	Comments
Value	Units	Value	Units		
6000	#/g	100	g	600 000	Reported concentration and amount consumed

Exposed	Response	Attack rate		Comments
13	11	84.62%	Reported attack rate	

Comments

In order to incorporate uncertainty in the dose, the concentration and amount consumed were assumed to have a potential range of 25% of the one reported. The lower and upper bounds for the dose were estimated to be 337 500 (4500 CFU/g × 75 g) and 937 500 (7500 CFU/g × 125 g), respectively. Since the size of the exposed population was reasonably small, it can be assumed that the uncertainty associated with the exposed and positive populations is minimal.

Outbreak parameter uncertainty

Dose		Exposed population	Positive
Uniform distribution			
Min	Max	Value	Value
337 500	937 500	13	11

—— § ——

Case Number: 24
Reference: Ministry of Health and Welfare [Japan], 1999
Serovar: *S.* Enteritidis (PT1)
Setting: School lunch
Medium: Peanut sauce

Concentration		Amount ingested		Dose	Comments
Value	Units	Value	Units		
1.4	#/g	35	g	49	Reported concentration and amount consumed

Exposed	Response	Attack rate	Comments
5320	644	12.11%	Reported attack rate

Comments

The attack rate that was reported for this outbreak was based on exposure of the entire school population that received lunch from the central kitchen. With such a large exposed population, which can be highly uncertain, the estimated attack rate can vary widely. It is highly unlikely that the entire reportedly exposed population was actually exposed to the contaminated food. Unlike the prior school outbreak (Case Number 20), there was no indication in this case of some schools being more likely to have been exposed than others. As a result, it was assumed that only a proportion, ranging down to 1/2 of the total potentially exposed population, were actually exposed. There could also be uncertainty in the number of positives, but given the size of the denominator (exposed population) and the size of the numerator (positives), incorporating a 5% allowance for under- and over-reporting has minimal effect on the attack rate uncertainty range.

In order to incorporate uncertainty in the dose, the concentration and amount consumed were assumed to have a potential range of 25% of the one reported. The lower and upper bounds for the dose were estimated to be 28 (1.05 CFU/g × 26.25 g) and 77 (1.75 CFU/g × 43.75 g), respectively.

Outbreak parameter uncertainty

Dose		Exposed population			Positive
Uniform distribution		Pert Distribution			
Min	Max	Min	ML	Max	Value
28	77	2660	3990	5320	644

—— § ——

Case Number: 25
Reference: Ministry of Health and Welfare [Japan], 1999
Serovar: *S.* Enteritidis
Setting: Day care
Medium: Cooked chicken and egg

Concentration		Amount ingested		Dose	Comments
Value	Units	Value	Units		
27	#/g	150	g	4050	Reported concentration and amount consumed

Exposed	Response	Attack rate	Comments
16	3	18.75%	Exposed and positive adults at day care
117	50	42.74%	Exposed and positive children at day care
133	53	39.85%	Exposed and positive population at day care

Comments

The food was a rice dish covered with cooked chicken and eggs. Of 133 exposed people, 16 were adults (3 became ill) and 117 were children (50 became ill). Day care-aged children were assumed to be of increased potential susceptibility to foodborne pathogens. Because of the outbreak setting (day care), the exposed and positive populations were assumed to be well characterized in this case.

In order to incorporate uncertainty in the dose, the concentration and amount consumed were assumed to have a potential range of 25% of the one reported. The lower and upper bounds for the dose were estimated to be 2278 (20.25 CFU/g × 112.5 g) and 6328 (33.75 CFU/g × 187.5 g), respectively.

Outbreak parameter uncertainty

Dose		Exposed population	Positive
Uniform distribution			
Min	Max	Value	Value
2 278	6 328	16	3
2 278	6 328	117	50
2 278	6 328	133	53

— § —

Case Number: 26
Reference: Ministry of Health and Welfare [Japan], 1999
Serovar: *S.* Enteritidis (PT1)
Setting: School lunch
Medium: Peanut sauce

Concentration		Amount ingested		Dose	Comments
Value	Units	Value	Units		
<100	#/g	80	g	8000	Reported concentration and amount consumed

Exposed	Response	Attack rate	Comments
2 267	418	18.44%	Reported exposed and positive population

Comments

The attack rate that was reported for this outbreak was based on exposure of the entire school population that received lunch from the central kitchen. With such a large exposed population, which can be highly uncertain, the estimated attack rate can vary widely. It is highly unlikely that the entire reportedly exposed population was actually exposed to the contaminated food. In addition, the reported concentration per gram of food was less than 100 CFUs, which introduces a second significant uncertain parameter.

—— § ——

Case Number: 27
Reference: Ministry of Health and Welfare [Japan], 1999
Serovar: *S.* Enteritidis
Setting: Hospital
Medium: Raw egg in *natto*

Concentration		Amount ingested		Dose	Comments
Value	Units	Value	Units		
1.20E+06	#/g	50	g	6.00E+07	Reported concentration and amount consumed

Exposed	Response	Attack rate	Comments
191 ?	45	23.56%	Reported exposed and positive population

Comments

Eggs were pooled in the preparation of this food. The number exposed was the number of people who were served with this dish. Of the 191 served, 128 answered the food-intake questionnaire. Some of the hospital patients could not talk. Among 128 responses, 36 did not actually consume this dish. Among the 45 cases, 2 were tuberculosis (TB) patients and apparently had taken antibiotics. The number of TB patients in the actual exposed population is unknown. This outbreak is highly unusual because the dose is very high but the attack rate is very low. In addition, the outbreak is reported to have occurred in a hospital, an environment in which one might expect, depending on the circumstances, the exposed population to be more susceptible than the overall population. Because of the uncertainties in these data and the potential confounding factors, this outbreak was not included for further analysis.

—— § ——

Case Number: 28
Reference: Ministry of Health and Welfare [Japan], 1999
Serovar: *S.* Enteritidis (PT4)
Setting: Hospital
Medium: Grated yam diluted with soup

Concentration		Amount ingested		Dose	Comments
Value	Units	Value	Units		
2400	#/g	60	g	144 000	Reported concentration and amount consumed

Exposed	Response	Attack rate	Comments
343 ?	75	21.87%	

Comments

This outbreak is unusual, like the previous hospital-associated outbreak (Number 27). Eggs were pooled and mixed well in preparing this dish. The actual number of individuals exposed is suspected to be lower than originally reported. The reported attack rate is lower than would be expected at this high dose level. It should be noted that some of the patients had antibiotic treatment, which may be a confounding factor in interpretation of these data.

—— § ——

Case Number: 29
Reference: Ministry of Health and Welfare [Japan], 1999
Serovar: *S.* Enteritidis (PT1)
Setting: Hospital
Medium: Tartar sauce

Concentration		Amount ingested		Dose	Comments
Value	Units	Value	Units		
100	#/g	36	g	3600	

Exposed	Response	Attack rate	Comments
126	36	28.57%	

Comment

This outbreak is also unusual, similar to the previous two hospital outbreaks, although in this case the dose is not as high as reported in Numbers 27 and 28. Information about confounding factors in these hospital outbreaks, such as diagnoses and treatments that patients were undergoing, was not available. Therefore, the three Japanese hospital outbreaks were not included in further analysis.

——— § ———

Case Number: 30
Reference: Ministry of Health and Welfare [Japan], 1999
Serovar: *S.* Enteritidis (PT1)
Setting: Restaurant
Medium: Cooked egg

Concentration		Amount ingested		Dose	Comments
Value	Units	Value	Units		
200	#/g	30	g	6000	Reported concentration and attack rate and average amount consumed

Exposed	Response	Attack rate	Comments
885	558	63.05%	

Comment

In order to incorporate uncertainty in the dose, the concentration and amount consumed were assumed to have a potential range of 25% of the values reported. The lower and upper bounds for the dose were estimated to be 3375 (150 CFU/g × 22.5 g) and 9375 (250 CFU/g × 37.5 g), respectively. The exposed and positive populations in this case were also potentially uncertain. Since the degree of uncertainty is unknown, it was assumed that the reported exposed population could not have been exceeded, and it was assumed that there could have been 10% fewer people actually exposed. The number of positives reported was assumed to represent the most likely number, but 5% under- and over-reporting was allowed for.

Outbreak parameter uncertainty

Dose		Exposed population			Positive		
Uniform distribution		Pert distribution			Pert distribution		
Min	Max	Min	ML	Max	Min	ML	Max
3 375	9 375	797	884	885	530	558	586

——— § ———

Case Number: 31
Reference: Ministry of Health and Welfare [Japan], 1999
Serovar: *Salmonella* Enteritidis (PT4)
Setting: Confectionery
Medium: Cake

Concentration		Amount ingested		Dose	Comments
Value	Units	Value	Units		
14	#/g	30	g	420	Reported concentration and amount consumed

Exposed	Response	Attack rate	Comments
5 103	1 371	26.87%	

Comment

In order to incorporate uncertainty in the dose, the concentration and amount consumed were assumed to have a potential range of 25% of the values reported. The lower and upper bounds for the dose were estimated to be 236 (11 CFU/g × 22.5 g) and 656 (18 CFU/g × 37.5 g), respectively. The exposed and positive populations in this case were also potentially uncertain. Since the degree of uncertainty is unknown, it was assumed that the reported exposed population could not have been exceeded, and it was assumed that there could have been 10% fewer people actually exposed. The number of positives reported was assumed to represent the most likely number, but 5% under- and over-reporting was allowed for.

Outbreak parameter uncertainty

Dose		Exposed population			Positive		
Uniform distribution		Pert distribution			Pert distribution		
Min	Max	Min	ML	Max	Min	ML	Max
236	656	4 593	5 102	5 103	1 302	1 371	1 440

——— § ———

Case Number: 32
Reference: Ministry of Health and Welfare [Japan], 1999
Serovar: *S.* Enteritidis
Setting: Day care
Medium: Egg salad

Concentration		Amount ingested		Dose	Comments
Value	Units	Value	Units		
0.78	#/g	30	g	23.4	Reported concentration and amount consumed

Exposed	Response	Attack rate	Comments
156	42	26.92%	

Comment

This outbreak was assumed to represent a susceptible population since the outbreak occurred in a day care facility. In order to incorporate uncertainty in the dose, the concentration and amount consumed were assumed to have a potential range of 25% of the values reported. The lower and upper bounds for the dose were estimated to be 13 (0.59 CFU/g × 22.5 g) and 37 (0.98 CFU/g × 37.5 g), respectively. The exposed and positive populations were assumed to be well characterized in this case because of the outbreak setting (day care).

Outbreak parameter uncertainty

Dose		Exposed population	Positive
Uniform distribution			
Min	Max	Value	Value
13	37	156	42

—— § ——

Case Number: 33
Reference: Ministry of Health and Welfare [Japan], 1999
Serovar: *S.* Oranienburg
Setting: Hotel
Medium: Grated yam diluted with soup

Concentration		Amount ingested		Dose	Comments
Value	Units	Value	Units		
5.00E+07	#/g	150	g	7.50E+09	Reported concentration and amount consumed

Exposed	Response	Attack rate	Comments
11	11	100.00%	

Comment

In order to incorporate uncertainty in the dose, the concentration and amount consumed were assumed to have a potential range of 25% of the one reported. The lower and upper bounds for the dose were estimated to be 4.22E+9 (3.75E+7 CFU/g × 112.5 g) and 1.17E+10 (6.25E+7 CFU/g × 187.5 g), respectively. Since the size of the exposed population was reasonably small, it can be assumed that the uncertainty associated with the exposed and positive populations is minimal.

Outbreak parameter uncertainty

Dose		Exposed population	Positive
Uniform distribution			
Min	Max	Value	Value
4.22E+9	1.17E+10	11	11

———— § ————

3.5.2 Epidemiological data summary and analysis

Of the 33 outbreak reports collected from the published literature and from unpublished data received by FAO and WHO following the call for data, 23 contained sufficient information on the number of people exposed, the number of people that became ill, and the number of organisms in the implicated food to enable calculation of a dose-response relationship. Of the 23 outbreaks, 3 were excluded because the immune status of the persons exposed could not be determined. The remaining 20 outbreaks comprise the database used to calculate a dose-response relationship.

Of the 20 outbreaks in the database, 11 occurred in Japan and 9 occurred in North America. Several serotypes were associated with the outbreaks, including Enteritidis (12 outbreaks), Typhimurium (3), and in single outbreaks, Heidelberg, Cubana, Infantis, Newport and Oranienburg. Several vehicles were implicated, including food (meat, eggs, dairy products and others), water, and a medical dye capsule (carmine dye).

Reports provided by the Ministry of Health and Welfare of Japan (1999) represent a valuable source of information on the real-world dose-response relationship and expand our database of *Salmonella* pathogenicity considerably. The data in these reports are generated as part of the epidemiological investigations that take place in Japan following any outbreak of foodborne illness. In accordance with a Japanese notification released on March 1997, large-scale cooking facilities that prepare more than 750 meals per day or more than 300 dishes of a single menu at a time are advised to save food for future possible analysis in the event of an outbreak. The notification is also applicable to smaller-scale kitchens with social responsibility, such as those in schools, day care centres and other child-welfare and social-welfare facilities. Thus, 50-g portions of each raw food ingredient and each cooked dish are saved for more than 2 weeks at a temperature below -20°C. Although this notification is not mandatory, the level of compliance is high, and some of the local governments in Japan also have local regulations that require food saving, but the duration and the storage temperature requirements vary.

The doses, attack rates, serovars and characteristics of the exposed populations derived from the outbreak reports described in the preceding section are summarized in Table 3.14 and Figure 3.10. The analysis of the epidemiological data was intended to serve three purposes:

[1]   To determine if there is any epidemiological evidence for greater attack rates in susceptible vs normal populations.

[2] To determine if there is any epidemiological evidence for different attack rates for *S*. Enteritidis compared with other *Salmonella* serotypes.

[3] To compare the epidemiological data for dose and attack rate with the estimates generated by the dose-response models.

Table 3.14. Summary of outbreak data.

Case no.	Serovar	Food	Popn.[1]	Dose[2] Log CFU	Attack Rate[2](%)	Reference(s)
1	*S*. Typhimurium	Water	N	2.31	10.63%	Boring, Martin and Elliott, 1971
	S. Typhimurium	Water	S	2.31	18.91%	
2	*S*. Heidelberg	Cheddar cheese	N	2.22	32.76%	Fontaine et al., 1980
3	*S*. Cubana	Carmine dye	S	4.57	70.93%	Lang et al., 1967
4	*S*. Infantis	Ham	N	6.46	100.00%	Angelotti et al., 1961
5	*S*. Typhimurium	Imitation ice cream	N	3.79	55.00%	Armstrong et al.,1970
7	*S*. Newport	Hamburger	N	1.23	1.07%	Fazil., 1996 Fontaine et al., 1978
11	*S*. Enteritidis	Hollandaise sauce	N	4.74	100.00%	Levy et al., 1996; USDA-FSIS., 1998
12	*S*. Enteritidis	Ice cream	N	2.09	6.80%	Vought and Tatini, 1998; Hennessy et al., 1996
13	*S*. Typhimurium	Ice cream	N	8.70	100%	Taylor et al., 1984
	S. Typhimurium	Ice cream	S	8.00	100%	
18	*S*. Enteritidis	Roasted beef	N	5.41	60.00%	Ministry of Health and Welfare, Japan, 1999
19	*S*. Enteritidis	Grated yam with soup	N	6.31	93.93%	
20	*S*. Enteritidis	Beef and bean sprouts	N	2.97	26.86%	
22	*S*. Enteritidis	Scallop with egg yolk	N	6.30	56.01%	
23	*S*. Enteritidis	Cake	N	5.80	84.62%	
24	*S*. Enteritidis	Peanut sauce	N	1.72	16.41%	
25	*S*. Enteritidis	Chicken and egg	N	3.63	18.75%	
25	*S*. Enteritidis	Chicken and egg	S	3.63	42.74%	
30	*S*. Enteritidis	Cooked egg	N	3.80	64.18%	
31	*S*. Enteritidis	Cake	N	2.65	27.33%	
32	*S*. Enteritidis	Egg salad	S	1.40	26.92%	
33	*S*. Oranienburg	Grated yam with soup	N	9.90	100%	

NOTES: (1) Popn. = population exposed, where N = Normal population and S = Susceptible population.
(2) Expected value based on defined uncertainty ranges and distributions.

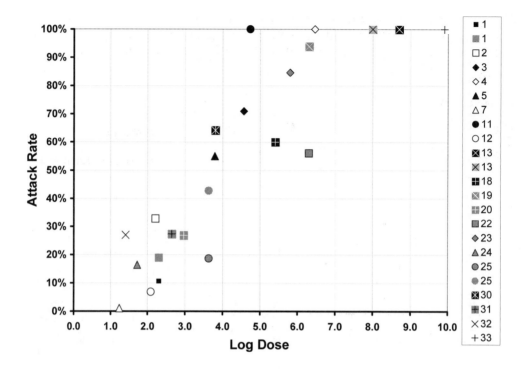

Figure 3.10. Summary of epidemiological data. Legend numbers indicate outbreak number given in text.

The data in Table 3.14 and Figure 3.10 are coded according to the outbreak number assigned in this document. If additional information related to a specific data point is required, for example the assignment of two data points, the details of the outbreak can be referred to in the previous section. The related assumptions for inclusion, exclusion or multiple data points are certainly issues for discussion and debate, and therefore included in the summary of reported outbreaks.

The data shown in Figure 3.10 appear to reflect our theoretical assumptions regarding the increasing trend in attack rates as dose increases. In addition, although there is a degree of clustering in some of the data points, a dose-response relationship is visually evident.

As noted earlier, some data were excluded from this summary and further analysis. For example, outbreaks numbers 27, 28 and 29 were attributed to *S.* Enteritidis in a hospital setting, where the exposed population would be expected to be more susceptible. The characteristics of the individuals that were exposed to the food is highly uncertain, so it may in fact be the case that the condition for which they were hospitalized is such that their immunity was not compromised. However, even if they are assumed to have normal susceptibility, these outbreaks were still distinctly different from outbreaks with a similar dose level, if the reported exposures were accurate. Alternative explanations for these data sets are that the individuals served the meal did not actually consume the implicated food, or that concurrent antibiotic therapy prevented the ingested *Salmonella* from colonization and illness production.

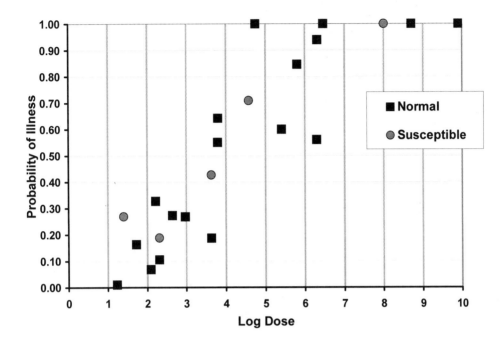

Figure 3.11. Attack rates corresponding to dose for "Normal" and "Susceptible" populations in reported outbreaks.

Susceptible vs Normal Populations

The observed outbreak data were used to gain some insight into the potential differences that may exist between "susceptible" and "normal" populations. The database of quantitative outbreak information collected during the course of this work includes several outbreaks that could be associated with "susceptible" and "normal" populations. Unfortunately, limited data allowed a comparison to be made based only on age. Susceptibility in this analysis was therefore limited to outbreak data for individuals less than 5 years old being classified as "susceptible", with other outbreak data representing a "normal" population. This was the case for all but one of the "susceptible" data points (estimated 85% attack rate, approximately 4.5 log dose), that occurred in a hospital and was attributed to carmine dye capsules. The "susceptible" and "normal" outbreak data were compared on the basis of reported attack rate corresponding with reported dose. Given the potential range in the observed data (dose and attack rate could vary based on the nature of the epidemiological investigation), the comparison was intended to look for overall trends first, and then, if necessary, additional analysis could be done. A plot of dose against attack rate for the "susceptible" and "normal" populations is shown in Figure 3.11.

Similarly, at other dose intervals there are outbreaks attributed to "normal" populations with attack rates either very similar to or higher than outbreaks involving "susceptible" populations. Given the data that currently exists from outbreaks, there is **insufficient** evidence to conclude that "susceptible" individuals, as defined in this database, have a higher probability of illness compared with the "normal" population.

It should be noted that, within the database of outbreaks, there are two outbreaks in which a "susceptible" and "normal" population were identified in the same outbreak with differing attack rates. The "susceptible" definition in these cases was again based on an age criteria (<5 years old and >5 years old). In these two outbreaks, shown in Figure 3.12, the attack rate was clearly reported to be higher for the susceptible population compared with the normal population. Taken in isolation, it could be concluded from this information that there is clearly a higher probability of illness for the susceptible population compared with the normal population. However, if we look at the whole picture, we can see other outbreaks involving a "normal" population with higher attack rates at similar doses.

Given the outbreak data that are currently available, it is not possible to conclude that some segments of the population are more susceptible to becoming ill upon exposure to *Salmonella* than are other segments. Furthermore, it is impossible to derive a quantitative estimate of the increased probability of illness for some segments of the population compared with others. The dose-response relationship for the probability of illness for different segments of the population was therefore assumed to be the same.

The key distinction that needs to be made in this conclusion is that the probability of illness is assumed indistinguishable, given the current data and the susceptible populations defined in the database. It is important to recognize that even if the probability of becoming ill, defined in the dose-response assessment as any degree of gastroenteritis, the *severity* of the illness may be markedly different for certain segments of the population. To quantify the probabilities of different outcomes, information is needed in the form of quantitative patient follow-up and data on physician visits, hospitalizations, death or other chronic outcomes.

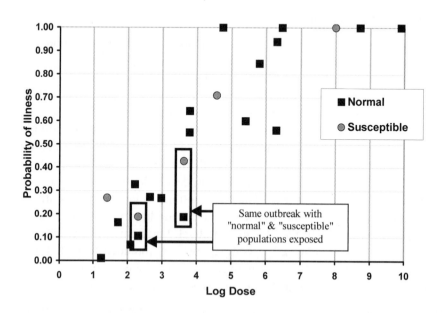

Figure 3.12. Attack rates for two outbreaks in which different populations in the same outbreak were identified.

S. *Enteritidis vs other* Salmonella *serovars*

In a similar manner to the comparisons made for susceptible and normal populations, the attack rates in outbreaks associated with *S.* Enteritidis were compared with outbreaks associated with other *Salmonella* serovars. This information is summarized in Figure 3.13.

The attack rates observed in outbreaks associated with other *Salmonella* serovars are indistinguishable from outbreaks associated with *S.* Enteritidis. At some dose ranges, the highest attack rate reported is for *S.* Enteritidis, while at others the highest attack rate is for other serovars. Based on this information, *S.* Enteritidis and other serovars were treated as equivalent for the purposes of the dose-response relationship. It is acknowledged, however, that less virulent strains may infrequently be the cause of foodborne outbreaks and hence would not be captured in this database.

In summary, it was concluded that for the purposes of the current assessment and based upon the existing observed evidence:

(1) a single dose-response relationship for the probability of illness would be used for all members of the population; and

(2) *S.* Enteritidis and other *Salmonella* serovars are assumed to have a similar probability of initiating illness at the same dose.

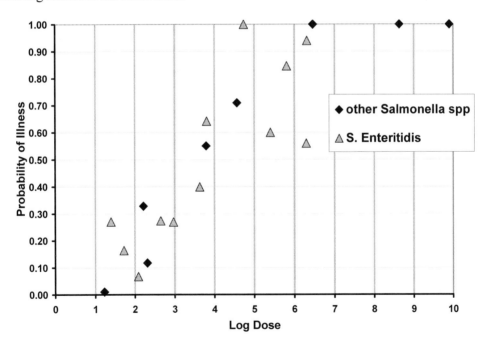

Figure 3.13. Attack rates corresponding to dose for *S.* Enteritidis and other *Salmonella* in reported outbreaks.

Comparison of outbreak data with existing Salmonella *dose-response models*

Three dose-response models for *Salmonella* exist in the literature. The first (Fazil, 1996) is the beta-Poisson model (Haas, 1983) fitted to the human feeding trial data for *Salmonella* infection (McCullough & Eisele, 1951a, c, d). The second model was proposed in the US SE RA (USDA-FSIS, 1998) and was based on the use of a surrogate pathogen to describe the dose-response relationship. This model assumed a shift in the dose-response model for "susceptible" and "normal" populations. The third model was introduced in a *Salmonella* Enteritidis risk assessment done by Health Canada (Health Canada, 2000, but unpublished), which was based on a Weibull dose-response relationship that was updated to reflect selected outbreak information using Bayesian techniques. Similar to the US SE RA model, this one also assumed a higher probability of illness for susceptible populations. The models and their comparison with the outbreak data are shown in Figures 3.14 to 3.16, and discussed in the following sections.

Naive human feeding trial data (beta-Poisson model)

The model suffers from the nature of the feeding trial data (i.e. the subjects used were healthy male volunteers) and may not reflect the population at large. The model also tends to greatly underestimate the probability of illness as observed in the outbreak data (Figure 3.14), even under the extremely conservative assumption that infection, as measured in the dose-response curve, equates to illness.

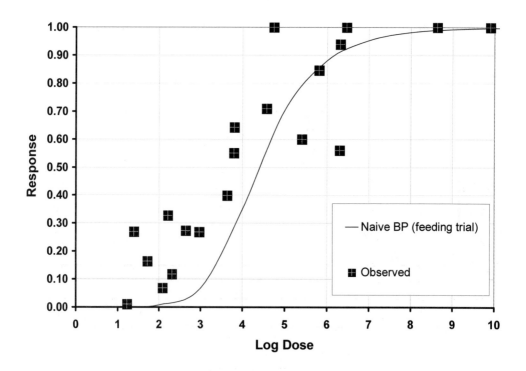

Figure 3.14. Beta-Poisson dose-response model fitted to naive human feeding trial data compared with reported outbreak data.

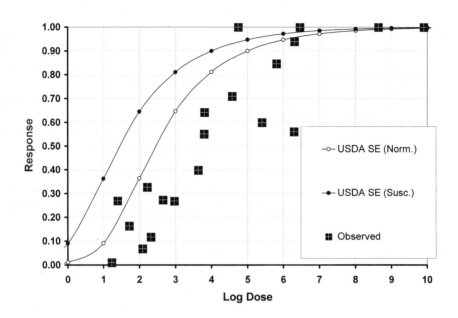

Figure 3.15. US SE RA dose-response model compared with reported outbreak data.

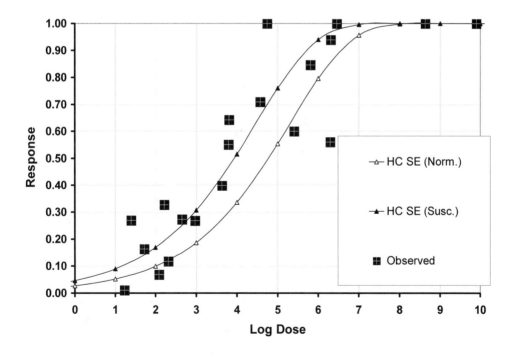

Figure 3.16. Health Canada *Salmonella* Enteritidis dose-response model compared with reported outbreak data.

US SE RA (beta-Poisson model)

The model uses human feeding trial data for *Shigella dysenteriae* as a surrogate pathogen, with illness as the measured endpoint in the data. The appropriateness of using *Shigella* as a surrogate for *Salmonella* is questionable given the nature of the organisms in relation to infectivity and disease. Compared with the outbreak data (Figure 3.15), and on a purely empirical basis, this curve tends to capture the upper range of the data, but overestimates the probability of illness that is observed in the outbreak data.

Health Canada Salmonella *Enteritidis (Weibull-Gamma model)*

To date, this model has not been fully documented and lacks transparency. The model uses data from many different bacterial-pathogen-feeding trials and combines this information with key *Salmonella* outbreak data using Bayesian techniques. Using data from many bacterial-feeding trials and the current lack of transparency regarding their influence is a point of caution. Empirically, the curve describes the outbreak data (Figure 3.16) at the low dose well but tends towards the lower range of response at higher doses.

Dose-response model based on outbreak data

The availability of a reasonably large data set representing real-world observations for the probability of illness upon exposure to *Salmonella* (outbreak data) allowed a unique opportunity to attempt to develop a dose-response relationship based upon this data. The beta-Poisson model (Equation 1) was used as the mathematical form for the relationship, and this was fitted to the outbreak data.

$$P_{ill} = 1 - \left(1 + \frac{Dose}{\beta}\right)^{-\alpha} \qquad \textbf{Equation 3.1}$$

The maximum likelihood technique was used as the basis for generating the best fitting curve to the data. The fit was optimized using an iterative technique that minimized the deviance statistic, based upon a binomial assumption (Haas, 1983).

The outbreak data have merits as real-world observations of the probability of illness upon exposure to a dose, but there are also some drawbacks in the data. Specifically, it should be recognized that there is a degree of uncertainty in the outbreak data, primarily due to the uncontrolled settings under which the information and data were collected. In some cases, the actual dose ingested can be uncertain, while in other cases the true number of people exposed or ill during the outbreak can be under- or over-estimated.

The uncertainty in the outbreak data set was incorporated into the fitting routine by reviewing the outbreak information and assigning an uncertainty distribution on observed variables that were potentially uncertain. A detailed summary of the assumptions associated with each outbreak and the estimation for the range of uncertainty for each of the variables were described in Section 3.2.2. A summary of the data set, with uncertainty for the variables, is given in Table 3.15.

Table 3.15. Uncertainty ranges assigned to variables in reported outbreak data

Case no.	Serovar	Log Dose (Uncertainty)		Response [Attack Rate] (Uncertainty)	
		Min	Max	Min	Max
1	*S.* Typhimurum	1.57	2.57	11.20%	12.36%
2	*S.* Heidelberg	1.48	2.48	28.29%	36.10%
3	*S.* Cubana	4.18	4.78	60.00%	85.71%
4	*S.* Infantis	6.06	6.66	100.00%	100.00%
5	*S.* Typhimurium	3.05	4.05	52.36%	57.64%
7	*S.* Newport	0.60	1.48	0.54%	2.59%
11	*S.* Enteritidis	4.00	5.00	100.00%	100.00%
12	*S.* Enteritidis	1.00	2.37	6.42%	7.64%
13	*S.* Typhimurium	8.00	8.88	100.00%	100.00%
18	*S.* Enteritidis	5.13	5.57	60.00%	60.00%
19	*S.* Enteritidis	6.03	6.48	87.70%	103.51%
20	*S.* Enteritidis	2.69	3.14	18.61%	36.41%
22	*S.* Enteritidis	6.02	6.47	52.17%	61.32%
23	*S.* Enteritidis	5.53	5.97	84.62%	84.62%
24	*S.* Enteritidis	1.45	1.89	12.19%	23.96%
25	*S.* Enteritidis	3.36	3.80	39.85%	39.85%
30	*S.* Enteritidis	3.53	3.97	60.14%	70.90%
31	*S.* Enteritidis	2.37	2.82	25.62%	30.04%
32	*S.* Enteritidis	1.11	1.57	26.92%	26.92%
34	*S.* Oranienburg	9.63	10.07	100.00%	100.00%

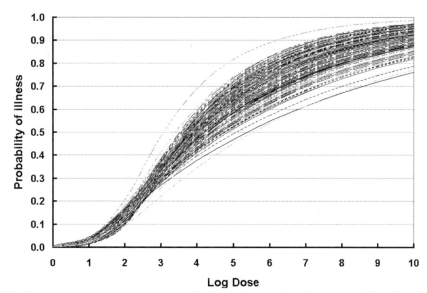

Figure 3.17. Dose-response curves generated by fitting to samples from uncertain outbreak observations.

In order to fit the dose-response model to the uncertain outbreak data, the data were re-sampled based on the uncertainty distributions, generating a new data set at each sample. The dose-response model was then fitted to each of the re-sampled data sets. This procedure was repeated approximately 5000 times, generating 5000 dose-response data sets, to which 5000 dose-response curves were fitted. The fitting procedure used (Haas, 1983) places a greater emphasis on fitting the curve through the larger-scale outbreaks compared with the smaller outbreaks. This is primarily a result of the binomial assumption and the greater variance associated with data from a small observation compared with a large one. Figure 3.17 shows an example of the dose-response curves that are generated by fitting to the uncertain data. The observed outbreak data were found to be over-dispersed compared with what would be expected from the binomial assumption inherent in the deviance statistic that is minimized during fitting. As a result, it was not possible to get a statistically significant single "best fitting" curve to the expected value of all the outbreak data points. However, the characterization of the observed outbreak data by the fitted dose-response model was better than that of the other dose-response models described previously. It is important to note that the range of possible responses at any one given dose shown in Figure 3.17 do not represent the statistical confidence bounds of the dose-response fit, but rather the best fit of the beta-Poisson model to different realizations of the observed data, given its uncertainties.

Figure 3.18 shows the comparison between the fitted curves and the expected value for the observed data. The upper bound, lower bound, expected value, 97.5th percentile and 2.5th percentile for the dose-response curves fitted to the 5000 data sets are also shown. The fitted dose-response range captures the observed outbreak data quite well, especially at the lower- and mid-dose range. The greater range at the high doses is due to the existence of several large-scale outbreaks at the lower- and mid-dose levels through which the curves attempt to pass, while the two high-dose data points are for relatively small-scale outbreaks that allow greater "elasticity" in the fit.

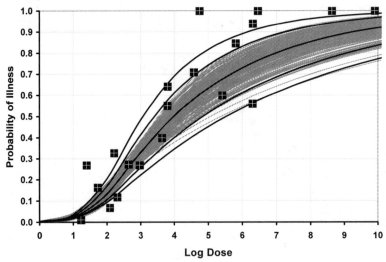

Figure 3.18. Uncertainty bounds for dose-response curves, compared with expected value for the outbreak data.

Since the fitting procedure generated a dose-response curve for each of the 5000 data sets, there are also 5000 sets of beta-Poisson dose-response parameters (alpha & beta). In order to apply the dose-response relationship in a risk assessment, the ideal approach would be to randomly sample from the **set** of parameters that are generated, thereby recreating the dose-response curves shown in Figures 3.17 and 3.18. As an alternative, it is also possible to use the upper, lower, expected value, 2.5^{th} percentile or 97.5^{th} percentile to represent the uncertainty ranges in the dose-response relationship, as opposed to a full characterization resulting from the sampling of the parameter sets. The parameters that generate dose-response curves that approximate the bounds shown in Figure 3.18 of the dose-response relationship are summarized in Table 3.16.

Table 3.16. Beta-Poisson dose-response parameters that generate the approximate bounds shown in Figure 3.18.

	Alpha	Beta
Expected Value	0.1324	51.45
Lower Bound	0.0763	38.49
2.5^{th} Percentile	0.0940	43.75
97.5^{th} Percentile	0.1817	56.39
Upper Bound	0.2274	57.96

Figure 3.19 summarizes all the dose-response models described so far, as well as the outbreak data. It also highlights the expected result of a better characterization of the outbreak data using the current model compared with the alternatives.

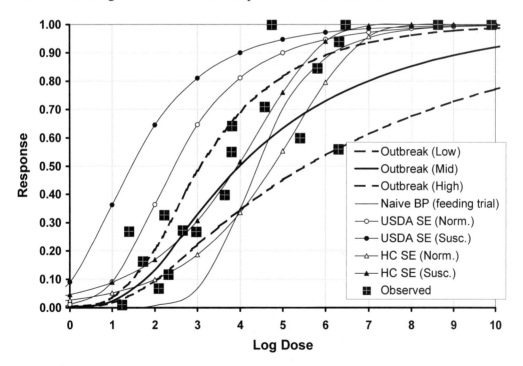

Figure 3.19. Comparison of all dose-response models with reported outbreak data.

In dose-response analysis, the critical region is the lower-dose region. These are the doses that are most likely to exist in the real world and this is also the region for which experimental data are mostly non-existent. The outbreak data extend to a much lower dose than is common in experimental feeding trials, and as such may offer a greater degree of confidence in the lower dose approximations generated by the outbreak dose-response model. Table 3.17 and Figures 3.20 and 3.21 summarize the low-dose estimates for the various dose-response models.

Table 3.17. Probability of illness, estimated by alternative dose-response models at selected low-mean-dose values.

	Mean Log Dose {Mean Dose}			
	0 {1 cell}	1 {10 cells}	2 {100 cells}	3 {1000 cells}
Outbreak (Mid)	0.25%	2.32%	13.32%	32.93%
Naive BP (feeding trial)	0.01%	0.08%	0.75%	6.77%
US SE RA (Susc.)	9.06%	36.27%	64.44%	81.08%
US SE RA (Norm.)	1.12%	9.14%	36.43%	64.54%
HC SE RA (Susc.)	4.65%	8.99%	16.97%	30.72%
HC SE RA (Norm.)	2.65%	5.16%	9.95%	18.72%

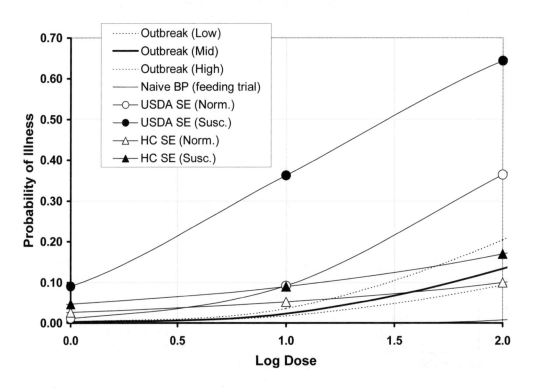

Figure 3.20. Comparison of alternative dose-response models in the 0 to 2.0 mean log dose interval

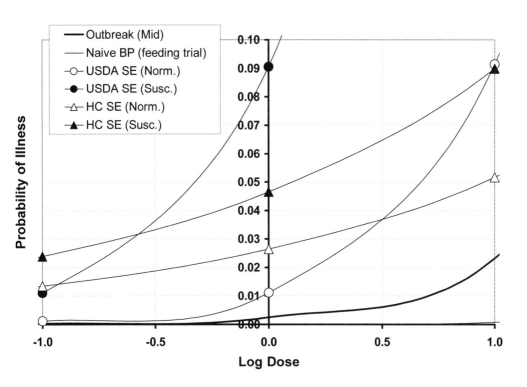

Figure 3.21. Comparison of alternative dose-response models in the -1.0 to 1.0 mean log dose interval.

There is a wide range of estimates generated by the dose-response models. At a dose of 1000 cells, the US SE RA model for the normal population estimates a 65% probability of illness, and an 81% probability for the susceptible population. The Health Canada *S.* Enteritidis model estimates a 31% probability for susceptible populations and 19% for normal populations, while the outbreak model estimates a probability of 33%. At a dose of 100 cells, the US SE RA model continues to be the most conservative, with estimates ranging from 37% to 64%, while the outbreak model estimates a probability of 13%, lying within the range (10–17%) estimated by the Health Canada *S.* Enteritidis model. Perhaps the most telling feature of low-dose estimates is the probability of illness estimated by the models upon ingestion of 1 cell. The US SE RA and Health Canada *S.* Enteritidis models for susceptible populations estimate 9% and 5% probabilities respectively. In the case of the normal population, the Health Canada *S.* Enteritidis model estimates a higher probability (2.7%) than the USDA model (1.1%). The outbreak model estimates the probability at 0.24%, approximately an order of magnitude lower than the Health Canada model for normal populations.

In conclusion, the dose-response model based upon the observed outbreak data provides an estimate for the probability of illness that is based on real-world data. Given the assumptions associated with some of the other models – surrogate pathogens; infection response with healthy male volunteers; and lack of transparency with non-linear low-dose extrapolation – the outbreak model offers the best current alternative for estimating the probability of illness upon ingestion of a dose of *Salmonella.*

3.6 DISCUSSION AND CONCLUSIONS

It has been postulated that some strains of *S.* Enteritidis, particularly the phage types isolated from the increased number of egg-related outbreaks seen in recent years, may be more virulent than other serovars of *Salmonella*. From the outbreak data used to examine the dose-response relationship, there was no evidence that the likelihood of *S.* Enteritidis producing illness differed from other serovars. In total, 12 sets of data were evaluated for *S.* Enteritidis, against 8 sets of data for other serovars. However, increased severity of illness once infected was not evaluated.

It was concluded that there is insufficient evidence in the current outbreak database to conclude that some segments of the population have a higher probability of illness compared with others. There was some indication in two instances, in which two populations potentially exposed to *Salmonella* in the same outbreak exhibited different attack rates. There is therefore a possibility that the probability of illness upon exposure may be different for some members of the population compared with others. However, in the absence of additional information, the probability of illness could be assumed the same for all members of the population, although the severity of the illness could be potentially different.

This document did not consider a quantitative evaluation of secondary transmission (person-to-person) or chronic outcomes. In addition, the impact of the food matrix was not incorporated into the assessment. These may be considerations for future document development.

The dose-response model fitted to the outbreak data offers a reasonable estimate for the probability of illness upon ingestion of a dose of *Salmonella*. The model is based on observed real-world data, and as such is not subject to some of the flaws inherent in using purely experimental data. Nevertheless, the current outbreak data also have uncertainties associated with them and some of the outbreak data points required assumptions to be made. Overall, the dose-response model generated in the current exercise can be used for risk assessment purposes, and generates estimates that are consistent with those that have been observed in outbreaks.

3.7 REFERENCES CITED IN CHAPTER 3

Angulo, F.J., & Swerdlow, D.L. 1995. Bacterial enteric infections in persons infected with human immunodeficiency virus. *Clinical Infectious Diseases*, **21**(Suppl. 1): S84–S93.

Angelotti, R., Bailey, G.C., Foter, M.J., & Lewis, K.H. 1961. Salmonella infantis isolated from ham in food poisoning incident. *Public Health Reports*, **76**: 771–776.

Armstrong, R.W., Fodor, T., Curlin, G.T., Cohen, A.B., Morris, G.K., Martin, W.T., & Feldman, J. 1970. Epidemic *salmonella* gastroenteritis due to contaminated imitation ice cream. *American Journal of Epidemiology*, **91**: 300–307.

Banatvala, N., Cramp, A., Jones, I.R., & Feldman, R.A. 1999. Salmonellosis in North Thames (East), UK: associated risk factors. *Epidemiology and Infection*, **122**: 201–207.

Bearson, S., Bearson, B., & Foster, J.W. 1997. Acid stress responses in enterobacteria. *FEMS Microbiology Letters*, **147**: 173–180.

Beck, M.D., Muñoz, J.A., Scrimshaw, N.S. 1957. Studies on the diarrheal diseases in Central America. I. Preliminary findings on the cultural surveys of normal population groups in Guatemala. *American Journal of Tropical Medicine and Hygiene*, **6**: 62–71.

Bellido Blasco, J.B., Gonzalez Cano, J.M., Galiano, J.V., Bernat, S., Arnedo, A., & Gonzalez, M.F. 1998. Factores asociados con casos esporádicos de salmonelosis en niños de 1 a 7 años. *Gaceta Sanitaria*, **12**: 118–125.

Bellido Blasco, J.B., Gonzalez, M.F., Arnedo, P.A., Galiano Arlandis, J.V., Safont, A.L., Herrero, C.C., Criado, J.J., & Mesanza, D.N. 1996. I. Brote de infección alimentaria por *Salmonella enteritidis*. Posible efecto protector de las bebidas alcohólica. *Medicina Clínica (Barcelona)*, **107**: 641–644.

Blaser, M.J., & Feldman, R.A. 1981. Salmonella bacteraemia: reports to the Centers for Disease Control, 1968–1979. *Journal of Infectious Diseases*, **143**: 743–746.

Blaser, M.J., & Newman, L.S. 1982. A review of human salmonellosis: I. Infective dose. *Reviews of Infectious Diseases*, **4**: 1096–1106.

Boring, J.R., III, Martin, W.T., & Elliott, L.M. 1971. Isolation of *Salmonella typhimurium* from municipal water, Riverside, California, 1965. *American Journal of Epidemiology*, **93**: 49–54.

Bruch, H.A., Ascoli, W., Scrimshaw, N.S., & Gordon, J.E. 1963. Studies of diarrheal disease in Central America. V. Environmental factors in the origin and transmission of acute diarrheal disease in four Guatemalan villages. *American Journal of Tropical Medicine and Hygiene*, **12**: 567–579.

Buzby, J.C. 2001. Children and microbial foodborne illness. *Food Review*, 24(2): 34–37.

Cash, R.A., Music, S.I., Libonati, J.P., Snyder, M.J.J., Wenzel, R.P., & Hornick, R.B. 1974. Response of man to infection with Vibrio cholerae. I. Clinical, serologic, and bacteriologic responses to a known inoculum. *Journal of Infectious Diseases*, **129**: 45–52.

Chiu, C.H., Lin, T.Y., & Ou, J.T. 1999. Prevalence of the virulence plasmids of nontyphoid *Salmonella* in the serovars isolated from humans and their association with bacteremia. *Microbiology and Immunology*, **43**: 899–903.

Coleman, M., & Marks, H. 1998. Topics in Dose-Response Modelling. *Journal of Food Protection*, **61**: 1550–1559.

Cowden, J.M., & Noah, N.D. 1989. Annotation: Salmonellas and eggs. *Archives of Disease in Childhood*, **64**: 1419–1420.

Craven, P.C., Mackel, D.C., Baine, W.B., Barker, W.H., Gangarosa, E.J., Goldfield, M., Rosenfeld, H., Altman, R., Lachapelle, G., Davies, J., & Swanson, R. 1975. International outbreak of *S. eastbourne* infection traced to contaminated chocolate. *The Lancet*, 5 April 1975: 788–792.

D'Aoust, J.Y. 1985. Infective dose of *Salmonella typhimurium* in cheddar cheese. *American Journal of Epidemiology*, **122**: 717–720.

D'Aoust, J.Y. 1991. Pathogenicity of foodborne Salmonella. *International Journal of Food Microbiology*, **12**: 17–40.

D'Aoust, J.Y. 1997. *Salmonella* species. *In:* M.P. Doyle, L.R. Beuchat and T.J. Montville (eds). *Food microbiology: Fundamentals and frontiers*. Washington, DC: American Society for Microbiology Press.

D'Aoust, J.Y., Warburton, D.W., & Sewell, A.M. 1985. *S. typhimurium* phage type 10 from cheddar cheese implicated in a major Canadian foodborne outbreak. *Journal of Food Protection,* **48**: 1062–1066.

D'Aoust, J.Y., Aris, B.J., Thisdele, P., Durante, A., Brisson, N., Dragon, D., Lachapelle, G., Johnston, M., & Laidely, R. 1975. *S. eastbourne* outbreak associated with chocolate. *Journal Institut Canadien de Technologie Alimentaire,* **8**: 181–184.

Davis, R.C. 1981. Salmonella sepsis in infancy. *American Journal of Diseases of Children,* **135**: 1096–1099.

Delarocque-Astagneau, E., Desenclos, J.C., Bouvet, P., Grimont, P.A. 1998. Risk factors for the occurrence of sporadic *Salmonella enterica* serotype *enteritidis* infections in children in France: a national case-control study. *Epidemiology and Infection,* **121**: 561–567.

Fazil, A.M. 1996. A quantitative risk assessment model for salmonella. Drexel University, Philadelphia PA. [Dissertation].

Fontaine, R.E., Arnon, S., Martin, W.T., Vernon, T.M. Jr, Gangarosa, E.J., Farmer, J.J. III, Moran, A.B., Silliker, J.H., & Decker, D.L. 1978. Raw hamburger: an interstate common source of human salmonellosis. *American Journal of Epidemiology,* **107**: 36–45.

Fontaine, R.E., Cohen, M.L., Martin, W.T., Vernon, T.M. 1980. Epidemic salmonellosis from cheddar cheese: Surveillance and Prevention. *American Journal of Epidemiology,* **111**: 247–253.

George, R.H. Small infectous dose of Salmonella. *The Lancet,* 24 May 1976: 1130.

Glynn, J.R., & Palmer, S.R. 1992. Incubation period, severity of disease, and infecting dose: evidence from a Salmonella outbreak. *American Journal of Epidemiology,* **136**: 1369–1377.

Greenwood, M.H., & Hooper, W.L. 1983. Chocolate bars contaminated with *S. napoli:* an infectivity study. *British Medical Journal,* **286**: 1394.

Haas, C.N. 1983. Estimation of risk due to low doses of microorganisms: a comparison of alternative methodologies. *American Journal of Epidemiology,* **118**: 573–582.

Haas, C.N., Rose, J.B., Gerba, C., Regli, S. 1993. Risk assessment of virus in drinking water. *Risk Analysis,* **13**: 545–552.

Health Canada. [2000]. Risk assessment model for *Salmonella* Enteritidis. Unpublished Health Canada document.

Hennessy, T.W., Hedberg, C.W., Slutsker, L., White, K.E., Besser-Wiek, J.M., Moen, M.E., Feldman, J., Coleman, W.W., Edmonson, L.M., MacDonald, K.L., & Osterholm, M.T. 1996. A national outbreak of *Salmonella enteritidis* infections from ice cream. *New England Journal of Medicine,* **334**: 1281–1286.

Hormaeche, E., Peluffo, C.A., & Aleppo, P.L. 1939. Nuevo contrabucion al estudio etiologico de las "Diarreas infantiles de Verano". *Archivas Uruguayos de Medicina, Cirurgia y Especialiadades,* **9**: 113–162.

Hornick, R.B., Geisman, S.E., Woodward, T.E., DuPont, H.L., Dawkins, A.T., & Snyder, M.J. 1970. Typhoid fever: pathogenesis and immunologic control. *New England Journal of Medicine,* **283**: 686–691.

Jaykus, L.A., Morales, R.A., & Cowen, P. 1997. Development of a risk assessment model for the evaluation of HACCP-based quality assurance programs for human infection from *Salmonella* Enteritidis: Preliminary estimates. *Epidemiolgie et Santé Animale*, **31–32**: 06.11.

Kapperud, G., Lassen, J., & Hasseltvedt, V. 1998. Salmonella infections in Norway: descriptive epidemiology and a case-control study. *Epidemiology and Infection*, **121**: 569–577.

Kapperud, G., Stenwig, H., & Lassen, J. 1998. Epidemiology of *Salmonella typhimurium* O:4-12 infection in Norway: evidence of transmission from an avian wildlife reservoir. *American Journal of Epidemiology*, **147**: 774–782.

Kass, P.H., Farver, T.B., Beaumont, J.J., Genigeorgis, C., & Stevens, F. 1992. Disease determinants of sporadic salmonellosis in four northern California counties. A case-control study of older children and adults. *Annuals of Epidemiology*, **2**: 683–696.

Khan, M.A. 1995. HLA-B27 and its subtypes in world populations. *Current Opinion in Rheumatology*, **7**: 263–269.

Khan, M.A. 1996. Epidemiology of HLA-B27 and arthritis. *Clinical Rheumatology*, **15** (Suppl. 1): 10–12.

Kourany, M., & Vasquez, M.A. 1969. Housing and certain socioenvironmental factors and prevalence of enteropathogenic bacteria among infants with diarrheal disease in Panama. *American Journal of Tropical Medicine and Hygiene*, **18**: 936–941.

Lang, D.J., Kunz, L.J., Martin, A.R., Schroeder, S.A., & Thomson, L.A. 1967. Carmine as a source of noscomial salmonellosis. *New England Journal of Medicine*, **276**: 829–832.

Le Bacq, F., Louwagie, B., Verhaegen, J. 1994. *Salmonella typhimurium* and *Salmonella enteritidis*: changing epidemiology from 1973 until 1992. *European Journal of Epidemiology*, **10**: 367–371.

Lee, L.A., Puhr, N.D., Maloney, E.K., Bean, N.H., & Tauxe, R.V. 1994. Increase in antimicrobial-resistant Salmonella infections in the United States, 1989–1990. *Journal of Infectious Diseases*, **170**: 128–134.

Levine, M.M., & DuPont, H.L. 1973. Pathogenesis of *Shigella dysenteriae* 1 (Shiga) dysentery. *Journal of Infectious Diseases*, **127**: 261–270.

Levy, M., Fletcher, M., Moody, M., and 27 others. 1996. Outbreaks of *Salmonella* serotype enteritidis infection associated with consumption of raw shell eggs – United States, 1994–1995. *Morbidity and Mortality Weekly Report*, **45**: 737–742.

Lipson, A. 1976. Infecting dose of salmonella. *The Lancet*, 1 May 1976: 969.

Mackenzie, C.R., & Livingstone, D.J. 1968. Salmonellae in fish and foods. *South African Medical Journal*, **42**: 999–1003.

McCullough, N.B., & Eisele, C.W. 1951a. Experimental human salmonellosis. I. Pathogenicity of strains of *Salmonella* Meleagridis and *Salmonella anatum* obtained from spray-dried whole egg. *Journal of Infectious Diseases*, **88**: 278–289.

McCullough, N.B., & Eisele, C.W. 1951b. Experimental human salmonellosis. II. Immunity studies following experimental illness with *Salmonella* Meleagridis and *Salmonella anatum*. *Journal of Immunology*, **66**: 595–608.

McCullough, N.B., & Eisele, C.W. 1951c. Experimental human salmonellosis. III. Pathogenicity of strains of *Salmonella* Newport, *Salmonella derby*, and *Salmonella* Bareilly obtained from spray dried whole egg. *Journal of Infectious Diseases*, **89**: 209–213.

McCullough, N.B., & Eisele, C.W. 1951d. Experimental human salmonellosis. IV. Pathogenicity of strains of *Salmonella pullorum* obtained from spray-dried whole egg. *Journal of Infectious Diseases*, **89**: 259–266.

Ministry of Health and Welfare, Japan. 1999. Research on micriobiological risk assessment. Prepared by Susumu Kumagai and Shigeki Yamamoto, *in:* Report of Grants for Health Science of the Ministry of Health and Welfare, Japan.

Morales, R.A., Jaykus, L.A., & Cowen, P. 1995. Preliminary risk assessments for the transmission of *Salmonella* Enteritidis in shell eggs. Abstract. Proceedings 76th Conference of Research Workers in Animal Diseases. Chicago, Illinois, 13-14 November 1995.

Mossel, D.A., & Oei, H.Y. 1975. Letter: Person to-person transmission of enteric bacterial infection. *Lancet*, 29 March 1975 1: 751.

Murray, M.J. 1986. Salmonella: virulence factors and enteric salmonellosis. *Journal of the American Veterinary Medical Association*, **189**: 145–147.

Narain, J.P., & Lofgren, J.P. 1989. Epidemic of restaurant-associated illness due to *S.* Newport. *Southern Medical Journal*, **82**: 837–840.

Nguyen, B.M., Lanata, C.F., Black, R.E., Gil, A.I., Karnell, A., & Wretlind, B. 1998. Age-related prevalence of Shigella and Salmonella antibodies and their association with diarrhoeal diseases in Peruvian children. *Scandinavian Journal of Infectious Diseases*, **30**: 159–164.

Olsen, S.J., Bishop, R., Brenner, F.W., Roels, T.H., Bean, N., Tauxe, R.V., & Slutsker, L. 2001. The changing epidemiology of *Salmonella*: Trends in serotypes isolated from humans in the United States, 1987–1997. *The Journal of Infectious Diseases,* **183**: 753–761.

Pavia, A.T., Shipman, L.D., Wells, J.G., Puhr, N.D., Smith, J.D., McKinley, T.W., & Tauxe, R.V. 1990. Epidemiologic evidence that prior antimicrobial exposure decreases resistance to infection by antimicrobial-sensitive *Salmonella*. *Journal of Infectious Diseases*, **161**: 255–260.

Regli, S., Rose, J.B., Haas, C.N., Gerba, C. 1991. Modelling risk for pathogens in drinking water. *Journal American Water Works Association*, **83**: 76–84.

Reitler, R., Yarom, D., & Seligmann, R. 1960. The enhancing effect of staphylococcal enterotoxin on *Salmonella* infection. *The Medical Officer*, **104**: 181.

Riley, L.W., Cohen, M.L., Seals, J.E., Blaser, M.J., Birkness, K.A., Hargrett, N.T., Martin, S.M., Feldman, R.A. 1984. Importance of host factors in human salmonellosis caused by multiresistant strains of Salmonella. *Journal of Infectious Diseases*, **149**: 878–883.

Rejnmark, L., Stoustrup, O., Christensen, I., & Hansen, A. 1997. Impact of infecting dose on severity of disease in an outbreak of food-borne *Salmonella* Enteritidis. *Scandinavian Journal of Infectious Diseases*, **29**: 37–40.

Ryder, R.W., Merson, M.H., Pollard, R.A., & Gangarosa, E.J. 1976. From the Center for Disease Control: salmonellosis in the United States, 1968–1974. *Journal of Infectious Diseases*, **133**: 483–486.

Schliessmann, D.J., Atchley, F.O., Wilcomb, M.J., & Welch, S.F. 1958. Relation of environmental factors to the occurrence of enteric disease in areas of eastern Kentucky. *Public Health Service Monographs*, **54**.

Schmid, H., Burnens, A.P., Baumgartner, A., & Oberreich, J. 1996. Risk factors for sporadic salmonellosis in Switzerland. *European Journal of Clinical Microbiology and Infectious Diseases*, **15**: 725–732.

Slauch, J., Taylor, R., & Maloy, S. 1997. Survival in a cruel world: how Vibrio cholerae and Salmonella respond to an unwilling host. *Genes and Development*, **11**: 1761–1774.

Smith, J. 2002. *Campylobacter jejuni* infection during pregnancy: Long-term consequences of associated bacteraemia, Guillain Barré Syndrome, and reactive arthritis. *Journal of Food Protection*, **65**: 696–708.

Smith, P.D., Lane, H.C., Gill, V.J., Manischewitz, J.F., Quinnan, G.V., Fauci, A.S., & Masur, H. 1988. Intestinal infections in patients with the acquired immunodeficiency syndrome (AIDS). Etiology and response to therapy. *Annals of Internal Medicine*, **108**: 328–333.

Sprinz, H., Gangarosa, E.J., Williams, M., Hornick, R.B., & Woodward, T.B. 1966. Histopathology of the upper small intestines in typhoid fever. *American Journal of Digestive Diseases*, **11**: 615–624.

Sprong, R.C., Hulstein, M.F., & Van der Meer, M.R. 1999. High intake of milk fat inhibits intestinal colonization of Listeria but not of Salmonella in rats. *Journal of Nutrition*, **129**: 1382–1389.

Tacconelli, E., Tumbarello, M., Ventura, G., Leone, F., Cauda, R., & Ortona, L. 1998. Risk factors, nutritional status, and quality of life in HIV-infected patients with enteric salmonellosis. *Italian Journal of Gastroenterology and Hepatology*, **30**: 167–172.

Taylor, D.N., Bopp, C., Birkness, K., & Cohen, M.L. 1984. An outbreak of salmonellosis associated with a fatality in a healthy child. *American Journal of Epidemiology*, **119**: 907–912.

Telzak, E.E., Greenberg, M.S., Budnick, L.D., Singh, T., & Blum, S. 1991. Diabetes mellitus: A newly described risk factor for infection from *Salmonella enteritidis*. *Journal of Infectious Diseases*, **164**: 538–541.

Teunis, P.F.M., Van der Heijden, O.G., Van der Giessen, J.W.B., & Havelaar, A.H. 1996. The dose response relation in human volunteers for gastro-intestinal pathogens. National Institute of Public Health and the Environment, Bilthoven, The Netherlands. RIVM Report No. 284550002.

Travers, K., & Barza, M. 2002. Morbidity of infections caused by antimicrobial-resistant bacteria. *Clinical Infectious Diseases*, **34**(Suppl. 3): S131–S134.

USDA-FSIS. 1998. *Salmonella* Enteritidis Risk Assessment. Shell Eggs and Egg Products. Final Report. Prepared for FSIS by the *Salmonella* Enteritidis Risk Assessment Team. 268 pp. Available on Internet as PDF document at: www.fsis.usda.gov/ophs/risk/contents.htm.

Varela, G., & Olarte, J. 1942. Infecion experimental del Hombre con Salmonella anatum. *Medicina Revista Mexicana*, **22**: 57–58.

Vought, K.J., & Tatini, S.R. 1998. *S. enteritidis* contamination of ice cream associated with a 1994 multistate outbreak. *Journal of Food Protection*, **61**: 5–10.

Waterman, S.R., & Small, P.L. 1998. Acid-sensitive enteric pathogens are protected from killing under extremely acidic conditions of pH 2.5 when they are inoculated onto certain solid food sources. *Applied Environmental Microbiology*, **64**: 3882–3886.

Wong, S.S., Yuen, K.Y., Yam, W.C., Lee, T.Y., & Chau, P.Y. 1994. Changing epidemiology of human salmonellosis in Hong Kong, 1982–93. *Epidemiology and Infection*, **113**: 425–434.

Woodward, W.E. 1980. Volunteer studies of typhoid fever and vaccines. *Transactions of the Royal Society of Tropical Medicine and Hygiene*, **74**: 553–556.

4. EXPOSURE ASSESSMENT OF *SALMONELLA* ENTERITIDIS IN EGGS

4.1 SUMMARY

This section outlines the components of an exposure assessment of *Salmonella* Enteritidis in eggs for the purpose of estimating the probability that an egg serving is contaminated with a certain number of the pathogen. Firstly, resource documents and currently available information are reviewed. These include those used in previously completed exposure assessments, and international data collected during this risk assessment, covering the flow of eggs from farm to consumption. Each input parameter considered in this section is critically reviewed, both regarding its uncertainty and from the viewpoint of how it was modelled in previously completed exposure assessments. This exposure assessment model considers contamination in yolk, and growth of *Salmonella* Enteritidis in eggs prior to processing for egg products. Some data and the associated modelling methods are country or region specific, while others are common to the world. This exposure assessment is itself not representative of any particular country or region.

4.2 REVIEW OF LITERATURE, DATA AND EXISTING MODELS

4.2.1 Introduction

Purpose

The practice of risk assessment will be advanced through critical review of existing models. Discussions between the *Salmonella* Enteritidis in Eggs drafting group and the FAO/WHO Secretariat determined the need for a comparison of existing exposure assessments in order to characterize the state of the art in the practice of risk assessment. Such a comparison would identify similarities and differences between existing models and provide the basis for the exposure model developed for the purposes of this work (Section 4.3, below). It is hoped that this critique of existing models will also be useful in further advancing methodologies for future exposure assessments of this product-pathogen combination.

The purpose of this section is to explain existing techniques and practices used to construct an exposure assessment of *Salmonella* Enteritidis in eggs. Three previously completed exposure assessments serve as case studies for this analysis.

This review intends to identify those methods that are most successful in previous exposure assessments, and to recognize the weaknesses of those assessments resulting from inadequate data or methodology. This report does not provide detailed instructions on constructing an exposure assessment. It also does not simply reproduce the contents of previously written reports. Instead, it was intended to highlight practices, techniques and inputs that are common to most, if not all, quantitative exposure assessments of *Salmonella* Enteritidis in eggs. Specific models are often designed for specific objectives, so each model may be different in important ways. Nevertheless, it is believed that the components and inputs presented in this section are useful to any exposure assessment modelling of this product-pathogen pair. Those wishing to complete such analysis, however, should refer to the original reports cited here, as well as texts on risk analysis.

The scope of this analysis is limited to human exposure risk associated with eggs that are internally contaminated with *Salmonella* Enteritidis. The problem of *Salmonella* Enteritidis in fresh shell eggs is a specific public health hazard that is unique within the general problem of human salmonellosis. This hazard is a food safety priority for public health officials in many countries.

The analysis and conclusions presented here apply only to currently understood mechanisms and variables, as incorporated in previous exposure assessments. Therefore caution should be exercised in interpreting this report in the context of data that has become available since these models were completed.

Organization

This section outlines the components of an exposure assessment of *Salmonella* Enteritidis in eggs. Exposure assessments depend on data. Therefore this report also summarizes data used in previously completed exposure assessments, as well as some of the international data pertaining to *Salmonella* Enteritidis in eggs. Because those previously reported risk assessments were conducted in North American countries, data used in this model are also mostly from those countries, but it is not the intention to focus this risk assessment on that region. Rather, those data are just examples for demonstrating how the data is used in a model.

While components of individual models may differ, an endeavour is made to explain the similarities among the models. For example, this report is structured in basic model stages that are common to any farm-to-table exposure assessment. Data used as model inputs may differ depending on the particular situation (e.g. country or product), but the form of the data and modelling are generally similar across models.

Some inputs to a *S.* Enteritidis in Eggs exposure assessment may be common between different countries. These common inputs are described, together with discussion of how they have been modelled in previous analyses. In addition, an extensive annotated bibliography was prepared of literature relevant for each stage of the model.

Components of an exposure assessment

A generic outline for quantitative exposure assessments of foodborne pathogens includes:

- prevalence of the pathogen in raw food ingredients,
- changes in the organisms per volume or weight of material subsequent to production, and
- preparation and consumption patterns among consumers.

Similarly, an exposure assessment of *Salmonella* Enteritidis in shell eggs consists of three main components: production; distribution and storage; and preparation and consumption. If the exposure assessment is concerned with commercially packaged liquid or dried egg products, then the analysis should have this additional component (Figure 4.1).

Figure 4.1. Schematic diagram showing the four general stages forming a farm-to-table exposure assessment of *Salmonella* Enteritidis in eggs.

Production   The production stage models the frequency of contaminated eggs at the time of lay and the level of bacteria initially present in contaminated eggs.

Distribution and storage   The distribution and storage stage models growth in the number of *Salmonella* Enteritidis   organisms between the laying of a contaminated egg and its preparation for consumption. Times and temperatures during storage and transportation can affect the microbe numbers within contaminated eggs.

Egg products processing   The egg products stage models the occurrence and concentration of *Salmonella* Enteritidis in egg products.

Preparation and consumption   The preparation and consumption stage models the effects of meal preparation and cooking on the number of *Salmonella* Enteritidis in meals containing egg. Eggs may travel different pathways depending on where they are used, how they are used, whether they are cooked, and to what extent they are cooked. Each of these pathways is associated with a frequency of occurrence and a variable number of servings. In addition, environmental conditions may differ for each pathway.

Previous exposure assessments

The drafting group identified five risk assessments previously conducted for *Salmonella* Enteritidis in eggs. They are briefly summarized below.

Salmonella *Enteritidis and eggs: assessment of risk (Morris, 1990)*

This simple analysis was conducted soon after the identification of the *Salmonella* Enteritidis epidemic in the United States of America. Data revealed that less than 1 in 1000 eggs from infected flocks were contaminated. An infected hen laid one contaminated egg in every 200, leading to an overall prevalence in endemic areas of 1 in 10 000 to 14 000 eggs produced. Approximately 0.9% of eggs were eaten without cooking.   This report summarized *Salmonella* Enteritidis outbreaks and contributing risk factors for human infection. These risk factors included poor refrigeration practices, improper storage of pooled eggs, use of raw eggs, substantial time and temperature abuse of eggs, and exposure of highly susceptible individuals. The report also includes pertinent facts about S*almonella* Enteritidis and eggs. Among the critical facts listed are that *Salmonella* Enteritidis has usual sensitivity to heat and is destroyed by pasteurization and cooking. Organisms may grow rapidly in egg mixtures (up to one log per hour), and warm summer temperatures may allow *Salmonella* Enteritidis to grow within shell eggs.

The analysis concluded by separating humans into four risk groups:

[1]   *Healthy adults who usually eat fully cooked eggs*. The prevalence of contaminated eggs (i.e. 1 in 14 000) and the frequency of consuming raw eggs (0.9%) equated to

a risk of one in 1.6 million eggs consumed. If an individual consumed 250 eggs per year and lived to 80 years old, the risk was reportedly one in 80 lifetimes.

[2]   *Healthy adults who frequently eat fried, soft boiled, and other less thoroughly cooked eggs.* The risk for this group was not quantified, but thought to be higher than the first risk group.

[3]   *Healthy adults who eat eggs not fully cooked and frequently eat at restaurants and other places where pooling and abusive storage of eggs are possible.* The risk for this group was thought to be proportional to the number of eggs pooled. If 10 eggs were pooled, the risk was 10 times greater. A specific quantification of risk for this group was not provided.

[4]   *More-susceptible individuals who eat higher-risk products as for group [3].* These individuals included residents of nursing homes and hospitals. No quantification of their risk was provided, but they are likely to be the population most at risk.

The assertions in Morris' analysis are not supported by references to research or data, but the mechanics of the analysis should be transparent to most readers.

A farm-to-table exposure assessment should consider all possible scenarios where human illness results from *Salmonella* Enteritidis in eggs. However, Morris' analysis was limited in the scenarios it considered and was, for the most part, non-quantitative. Therefore it was not considered in the comparative evaluation of different risk assessments for this report.

Risk assessment of use of cracked eggs in Canada (Todd, 1996)

The objective of this analysis was to determine the probability of illness associated with consuming cracked shell eggs in Canada. Eggs with cracks in the shell are considered hazardous because their contents are potentially exposed to pathogens more readily than eggs with intact shells. The hazard identification evaluated the possible association of *Bacillus* spp., *Campylobacter* spp., *Salmonella*, *Staphylococcus aureus*, *Yersinia enterocolitica*, *Escherichia coli* O157:H7 and *Listeria monocytogenes* with cracked eggs and human illnesses. *Salmonella* was the only hazard conclusively linked to human illness in this assessment. Therefore *Salmonella* was the only hazard used in further analysis.

Research was cited demonstrating that *Salmonella* can penetrate the shells of intact eggs. Nevertheless, it was concluded that little growth of *Salmonella* organisms would occur unless these organisms gained access to the yolk. Research was cited demonstrating that between 1.3% and 6.3% of eggs examined in Canada were cracked. Risk factors noted to be associated with processing included washing and rapid cooling. Both of these factors were thought to reduce shell integrity and make cracks more likely to occur. Research was cited demonstrating that *Salmonella* was more likely to be isolated from cracked eggs than from intact eggs.

The number of cracked shell eggs was estimated by multiplying the fraction of all eggs that were cracked by the number of eggs produced annually in Canada. To determine the illness burden, 13 outbreaks involving shell eggs were analysed and five of the outbreaks were identified as associated with cracked eggs. Given the estimated ratio of cracked to uncracked eggs, and the ratio of outbreaks associated with cracked eggs to those associated with intact eggs, a relative risk of 23:1 was calculated. Uncertainty analysis suggested the relative risk might range from 3:1 to 93:1.

By using reported human cases, adjusted for underreporting, Todd (1996) estimated that 10 500 cases per year are associated with cracked eggs. The risk of illness was calculated as one case per 3800 cracked eggs consumed, using an estimated exposure to 40 million cracked eggs.

This exposure assessment is transparent and data based. It relies on human epidemiological data to determine the illness burden associated with cracked eggs. Substantial uncertainty attends the estimates, but these are probably more defensible than a mechanistic farm-to-table model based on limited evidence. At the same time, assumptions regarding correspondence of eggs to cases are problematic, because a single egg may contribute to many servings. This effect is not captured by the analysis. Furthermore, the lack of a mechanistic explanation of the chain of events leading to illness makes risk management options difficult to evaluate. General policies, such as requiring all cracked eggs to be pasteurized, can be reasonably evaluated with this approach, but more subtle interventions, such as strict temperature-controlled storage requirements for cracked eggs, cannot be easily analysed without a mechanistic modelling approach.

Because this analysis was not a farm-to-table exposure assessment that incorporated quantitative data for each stage, it was not included in the comparative evaluation of exposure assessments. However, the approach used in this analysis is useful for certain types of exposure assessments that require rapid approximations of risk. In particular, preliminary assessments could be based on this approach to determine if a problem deems further, more time consuming, analysis.

Development of a quantitative risk assessment model for Salmonella *Enteritidis in pasteurized liquid eggs (Whiting and Buchanan, 1997)*

This farm-to-table quantitative risk assessment estimated the potential risks associated with consuming mayonnaise prepared from pasteurized liquid whole eggs. Although it does not consider all possible pathways that might lead to illness from pasteurized egg products, it comprises many of the components of a production-to-consumption exposure assessment. It was therefore included in this comparative evaluation of exposure assessments.

The exposure assessment model includes inputs on the proportion of commercial flocks that are affected by *Salmonella* Enteritidis (i.e. contain infected birds), the frequency that infected flocks produce contaminated eggs, the numbers of *Salmonella* Enteritidis in contaminated eggs, and the influence of time and temperature abuse on growth of *Salmonella* Enteritidis before and after pasteurization. This model also includes an input that predicts the effectiveness of pasteurization when applied according to regulatory standards. Time, temperature and pH inputs are varied to demonstrate their influence on the number of *Salmonella* Enteritidis organisms that remain in a serving of home-made mayonnaise prepared using pasteurized egg product.

This assessment determined that pasteurization reduced consumer risk associated with a high prevalence of *Salmonella* Enteritidis infection in layer flocks. Reducing time and temperature abuse of contaminated eggs before pasteurization was also effective for risk reduction. However, inadequate pasteurization temperatures and temperature abuse during post-pasteurization storage were associated with increased risk of human *Salmonella* Enteritidis exposure and illness.

Salmonella *Enteritidis Risk Assessment: Shell Eggs and Egg Products (USDA-FSIS, 1998)*

This farm-to-table quantitative risk assessment model examined the human illness risk associated with *Salmonella* Enteritidis in shell eggs, covering an exhaustive number of consumption pathways. It also examined the levels of *Salmonella* Enteritidis in liquid egg products before and after pasteurization. It contains the components of an exposure assessment from production to consumption, and is included here in the comparative evaluation of such analyses.

The exposure assessment model estimated the unmitigated risk of exposures resulting from consumption of table eggs that were internally contaminated with *Salmonella* Enteritidis. In concert with a hazard characterization, the baseline exposure assessment was then used to identify target areas for risk reduction activities along the farm-to-table continuum. These target areas could be further evaluated to compare the public health benefits accruing from the mitigated risk of *Salmonella* Enteritidis egg-borne illness resulting from various intervention strategies. Furthermore, the exposure assessment was used to identify data gaps and guide future research efforts.

Example mitigations included reduction of storage times and temperatures, reduction in the prevalence of infected flocks, and diversion of contaminated eggs. These were examined to evaluate the proportional effect on estimated human cases per year. Diversion of contaminated eggs resulted in a direct reduction in human cases, as did a mitigation strategy that combined reduction of the prevalence of infected flocks with reduction in egg storage times. Other mitigation scenarios were less efficient, but the costs of achieving any of the intervention strategies were not considered. A specific policy requiring storage of eggs at an ambient temperature at or below 45°F [7.2°C] before and during processing resulted in an average 8–12% reduction in human cases per year.

Salmonella *Enteritidis in eggs risk assessment (Health Canada, 2000)*

This farm-to-table quantitative risk assessment focused on *Salmonella* Enteritidis in table eggs. The FAO/WHO drafting group members for the present report were given a copy of the spreadsheet model for review and analysis. The model consists of all components of an exposure assessment, with the exception of egg products processing. It was therefore included in most of the comparative evaluation of exposure assessments.

The Whiting and Buchanan (1997), USDA-FSIS (1998) and Health Canada (2000) exposure assessments are discussed and the data quality and biases are evaluated. The pathways modelled in these exposure assessments are also compared. Discussions included the issues of variability and uncertainty – concepts important in the field of risk assessment. Variability describes naturally occurring observable differences within or between populations, while uncertainty describes our confidence about the true value of some parameter, or the frequency distribution of some variable; in essence, our understanding of the system under investigation. Uncertainty can be reduced by the gathering of more data, but variability cannot be changed without some intervention in the physical world. The explicit separation of variability and uncertainty for model inputs and outputs is a goal of risk assessors. Such separation allows decision-makers to understand how model outputs might improve if uncertainty were reduced. However, accomplishing this separation is a daunting task, so model inputs are described as uncertain, variable or both. Methods are also introduced for separating uncertainty and variability for inputs, as well as for model outputs.

Exposure assessments should be transparent to decision-makers. Through discussion and critical review, it is hoped that an understanding of the exposure assessments examined in this section will be attained.

4.2.2 Production

The production component of a *Salmonella* Enteritidis exposure assessment will produce an output consisting of a distribution of contaminated eggs at varying levels of contamination. This distribution describes the frequency of eggs that contain *Salmonella* Enteritidis bacteria per unit time or per egg. Additional outputs might describe the fraction of *Salmonella* Enteritidis contaminated eggs by geographic region, by flock type (e.g. battery or free range), or by other factors that distinguish egg production facilities (e.g. flock size).

Inputs to a production component include the prevalence of infected flocks; the frequency at which infected flocks produce contaminated eggs; the number of *Salmonella* Enteritidis bacteria initially present at the time of lay (or soon thereafter); and possibly moulting practices. These data may be derived from several sources, including prevalence studies of *Salmonella* Enteritidis in layer flocks, epidemiological studies of risk factors, transmission study results, industry demographic data, and experimental or survey data concerning the concentration of organisms in, or on, infected animals or their products.

Prevalence data are usually adjusted for the sensitivity or specificity, or both, of the diagnostic assay used. In this context, sensitivity describes the frequency that truly infected hens or flocks are detected using surveillance or testing protocols. Specificity describes the frequency that truly non-infected hens or flocks are properly classified as non-infected. Because diagnostic tests for the presence of *Salmonella* Enteritidis are based on microbiological culture, most analysts assume that specificity is 100%. Surveys typically use diagnostic tests with imperfect sensitivity and do not sample all birds in the flock. Imperfect laboratory tests result in biased estimates of the number of infected hens in flocks. Sampling less than 100% of the birds in a flock can result in misclassification of infected flocks.

The availability of detailed epidemiological data provides better risk assessments. Increased detail provides information that is more precise for decision-making based on risk assessments. For example, the proportion of all eggs in a country or region that are contaminated can be calculated from: (1) an estimate of the proportion of flocks containing *Salmonella* Enteritidis-infected hens, and (2) the proportion of eggs laid by these flocks and which are contaminated. An estimate of the contaminated egg proportion could be derived from a random sampling across all egg production, but such an approach is extremely costly and not useful for analysis of mitigation of the risk to humans when the estimate is unattached to status of the producing flock. For example, if a random sample of eggs across the country estimated that 1 egg in 20 000 was contaminated, this information may be of little value to decision-makers without information about spatial and temporal clustering of infected flocks. One could not determine whether some flocks produce contaminated eggs more frequently than others (i.e. spatial clustering), nor could one determine if there were certain times when flocks produce more contaminated eggs (i.e. temporal clustering).

Many factors contribute to variability in the production of contaminated eggs. These include regional differences in flock prevalence – if egg marketing is regional – and flock age. Other factors (e.g. stage of infection in flock, season, control efforts by management) may also modulate within-flock prevalence and egg contamination frequency. Moult status of flocks is a proven risk factor that can influence flock-to-flock variability in egg production and egg contamination frequency.

Flock prevalence

By definition, flock prevalence is the proportion of flocks containing one or more birds infected with *Salmonella* Enteritidis. As contaminated eggs can only be produced by infected flocks, exposure assessments must concentrate on these flocks.

Flock prevalence data always represents apparent prevalence. Apparent prevalence is the observed prevalence without accounting for the effects of diagnostic test imperfections. For present purposes, apparent prevalence equals the true prevalence of infection times the sensitivity of the methods used to generate the observations.

Most evidence suggests that infected flocks remain infected for most of their productive life. Hens usually begin egg production at about 20 weeks of age. Flocks usually become infected soon after immature hens (i.e. pullets) are placed in laying houses. Carryover infection from a previously infected flock and rodent reservoirs in the environment of such flocks serve to perpetuate infection across flocks. Infection of flocks during pullet grow-out probably can occur via a previously contaminated environment. Infection at the hatchery is also possible.

Local trends in flock prevalence for a country or region might be inferred from surveillance data. Such an inference might suggest that the proportion of infected flocks in a country or region is increasing or decreasing over time. Nevertheless, these trends must be demonstrated across a sufficient period to be convincing. Cross-sectional surveys may imply seasonal patterns in flock prevalence, but this is not likely to be the case. Instead, observed seasonal differences in flock prevalence in cross-sectional studies are probably the result of changes in within-flock prevalence and the effect of increases or decreases in within-flock prevalence on the capacity of a survey to detect infected flocks. Therefore, unless local trends are clearly proven, it is generally best to model flock prevalence as an invariant, fixed value. The methods used to model flock prevalence should incorporate all uncertainty regarding the true fixed value.

Data

Table 4.1 summarizes data on flock prevalence. The three quantitative exposure assessments have used some of these data to estimate flock prevalence.

Table 4.1. Studies to determine the proportion of layer flocks that contain one or more infected hens.

Flocks positive	Flocks sampled	Hens sampled per flock	Apparent flock prevalence	Country and source
247	711	300	35%	USA. Hogue et al., 1997
8	295	60 faecal, 12 egg belts	3%	Canada. Poppe et al., 1991
2	37	20	5%	Japan. Sunagawa et al., 1997
10	422	100	2%	Denmark. Gerner-Smidt and Wegener, 1999

The studies in Table 4.1 differ in the number of flocks sampled, the intensity of sampling within each flock and the test methodology, as well as in the reported apparent prevalence. Because flock prevalence is constant, the main interest becomes describing the uncertainty about the true value, so methods are described for modelling this uncertainty. Furthermore, apparent prevalence estimates are biased, so methods are described to correct this bias.

Methods

The Beta distribution is commonly used to model prevalence in quantitative exposure assessments. When using the @Risk® software (Palisade Corporation, Newfield, NY), this distribution is modelled as @RISKBETA(s+1,n-s+1), where s is the number of positives observed and n is the total number sampled. This distribution can be derived by applying Bayes Theorem to the binomial distribution, where p is the probability of a positive, or prevalence (Vose, 1996). The Beta distribution demonstrates the increased certainty in estimated prevalence resulting from increasing the number of samples collected. For example, Figure 4.2 illustrates how a probability density function becomes increasingly narrowed for increasing numbers of samples when the underlying prevalence of positives is fixed at 10%.

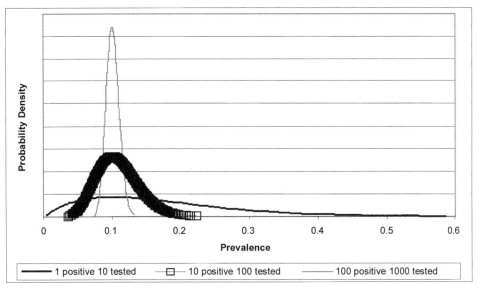

Figure 4.2. Illustrating the effect of increased sample size on certainty regarding prevalence.

Data similar to those presented in Table 4.1 are typically used to estimate regional or national flock prevalence. In the US SE RA, the data from Hogue et al. (1997) were used to estimate the United States of America flock prevalence. These data were generated from sampling hens at slaughter plants, and summarized at a regional level (Figure 4.3).

The proportion of all flocks sampled in these four regions did not match the proportion of regional production. Instead, more flocks were sampled in the high prevalence regions. Therefore the raw data were adjusted by calculating the expected value of prevalence:

$$\sum_{1}^{4}(p_i w_i)$$

where p_i was the observed prevalence in region i, and w_i was the proportion of production in region i. The total number of positive flocks was calculated as the product of this expected value and the observed number in Table 4.1. The total number of flocks sampled (i.e. 711) was not changed, so the uncertainty in the estimated prevalence was consistent with the level of sampling used. Figure 4.4 shows the effect of this adjustment on apparent flock prevalence.

Given apparent prevalence evidence like that shown in Table 4.1, exposure assessment models must make adjustments for false-negative results. No survey of flocks can definitively determine the status of flocks sampled. Given the limited number of flocks sampled in surveys, the limited sampling within flocks, and the imperfect nature of diagnostic tests applied to individuals – and our imperfect understanding of these imperfections – uncertainty about true flock prevalence can be substantial.

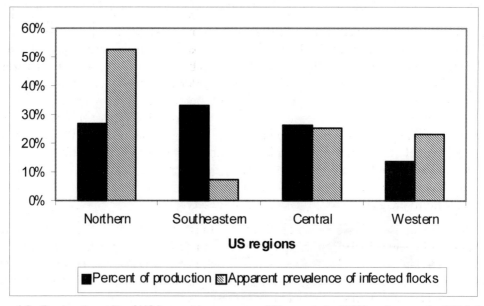

Figure 4.3. Regional results of USA spent hen surveys (Hogue et al., 1997) and percent of USA flocks by region. National estimates of flock prevalence should be adjusted for spatial bias.

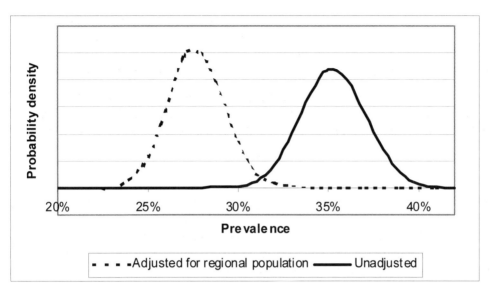

Figure 4.4. Illustration of the effect of weighting USA survey results (Hogue et al., 1997) for regional hen populations to estimate uncertainty regarding the national flock prevalence

Two factors influence the likelihood of false-negative results: the number of hens sampled per flock, and the underlying likelihood of detecting an infected hen given the methods used to test individuals.

Tables 4.2 and 4.3 show the results of sampling within infected flocks. The Poppe et al. (1992) study was a follow-up to the survey listed in Table 4.1. The original survey identified eight infected Canadian flocks, but was able to measure within-flock prevalence in only seven flocks. A variable number of hens were cultured in each of these flocks and, in four flocks, no infected hens were detected, despite previous positive hen or environmental test results, or both. The mean of the Beta distribution based on these results provided a non-zero point estimate for within-flock prevalence (Table 4.2). Table 4.3 summarizes the findings of the studies analysed by Hogue et al. (1997). In two different surveys, 247 positive flocks were detected. For each flock, 60 pooled caecal samples comprising five hens each were collected (i.e. caecae were collected from 300 hens per flock). Apparent within-flock prevalence was estimated by assuming that only one infected hen contributed to each positive pool. Such an assumption is reasonably unbiased (i.e. <5% difference between assumed and calculated within-flock prevalence) for those flocks with up to seven positive pools, but this negative bias increases with the number of positive pools. The average bias from this simplifying assumption is 5% across all observations.

Tables 4.2 and 4.3 provide evidence of the variability in number of *S.* Enteritidis infected hens between infected flocks. Both studies suggest that low within-flock prevalence is more frequent than high within-flock prevalence (Figure 4.5). Despite different populations sampled (e.g. Canadian vs United States of America layer flocks) and the dramatically different numbers of samples collected, the distributions are similar.

Table 4.2. Results of sampling known-infected layer flocks from Canadian study and mean of Beta distribution for predicting apparent within-flock prevalence

Number of flocks sampled	Positive hens	Hens sampled per flock	Apparent within-flock prevalence
4	0	60	1.6%
1	2	150	2.0%
1	0	40	2.4%
1	24	150	16.4%

SOURCE: Poppe et al., 1992

Table 4.3. Results of sampling known-infected layer flocks from USA studies. To calculate within-flock prevalence, it is assumed that a positive pool is equivalent to one positive hen, and 300 hens (60 pools of 5 hens) were sampled per flock.

Number of flocks sampled	Positive pools	Apparent within-flock prevalence
77	1	0.33%
39	2	0.67%
23	3	1.00%
18	4	1.33%
9	5	1.67%
6	6	2.00%
8	7	2.33%
7	8	2.67%
8	9	3.00%
4	10	3.33%
6	11	3.67%
4	12	4.00%
4	13	4.33%
2	14	4.67%
2	15	5.00%
6	16	5.33%
1	17	5.67%
3	18	6.00%
3	19	6.33%
2	21	7.00%
3	22	7.33%
1	23	7.67%
1	24	8.00%
1	25	8.33%
2	26	8.67%
2	27	9.00%
1	28	9.33%
1	36	12.00%
1	39	13.00%
1	42	14.00%
1	44	14.67%

SOURCE: Hogue et al., 1997

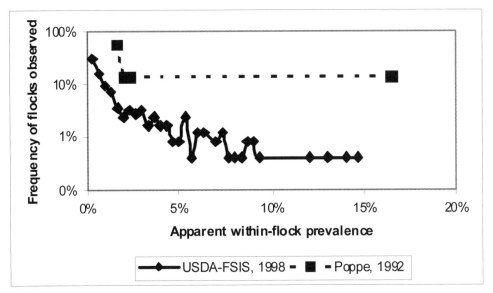

Figure 4.5. Comparison of evidence for within-flock prevalence from two studies that sampled multiple infected flocks.

Flock prevalence estimation methods have been proposed (Audige and Beckett, 1999; USDA-FSIS, 1998). These methods account for less-than-complete sampling within flocks. Given a fixed within-flock prevalence, infected flocks can be incorrectly classified as negative when a limited number of samples are collected in the flock (Martin, Meek and Willeberg, 1987). In practice, sample size is usually fixed in surveys, while within-flock prevalence varies between infected flocks. Therefore, collecting a number of samples sufficient to detect at least one positive hen in one infected flock (with reasonable likelihood) may not be a sufficient number in another infected flock.

In the US SE RA, the probability that a positive flock is detected given a fixed sample size is calculated as $1-(1-p)^n$, where p is apparent within-flock prevalence (i.e. proportion of detectable infected hens within an infected flock) and n equals the number of hens sampled per flock. Apparent within-flock prevalence was modelled as a cumulative distribution based on the survey evidence in Table 4.3. The cumulative distribution (Vose, 1996) converts within-flock prevalence data into a continuous probability function by specifying the minimum possible value (arbitrarily set at 0.001%, or 1 in 100 000 hens), the maximum value (arbitrarily set at 100% of hens), and the evidence in Table 4.3. Integrating $1-(1-p)^n$ across the distribution for apparent within-flock prevalence indicated that the sample size of 300 hens per flock used in the Hogue et al. (1997) surveys detected 76% of infected flocks. Integration was accomplished by simulating $1-(1-p)^n$, where p varied from iteration to iteration, and calculating the average of the simulated output.

For the US SE RA, the number of truly infected flocks in the Hogue et al. (1997) surveys was modelled using a Negative Binomial distribution. In the @Risk software language, the @RISKNEGBIN(s,p) function predicts the number of flocks missed given the number successfully detected, s, and the probability, $p = 0.76$, of detecting flocks (Vose, 1996). Adding the number of infected flocks misclassified in the survey to the number of infected flocks actually observed, then using this estimate with the total number of flocks sampled (i.e. 711) as inputs to a Beta distribution, provides the best description of uncertainty regarding true national prevalence.

An alternative to the method described for the US SE RA is to use a direct Bayesian methodology. In this case, Bayes Theorem is used to estimate the true flock prevalence:

$$f(\Phi \mid y) = \frac{f(y \mid \Phi) f(\Phi)}{\int\limits_0^1 f(y \mid \Phi) f(\Phi) d\Phi}$$

This depiction of Bayes Theorem is used to predict the distribution for flock prevalence (Φ), given the available evidence (y) (i.e. f(Φ|y)). In this case, the likelihood function, f(y|Φ), calculates the likelihood of observing a particular sampling result (e.g. 247 positive flocks in 711 flocks sampled) given that the true flock prevalence is Φ.

The likelihood function, f(y|Φ), determines the probability of the sampling evidence (i.e. apparent flock prevalence), given the true prevalence, Φ, and the sensitivity of the survey design. In this case,

$$Sens = 1 - \int (1 - p_i)^n \, f(p_i) \, dp$$

where p_i is the apparent within-flock prevalence in flock i, $f(p_i)$ is the likelihood of p_i occurring, and n is the number of samples collected in each flock. Operationally, the likelihood function is the binomial distribution,

$$f(y \mid \Phi) = \binom{N}{S} (Sens * \Phi)^S (1 - Sens * \Phi)^{N-S}$$

where N is the number of flocks sampled in a study, and S is the number found positive. Although this approach was not used in the US SE RA, it should give similar results to the negative binomial method previously described.

The Health Canada (2000) and Whiting and Buchanan (1997) exposure assessments did not adjust flock prevalence evidence for sensitivity. In the Health Canada assessment, the Poppe et al. (1991) data were modelled directly using a Beta distribution. In the Whiting and Buchanan (1997) assessment, two fixed values were used to model flock prevalence: 10% and 45%. These values were selected to approximate the regional variability observed in the United States of America surveys of slaughtered hens (Hogue et al., 1997).

Simply modelling apparent flock prevalence will result in a depiction of this parameter that differs from that found if true flock prevalence is modelled. Figure 4.6 illustrates the Beta distributions implied by the data in Table 4.1. In the case of the Hogue et al. (1997) data, the distribution that results from estimating true prevalence from apparent prevalence is illustrated. The effect of this adjustment is to shift the distribution towards higher flock prevalence levels, as well as slightly increasing the spread of the distribution.

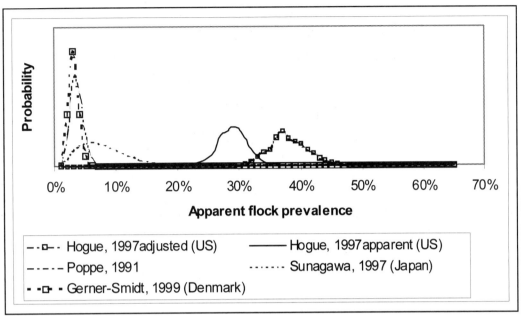

Figure 4.6. Implied distributions for apparent flock prevalence using the sampling evidence listed in Table 4.1. Evidence was modelled using Beta distributions. The Hogue et al. *(1997)* data are modelled for both apparent prevalence (using just the sampling evidence), and after adjusting the sampling evidence for false-negative flocks (i.e. true prevalence).

Egg contamination frequency

Ideally, egg-culture data would be available from flocks known to be infected. However, results from sampling eggs from infected flocks will show variability across time in the same flock, and between flocks. Variability is expected in any biological system. Seasonal variability in egg culturing results may also be observed, but previous analysis has not detected a consistent pattern (Schlosser et al., 1999).

Unfortunately, the logistics and cost of egg sampling limit the availability of such data. Furthermore, when egg sampling is conducted at the flock level, the number of eggs sampled is usually inadequate to calculate precise estimates of egg contamination frequency. In fact, the low apparent prevalence of contaminated eggs from infected flocks suggests that inadequate sampling of eggs will usually result in culture-negative results for all samples collected.

Sampling eggs is not a cost-effective surveillance method when the prevalence of egg contamination in infected flocks is low (Morales and McDowell, 1999). It is possible, however, that variability in egg contamination from flock to flock might be modelled using evidence concerning within-flock prevalence of infected hens. Evidence may come from the proportion of hens in a flock that are faecal shedders of *Salmonella* Enteritidis, or have organ or tissue samples that are culture-positive for *Salmonella* Enteritidis. Regardless of the endpoint measured, some estimate of the fraction of contaminated eggs laid by infected (or colonized) hens will allow the modelling of egg contamination frequency at the flock level. However, uncertainty regarding the variability in egg contamination frequency is greater using this approach than one that relies on direct egg culturing evidence. For that reason, this approach to estimating egg contamination frequency is not preferred when direct egg culturing evidence is available.

Data

Tables 4.4 and 4.5 summarized the egg sampling evidence from known infected flocks or hens used by the three quantitative exposure assessments. These data are used in the respective models to estimate egg contamination frequency.

For the US SE RA and Health Canada exposure assessments, two forms of data from the same field project are used. The Health Canada exposure assessment data are from a study of 43 positive flocks; the number of samples analysed was limited to the first 4000 eggs collected from these flocks. In contrast, the complete egg sampling results from the 43 flocks were summarized in the US SE RA. The flocks were stratified into high and low prevalence in the US SE RA analysis. The basis of this stratification was the finding that egg contamination frequency was correlated with environmental status and there was a bimodal pattern to environmental test results in infected flocks. Additional studies were included in each strata, based on the same criteria or similarity in results. The combined results from the US SE RA study suggest an overall egg contamination frequency of 0.03%; the same as the average based on the Health Canada exposure assessment data.

In the Whiting and Buchanan (1997) model, the egg contamination frequencies implied by 27 published studies were summarized (Table 4.5). Some of these studies were reportedly experimental. The relevance of experimental studies to populations of naturally infected hens is arguable. In particular, if a study reports the frequency at which a cohort of experimentally inoculated hens produced contaminated eggs, then these results need adjusting for the prevalence of naturally infected hens in a flock to be comparable to field-based evidence. The median frequency from this series of studies is between 0.6% and 0.9%.

Table 4.4. Summary of evidence used in two exposure assessments to model egg contamination frequency. The number positive eggs (s) and the total number of eggs sampled (n) are reported by study cited

Risk assessment	Flock type	s	n	Data source
USDA-FSIS, 1998	High prevalence	58	85 360	Kinde et al., 1996
		56	113 000	Schlosser at al., 1995
		41	15 980	Henzler et al., 1994
	Total	155	214 340	
	Low prevalence	22	381 000	Schlosser at al., 1995
		2	10 140	Henzler et al., 1994
	Total	24	391 140	
Health Canada (2000)		34	100 000	Schlosser at al., 1995
		2	16 560	Poppe et al., 1991
	Total	36	116 560	

Table 4.5. Summary of evidence used to model egg contamination frequency in the Whiting and Buchanan (1997) exposure assessment model

Frequency of culture-positive eggs	Number of studies
0.00%	5
0.06%	1
0.08%	1
0.10%	1
0.20%	1
0.30%	1
0.40%	2
0.60%	1
0.90%	1
1.00%	4
1.40%	1
1.90%	1
2.90%	2
4.30%	1
7.50%	1
8.10%	1
8.60%	1
19.00%	1
Total	27

Methods

A histogram relating egg contamination frequency with the number of infected flocks observed can be derived if enough eggs from enough flocks are sampled in a cross-sectional survey. Such a histogram provides an empirical description of the variability in egg contamination. This distribution may be skewed if most flocks express very low egg contamination frequencies and few flocks experience higher contamination frequencies.

In surveys that are prospective and cross-sectional in design, the data can be summarized using the average contamination frequency across all egg collections in each flock. Because individual egg collections usually involve an insufficient numbers of eggs (e.g. 1000), the most confident estimate is that applied to the entire period during which sampling was completed in that flock.

Figure 4.7 illustrates the distribution for egg contamination frequency found in one research project covering 60 infected flocks (Henzler, Kradel and Sischo, 1998). The reported frequencies are set equal to the mean of the Beta distribution to provide non-zero estimates for those flocks where no positive eggs were detected. A significant finding from this study was that a high proportion of the flocks with the lowest observed egg contamination frequency were also flocks with fewer numbers of positive environmental samples. Most of the flocks with higher egg contamination frequencies also had more positive environmental samples.

Egg contamination frequency evidence should be adjusted for uncertainty resulting from pooling of samples, and the sensitivity or specificity, or both, of laboratory culture techniques (Cowling, Gardner and Johnson, 1999). For pooled sample results, it is probably appropriate, when prevalence is low (e.g. <0.1%), to assume that individual prevalence is equivalent to *x/km*, where *x* is the number of positive pools, *k* is the size of pools (e.g. 10 or 20 eggs), and *m* is the number of pools sampled. Nevertheless, it is instructive to evaluate the probability theory of pooling.

Given some probability that an individual egg is positive, *p*, and a pool size of *k*, the probability of a pool being positive, *P*, equals $1 -(1 - p)^k$, where $(1 - p)^k$ calculates the probability of selecting *k* negative individuals. From available sampling evidence, we have *P*. Therefore we can solve this equation for *p*, and it equals $1 -(1 - P)^{1/k}$. Cowling, Gardner and Johnson (1999) present various methods for describing the uncertainty about this estimated probability. A simple method is to describe the uncertainty about *P* as a Beta(x + 1, m − x + 1) distribution and directly map the values for *p* to the probabilities predicted for the Beta distribution. When *p* is very small and *m* is very large, the need to incorporate uncertainty about the effect of pooling is insubstantial.

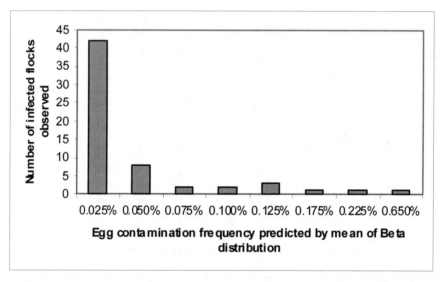

Figure 4.7. Results of a study in 60 United States of America infected flocks, showing variability in egg contamination frequency between infected flocks (Source: Henzler, Kradel and Sischo, 1998).

Evidence for the sensitivity of laboratory culture techniques comes from an experiment on the isolation of *Salmonella* Enteritidis in pooled eggs. In this study, pools of 10 eggs were spiked with approximately 2 CFU of *Salmonella* Enteritidis, and 24 out of 34 (70.6%) of the pools were detected as positive using standard culture techniques (Gast, 1993). Although Cowling, Gardner and Johnson (1999) argue that the best estimate of sensitivity for pooled egg culturing should be centred about 70.6%, this estimate may understate the likelihood of detection. Most contaminated eggs, unless cultured within a few hours of lay, contain many more than two *Salmonella* Enteritidis organisms (see section below). Therefore one must calculate the probability that laboratory culturing will detect a single organism and then apply that probability to the number of organisms expected to be found in contaminated egg pools.

If a pooled sample contains two *Salmonella* Enteritidis organisms, the probability of correctly classifying the sample as positive using culture techniques equals $1 -(1 - p)^2$, assuming a binomial process and *p* equal to the probability of detecting one organism. From Gast (1993), 70.6% of samples with two *Salmonella* Enteritidis organisms were found positive. Therefore one can solve for *p* to determine the probability of detecting one organism in a pooled sample. In this case, *p* equals 46%.

If the probability of detecting one organism in a pooled sample is 46% – and the probability of detecting two organisms is 70.6% – then one can calculate the sensitivity of pooled egg testing for any number of organisms contained in a sample. Figure 4.8 shows how the probability of a positive result increases as the number of organisms in the sample increases. At eight organisms (or more) in a pooled sample, the probability of a positive test result is essentially 100%. Given that the predicted mean number of *Salmonella* Enteritidis organisms per contaminated egg typically exceeds seven, these results suggest it is unlikely that sensitivity of egg testing is an important input to exposure assessments.

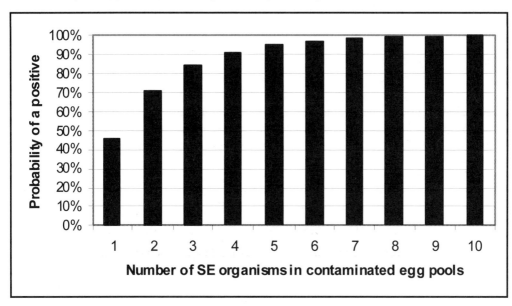

Figure 4.8. Predicted probability of a positive pooled egg sample when contaminated with varying numbers of *Salmonella* Enteritidis (SE) organisms and a probability of detecting just one organism equal to 46%.

The frequency that an infected flock produces contaminated eggs is modelled in the Health Canada exposure assessment by incorporating the data in Table 4.4 into a gamma distribution. In @Risk, this distribution is specified as: @RISKGAMMA(s,1/n), where *s* is the number of positive eggs and *n* is the total number of eggs sampled. The gamma distribution is a theoretical distribution for estimating uncertainty about the average of a Poisson process. In practice, the difference is insignificant between assuming egg contamination frequency follows either a binomial or a Poisson process. Therefore, either the gamma or the beta distribution would suffice for modelling these data.

One can model the data cited for the Whiting and Buchanan (1997) exposure assessment (Table 4.5) using a cumulative distribution. In @Risk, the cumulative distribution is specified as:

@RISKCUMULATIVE(min,max, {x}, {p})

where theoretical minima and maxima are estimated, {x} is an array of observed egg contamination frequencies, and {p} is an array of cumulative probability densities corresponding to values in {x}. Alternatively, these data could be modelled using discrete or histogram distributions.

In the US SE RA exposure assessment, egg contamination frequencies for high and low prevalence flocks are modelled using the gamma distribution. The egg contamination frequencies for each type of infected flock only apply to the fraction of infected flocks within these two strata. This exposure assessment also explicitly models moulted and non-moulted flocks. Therefore, egg contamination frequency from infected flocks is calculated as a weighted average across all infected flocks by prevalence strata and moult status.

The data in Table 4.5 on the frequency of *Salmonella* Enteritidis-positive eggs produced by positive flocks were from flocks that were typically detected via environmental sampling. Estimating egg contamination frequencies directly from these data can result in biased estimates, possibly introduced because environmental testing is more likely to detect infected flocks with high within-flock prevalence levels, compared with flocks with low within-flock prevalence levels. Therefore, egg-culturing evidence is disproportionately influenced by higher prevalence flocks relative to the actual egg contamination frequency in the total population of infected flocks. In US SE RA, the proportion of infected flocks classed as high prevalence was adjusted for the sensitivity of environmental testing to account for this phenomenon.

The effect of moulting on egg contamination frequency has been experimentally examined and found significant (Holt and Porter, 1992, 1993; Holt et al.,1994; Holt,1995, 1998). However, there is only one field study that examines this phenomenon (Schlosser et al., 1999). In that study, 31 of 74 000 (0.04%) eggs sampled from infected flocks that were within 20 weeks following moult were *Salmonella* Enteritidis-positive. In contrast, only 14 of 67 000 (0.02%) eggs sampled from infected flocks that were within 20 weeks prior to moult were *Salmonella* Enteritidis-positive. These results imply that moulting is associated with a nearly twofold increase in egg contamination in the 20 weeks following moult. Figure 4.9 shows the probability distributions for high and low prevalence flocks that are moulted or not moulted, using these data and those shown in Table 4.4. Figure 4.10 illustrates the different distributions for egg contamination frequency in infected flocks, generated by the three exposure assessments. The weighting of egg contamination frequencies for different types of flocks in the US SE RA has the effect of reducing the overall frequency of contaminated eggs from infected flocks. Nevertheless, the predicted distributions for the US SE RA and Health Canada exposure assessments are more similar to each other than either is to the distribution implied by Whiting and Buchanan (1997). The Whiting and Buchanan (1997) distribution is bimodal, with some flocks producing contaminated eggs at frequencies at or below 10^{-6} and many more flocks producing contaminated eggs at frequencies at or above 1%.

It should be noted that the US SE RA and Health Canada distributions in Figure 4.10 represent uncertainty about the true fraction of contaminated eggs produced by infected flocks. In contrast, the Whiting and Buchanan (1997) distribution is best characterized as a frequency distribution of the predicted proportion of infected flocks producing contaminated eggs at different frequencies. Only the reported frequencies, and not the number of positives and samples, were used to derive this distribution. Therefore uncertainty about the true egg contamination frequencies reported by each of the 27 studies used by Whiting and Buchanan (1997) is not incorporated into this analysis.

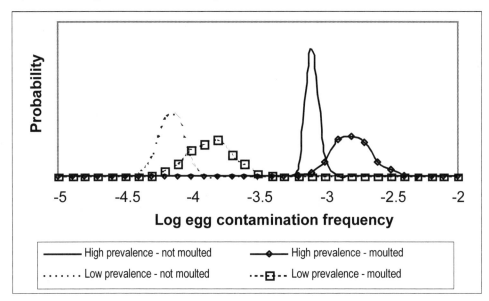

Figure 4.9. Comparison of uncertainty regarding egg contamination frequencies between so-called high and low prevalence flocks that are moulted or not moulted (Source: US SE RA).

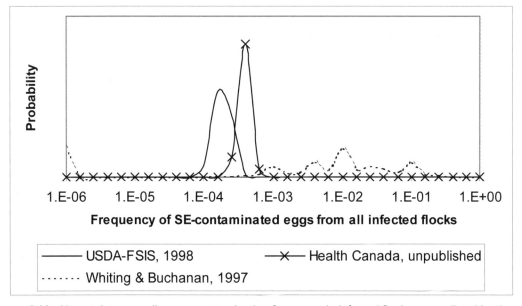

Figure 4.10. Uncertainty regarding egg contamination frequency in infected flocks as predicted by three published exposure assessments.

The proportion of flocks that are infected (i.e. flock prevalence) and the egg contamination frequency for infected flocks are combined to estimate the overall frequency of contaminated eggs among all eggs produced. Figure 4.11 shows these results for the three quantitative exposure assessments. For both the US SE RA and Health Canada exposure assessments, the predicted fraction of all eggs that are *Salmonella* Enteritidis-contaminated is most probably between 10^{-5} and 10^{-4}. The overall contamination frequency implied by Whiting and Buchanan (1997) is most probably between 10^{-3} and 10^{-2}.

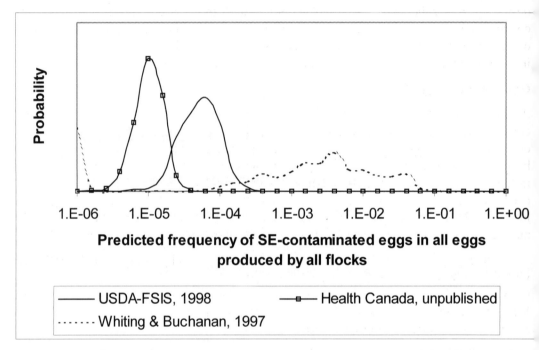

Figure 4.11. Uncertainty regarding frequency egg contamination by *Salmonella* Enteritidis (SE) among all eggs produced, regardless of flock status, as predicted by three published exposure assessments.

Methods for modelling egg contamination frequency varied among the three quantitative exposure assessments. Although the USDA-FSIS and Health Canada approaches are similar in some respects, the US SE RA explicitly modelled variability in egg contamination frequency by stratifying infected flocks into different categories (e.g. high and low prevalence). Disaggregation of infected flocks by degree of severity accomplishes two purposes: first, it allows one to explicitly model control interventions on a subset of the total population of infected flocks; and, second, it illustrates the relative importance of different types of flocks to the overall frequency of contaminated eggs. By modelling high and low prevalence flocks, as well as moulted or unmoulted subpopulations, the US SE RA may be a more useful tool for risk managers. Furthermore, the explicit delineation of flocks because of severity of their infection enabled the identification of a potential bias resulting from easier detection of the more severely affected flocks.

The methods used by Whiting and Buchanan (1997) may illustrate the possible bias introduced using experimental data. Egg contamination frequency in infected flocks is best estimated using field research results. Experimental studies cannot replicate the infectious dose and transmission characteristics that exist in naturally infected flocks. This distribution may also imply increased egg contamination frequencies because the underlying data reflect higher-virulence strains of *Salmonella* Enteritidis than typically occur in naturally infected populations (Gast, 1994).

Ideally, this exposure assessment input should describe the variability of egg contamination frequency between infected flocks. Although the US SE RA accomplishes this to a limited extent by stratifying flocks, a more continuous description of this input is desirable. Either the gamma or the beta distribution can be used to model uncertainty in egg contamination frequency based on the available data. Furthermore, given the numbers of organisms expected within contaminated eggs, and the low frequency of contaminated eggs, it seems unnecessary to adjust observed data for pool size or sensitivity of tests.

Organisms per egg at Lay

An exposure assessment must include the initial concentration of *Salmonella* Enteritidis in contaminated eggs. The number of *S.* Enteritidis in contaminated eggs varies from egg to egg. Available evidence suggests that most contaminated eggs have very few *S.* Enteritidis bacteria within them at the time of lay. It is the initial contamination level in an egg that is influenced by subsequent distribution and storage practices. If the egg is handled under conditions that allow growth of the bacteria in the egg, then the initial concentration will increase. Nevertheless, some contaminated eggs will arrive at the kitchen with the same number of bacteria within them that they contained at the time of lay.

Most experts believe the *S.* Enteritidis in eggs is initially limited to the albumen or vitelline (yolk) membrane. Nevertheless, it is possible that *S.* Enteritidis may gain access to the yolk of the egg before or just after the egg is formed. If this occurs, it is a rare event. While egg albumen is not conducive to *S.* Enteritidis multiplication, yolk nutrients will foster relatively rapid growth of these bacteria (Todd, 1996).

Immediately following lay, the pH of the interior contents of an egg begins to increase. Elevated pH suppresses growth of *S.* Enteritidis. It is estimated that about one log of growth can occur between the time of lay and stabilization of pH inside the egg (Humphrey, 1993). Because of this phenomenon, it is difficult to know whether the observed number of organisms in a fresh egg is the result of some initial growth or the actual inoculum present at lay.

Data

In a study of contaminated eggs produced by naturally infected hens, 32 positive eggs were detected (Humphrey et al., 1991). Enumeration of their contents found that 72% of these eggs contained less than 20 *S.* Enteritidis organisms. The calculated mean number of *S.* Enteritidis per contaminated egg was 7. However, there were a few eggs that contained many thousands of *S.* Enteritidis bacteria following >21 days of storage at room temperature.

In a study of experimentally infected hens, 31 *Salmonella* Enteritidis positive eggs were detected (Gast and Beard, 1992a). Enumeration of their contents found that the typical contaminated egg harboured about 220 *Salmonella* Enteritidis organisms. Yet, there were marked differences in levels depending on storage time and temperature. Four of the contaminated eggs contained more than 400 *Salmonella* Enteritidis organisms per egg.

Methods

Growth of bacteria within the egg is a function of temperature and time between lay and consumption. At the time of lay, the egg's internal temperature is essentially that of the hen (i.e.~42°C). This temperature equilibrates with the environment over time.

Conventionally, concentration of bacteria per unit volume is modelled using a lognormal distribution (Kilsby and Pugh, 1981). Such a distribution describes the frequency of variable numbers of bacteria in contaminated eggs. Assuming there is sufficient data to estimate a population distribution, then uncertainty in an exposure assessment model stems from the fitting procedure used to describe the distribution. Lacking evidence from a sufficient sample of eggs to warrant direct fitting of evidence to a distribution, alternative approaches include using expert opinion to develop a distribution, or representing the data with an empirical distribution.

In the Whiting and Buchanan (1997) exposure assessment, a two-stage distribution for concentration of organisms per contaminated egg is modelled. The majority of eggs are modelled as containing 0.5 organisms/ml. For a 60-ml volume egg, this equates to 30 organisms per egg. Development of this estimate is based on the Humphrey et al. (1991) enumeration data. Some eggs in the Whiting and Buchanan (1997) model, however, are assumed to be held for up to 21 days at room temperature, or for shorter times at higher temperatures. These eggs experience growth of *S.* Enteritidis and the resultant number of organisms per egg is modelled using the following probability distribution: 58% = 0.5 organisms/ml; 25% = 13 organisms/ml; 8% = 375 organisms/ml; and 8% = 3 000 organisms/ml. Because this model is concerned with eggs that are sent for pasteurization, the large concentrations predicted here are arguably appropriate.

The fraction of eggs that are time- or temperature-abused in the Whiting and Buchanan (1997) model is assumed to range from 2.5% to 10% of all eggs. Therefore contaminated eggs usually contain about 30 organisms, but for 2.5–10% of this model's iterations the eggs may contain from 30 to 180 000 organisms.

In the Health Canada exposure assessment, data from two studies are combined to depict initial concentration. An initial number of organisms is modelled as the output of a Poisson process, where the mean is 7 organisms per egg (Humphrey et al., 1991). To this initial number of organisms is added another 0 to 1.5 logs of bacteria to account for the period immediately after lay, when pH is increasing in the egg (Humphrey, 1993; Humphrey and Whitehead, 1993). This additional step is modelled using the @Risk function @RISKPERT(min, most likely, maximum),where the minimum is zero, the most likely value is 1 log, and the maximum is 1.5 logs.

In the US SE RA, the data from Humphrey et al. (1991), Humphrey (1993) and Gast and Beard (1992) are combined to derive a distribution for initial number of *S.* Enteritidis organisms per contaminated egg. A truncated exponential distribution is used to model this input. In @Risk, this distribution is specified as @RISKTEXPON(mean, min, max), where the minimum equals 1 organism, the maximum is 420 organisms (based on Gast and Beard, 1992), and the mean is 152 organisms per egg.

The effect of the different modelling approaches used in the three exposure assessments is shown in Figure 4.12. The curve predicted by Whiting and Buchanan (1997) shows a peak frequency at 30 organisms per egg; however, the instances of heavily contaminated eggs are not shown in this graph. Although heavily contaminated eggs are infrequently predicted, the expected value of this distribution (>900 organisms per egg) reflects these occasionally large values.

The expected values of the Health Canada and US SE RA distributions are 88 and 152 cells, respectively. The US SE RA distribution reflects the incorporation of the Gast and Beard (1992) evidence. This evidence is from experimentally infected hens that were inoculated with large doses of *S*. Enteritidis. Its relevance to naturally contaminated eggs is arguable. Nevertheless, the combined evidence from Humphrey et al. (1991) and Gast and Beard (1992) amounts to just 63 eggs. Therefore the USDA-FSIS distribution may be interpreted as containing elements of variability and uncertainty regarding the actual frequency distribution for initial contamination levels in eggs.

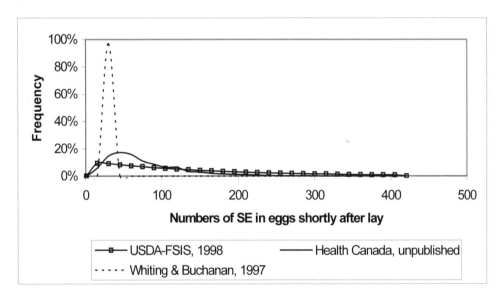

Figure 4.12. Comparison of varying levels of *Salmonella* Enteritidis (SE) within contaminated eggs, predicted by three published exposure assessments

Summary

The production component of an exposure assessment should include estimates of flock prevalence, egg contamination frequency in infected flocks, and number of *S*. Enteritidis per contaminated egg.

Flock prevalence is arguably a fixed value, but one for which uncertainty can be substantial. Apparent flock prevalence from surveys that use imperfect diagnostic assays should be adjusted for expected bias. Several methods are available to make these adjustments. Audige and Beckett (1999) have published a method that relies on the hypergeometric distribution. Such a method is particularly appropriate when the total population is small. The consequence of incorrectly assuming evidence was generated from a binomial distribution (that assumes sampling with replacement in a large population) is illustrated in Figure 4.13.

This example assumes that just one infected flock exists in a total population of 100 flocks. The probability of detection in this case is different if we assume a binomial distribution versus assuming the correct hypergeometric distribution. If we had a sample of 100 flocks and all were negative, the binomial distribution would imply that there was a 40% chance of such a result if the prevalence was 1%. In contrast, the hypergeometric distribution would tell us there was 0% probability of such a result if one infected flock existed. The probability of detection is used to adjust apparent flock prevalence, so it behoves the risk analyst to consider which distribution is appropriate when conducting an exposure assessment.

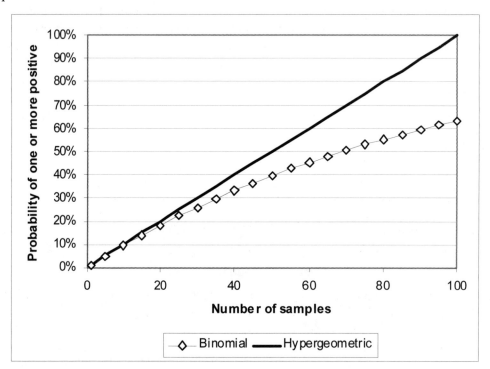

Figure 4.13. Illustration of the effect of assuming a binomial distribution when the population is limited to 100 and the prevalence is fixed at one infected flock (1%)

None of the exposure assessments explicitly accounted for the mechanisms by which flocks become infected. To assess pre-harvest interventions, more data is needed on the prevalence of *S.* Enteritidis in breeder and pullet flocks, as well as in feedstuffs. In particular, associations between the occurrence of *S.* Enteritidis in these pre-harvest steps and its occurrence in commercial layers should be quantified. The existing models, however, can be used to evaluate the effect of interventions that might reduce the risk of flocks becoming infected. The public health effect of such hypothetical interventions would be modelled by appropriately reducing the flock prevalence input of the existing models.

Egg contamination frequency should be a variable input to a *S.* Enteritidis exposure assessment. However, data are needed to accurately estimate the proportion of flocks with varying egg contamination frequencies. An alternative to modelling egg contamination frequency as a continuous distribution is to stratify flocks into two or more categories and model the estimated egg contamination frequency separately for each category. Such an approach provides more information to risk managers regarding control options. Nevertheless, stratifying infected flocks requires epidemiological evidence of differences among infected flocks. Without such evidence, use of available egg sampling evidence is a second-best approach.

The concentration of *S*. Enteritidis in contaminated eggs is also a variable input, yet few data are available to describe this variability. In the exposure assessments evaluated, estimation of the number of *S*. Enteritidis in contaminated eggs at, or soon after lay, was based on empirical data from, at most, 63 eggs. Evidence that associates modulation in numbers of *S*. Enteritidis per egg with causative factors (i.e. strain of *S*. Enteritidis, hen strain, environmental conditions) would provide analysts with better methods for modelling this input. Lacking such evidence, however, suggests that most *S*. Enteritidis exposure assessments will rely on the same evidence used by previous exposure assessments. Therefore, it is expected that this input will be common to most models. Of the methods used to model initial contamination, those used by Health Canada seem most intuitively appealing. The method used by US SE RA gives similar results but is potentially biased upwards.

4.2.3 Distribution and Storage

Once *S*. Enteritidis-contaminated eggs are produced, they undergo multiple stages of handling and storage before they are finally consumed. After the eggs are collected in the hen house, they may be processed, transported and stored for varying times under variable environmental conditions. The distribution and storage stage covers the period after the eggs are laid until the eggs arrive at a point where they are prepared, possibly cooked, and consumed.

Generally, the distribution and storage stage is concerned with two things: (1) the effect of time and temperature on the *S*. Enteritidis within contaminated eggs, and (2) the fraction of eggs that are marketed as fresh table eggs versus the fraction marketed in some other form. For example, some shell eggs may be sent to a pasteurization plant or cooked before sale.

Important inputs to include in this part of the exposure assessment model are algorithms for predicting the microbial dynamics within eggs; time and temperature distributions; and marketing fractions. Egg thermodynamics may also be explicitly modelled to simulate internal egg temperature as a function of ambient temperature.

The output of the distribution and storage stage should consist of a frequency distribution for the contamination levels in eggs just before preparation, cooking and consumption. Changing concentration of bacteria in eggs depends on the lag period before *S*. Enteritidis growth. Lag period is the period during which bacteria adjust to environmental conditions. Lag period inside an egg is a complex function of nutrient availability, pH, time and temperature (Humphrey, 1999).

In Monte Carlo simulations, the modelling of the paired occurrence of concentration and lag period requires that individual eggs are, to some extent, kept track of as they move through the various stages of distribution and storage. These individual eggs can then be carried over to the preparation and consumption stage for further consideration. In this manner, the attributes of organisms per egg and lag period remaining are paired at the individual egg.

Marketing fractions

As eggs move from the farm to table, there are two general users to consider: homes and food service institutions. Home users of eggs generally purchase their eggs from retail settings, while institutional users of eggs (e.g. restaurants and hospitals) get their eggs from wholesale distributors or directly from the producer.

If it is assumed that on average eggs are handled differently by these different users, then at least two growth pathways should be included in any *S.* Enteritidis exposure assessment model. In this case, the marketing route categorizes the outputs of the distribution and storage stage.

Shell eggs will also be marketed either as table eggs or as eggs to be broken and sold as processed egg products. It is likely that the marketing route influences the period (and possibly temperatures) that eggs experience in the distribution and storage stage. For example, some eggs are diverted to egg products processors following grading. These eggs may have a distribution of transport time post-processing that is different from eggs destined for table egg markets. Another example might be eggs that are directly marketed for egg products. Such eggs will probably spend less time in processing because they do not undergo candling or grading or sorting (so-called "nest run" eggs).

Marketing fractions are path probabilities that determine what fraction of contaminated eggs experience the time and temperature conditions of specific pathways. These are important inputs to an exposure assessment.

Data

Both the Health Canada and US SE RA exposure assessments determined that about 25% of all table eggs consumed were marketed to institutional users. These users consumed 18% of all eggs produced in the United States of America, but after adjusting for the 28% of eggs that are marketed as processed egg products, institutional users consume 25% of the table eggs (Table 4.6). The Whiting and Buchanan (1997) exposure assessment was not concerned with table eggs, so user fractions were not considered.

Table 4.6. Distribution of eggs by market outlet, as estimated for USA production

Egg market	Million cases[1] of eggs	Proportion of all eggs
Retail	94	53%
Egg products processing	49	28%
Food service	31	18%
Exported	3	2%

NOTE: (1) A standard USA case of eggs contains 360 eggs (30 dozen).
SOURCE: US SE RA

Time and temperature

Eggs may be stored before and after they are sold. They may be transported to and from wholesale or retail distributors. Times and temperatures during storage and transportation vary. Given the low frequency of contaminated eggs, it is reasonable to consider these eggs as independent from each other regarding the times and temperatures they experience.

An exposure assessment model must consider the different times and temperatures that contaminated eggs experience. Therefore, distributions of time and temperature must be included in the model. Such distributions should, for example, describe the proportion of eggs that experience different temperature at a given stage. Furthermore, a different distribution for the same stage must describe the proportion of eggs that are maintained in that stage for different times. It might be assumed that there is some negative correlation between time and temperature within a particular stage. It seems reasonable to expect that eggs held at higher temperatures are held for shorter times, but there is no evidence available to estimate such a correlation and it is possible that the correlation is only applicable over a portion of the temperature and time distribution.

For convenience, models of distribution and storage typically describe the passage of time in discrete steps. Clearly, an egg can experience a constantly changing environment from the point of lay until consumption. Describing distributions for ambient temperature within a discrete stage is nearly impossible without continual data collection. Using expert opinion and available evidence, however, it is possible to estimate a distribution for the length of time that eggs are stored (e.g. on the farm), and a distribution of average ambient temperatures that apply to that storage period.

Data

Times and temperatures for various stages of the farm-to-table continuum are not readily available from the published literature. An exposure assessment portrays the variability in times and temperatures that individual eggs experience between lay and consumption. Not all eggs are handled in the same way, and it is the combination of inordinately high temperatures and times that result in large amounts of *S.* Enteritidis growth in contaminated eggs.

In the absence of survey or sampling data, other types of information can be used. For instance, the recommended shelf life of eggs can serve as a surrogate for retail storage time. Such a measure has the added advantage of allowing measurement of mitigation effects by adjusting the level of compliance with recommended procedures.

Methods

Table 4.7 summarizes the time and ambient temperature inputs to the Health Canada and US SE RA exposure assessments. The averages portrayed here suggest that the underlying assumptions are similar between the models. In general, the stages modelled are the same between the models, with the exceptions that the Health Canada exposure assessment explicitly delineates pre-collection and wholesale storage stages, and the US SE RA explicitly delineates a post-cooking storage stage. The average cumulative time for an egg to pass through these stages is 429 hours (17.9 days) for the Health Canada and 565 hours (23.5

days) for the US SE RA. The average ambient temperature, weighted for time, is 9°C for the Health Canada and 8°C for the US SE RA. The scope of the Whiting and Buchanan (1997) exposure assessment was limited to pasteurized liquid whole egg, so farm-to-table stages were not modelled explicitly.

Table 4.7. Summary of average input times and ambient temperatures used to model farm-to-table pathways in exposure assessments. Probability distributions are used in these models and the central tendency of each distribution is presented here.

Stage	Health Canada		US SE RA	
	Av. temp. (°C)	Av. time (hours)	Av. temp. (°C)	Av. time (hours)
Pre-collection of eggs	26	7	N.A.	N.A.
Storage before transport from farm	13	35	13	48
Transport to grading or processing	13	3	13	1
Storage before grading or processing	13	13	20	5
Grading or processing	20	0.2	18	1
Storage after processing or grading	13	62	8	48
Transport to wholesale or retail	13	3	10	6
Wholesale storage	4	0	N.A.	N.A.
Retail storage	7	142	7	168
Consumer storage	7	164	7	288
Post-cooking storage	N.A.	N.A.	18	1

Note: N.A. = not available.

Growth is a function of time and temperature inputs to the exposure assessment models. These inputs can vary by pathway (e.g. home vs institution) as well as within the pathway. The average predicted temperature is greater for the US SE RA model (11°C) than for the Health Canada model (9°C) (Figure 4.14). The temperatures captured in this analysis are ambient temperature in the Health Canada model, and internal egg temperature in the US SE RA model. These parameters determine lag time and growth rates in the respective models. It is noteworthy that the average ambient temperature is actually lower in the US SE RA model (7°C). Nevertheless, accounting for cooling rates in eggs results in higher average internal egg temperatures.

As shown in Figure 4.14, there is much more variability in average temperature per egg for the US SE RA output. Increased variability implies that greater extremes in temperature are possible in this model when compared with the Health Canada model. Sustained higher temperatures result in shorter lag periods and faster growth rates (Humphrey, 1999).

The average time between lay and consumption is longer in the USDA model than in the Health Canada model. Figure 4.15 illustrates that this time is also more variable in the USDA model. Therefore individual contaminated eggs may spend longer in going from the producer to a prepared meal. As with higher temperatures, longer times can be associated with shorter lag periods and greater growth rates within contaminated eggs.

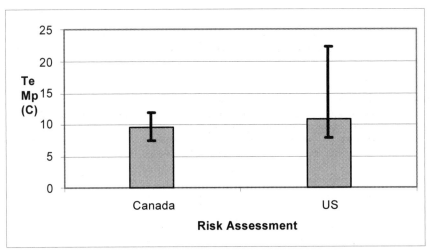

Figure 4.14. Average temperature between lay and consumption for all eggs in two exposure assessment models. In the US SE RA, temperature is internal egg temperature. Error bars depict 95% of the variability in each model.

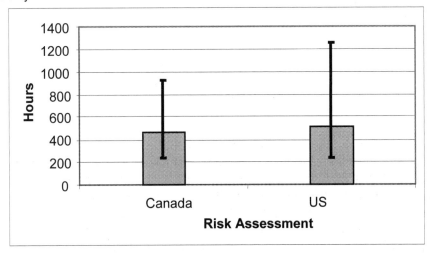

Figure 4.15. Average total time between lay and consumption for all eggs in two exposure assessment models. Error bars depict 95% of the variability in each model.

Microbial growth dynamics

Considering only *S.* Enteritidis bacteria that are inside the egg soon after lay, available evidence suggests that growth of the bacteria depends on an increase in the permeability of the vitelline (yolk) membrane. This increase allows the bacteria access to critical growth nutrients. However, the change in permeability of the yolk membrane is time and temperature dependent. The process may take three weeks or longer, depending on the temperature at which eggs are held. Until this process is complete, there is little or no growth of *S.* Enteritidis bacteria within the egg. Essentially, this period represents a lag phase for the bacteria.

Once there is yolk membrane permeability sufficient for *S.* Enteritidis to grow, multiplication of the bacteria can occur in the egg. The rate of growth of bacteria is also a function of time and temperature. Therefore, as the egg moves from the point of lay to the point of consumption, yolk membrane permeability and growth must be monitored continually with respect to temperature and time. In the US SE RA, the relevant temperature for yolk membrane permeability and microbial growth is the internal egg temperature. Therefore the thermodynamics of temperature equilibration between ambient and internal temperature must also be included in the model.

It has been argued the lag period in a discrete stage model should be modelled cumulatively (Zwietering et al., 1994). Therefore the fraction of lag period remaining for an individual egg should be monitored for each stage and accumulated across successive stages. For example, if 50% of an egg's lag period is expended during one step of the processing stage (e.g. pre-processing storage), then this should be subtracted from the available abuse time in the next step. Therefore, if the next step results in the use of 75% of an egg's lag period, there would actually be 25% of the time in that step when active growth of *S.* Enteritidis could occur.

Eggs produced in commercial flocks in many countries are usually processed. Processing can include candling, grading and sorting, washing, sanitizing and packaging. In general, egg processing does not result in any reduction in the number of bacteria present in contaminated eggs. Instead, processing either increases the number of bacteria in a contaminated egg or leaves the concentration unchanged.

Processing may detect some contaminated eggs, thereby preventing these eggs from reaching the table egg market. Candling and grading of eggs are activities that evaluate quality characteristics of eggs. Candling will identify blood spots and defective shells. Grading eggs involves valuing the qualities of the eggs based on the outcome of candling. Sorting eggs basically groups the eggs dependent on their grades. There is reportedly an association between blood spot defects and the likelihood of these eggs being internally contaminated with *S.* Enteritidis (Schlosser et al., 1999), so, if blood-spot eggs are less likely to be marketed as table eggs, then sorting of eggs may result in a lower proportion of contaminated eggs in that market.

Data

To model the growth of *S.* Enteritidis in eggs, mathematical models are used to account for the lag and growth dynamics of this product and pathogen. Ideally, studies are conducted that closely mimic the range of conditions that commercially contaminated eggs experience. Such studies should provide sufficient statistical rigour to estimate the length of time before *S.* Enteritidis begins to grow at a given temperature, and the rate it grows once multiplication commences. Furthermore, evidence would ideally explain the effect of dynamic ambient temperatures on the time until yolk membrane permeability is sufficient for *S.* Enteritidis growth, as well as the effect of varying temperature before growth on the subsequent growth rate.

Research on *S.* Enteritidis growth has focused on the effects of storage time and temperature. This research varies in methodology but, in most cases, eggs are artificially inoculated with *S.* Enteritidis bacteria. In some cases, the inoculum is very high (Hammack et al., 1993; Schoeni et al., 1995). In other cases, the inoculum is placed into the yolk. The most relevant research involves inoculation of numbers of *S.* Enteritidis consistent with those observed in naturally contaminated eggs (Humphrey, 1999), and placement of the inoculum outside of the yolk (Humphrey, 1993). Nevertheless, incorporation of these research findings into an exposure assessment should account for uncertainties that result from the experimental (versus field observational) nature of this research.

In the US SE RA, lag period duration and exponential growth rate equations were estimated from data provided by Dr T. Humphrey (Exeter, UK, personal communication), as well as from published reports (e.g. Schoeni et al., 1995). The lag period duration was denoted as yolk membrane breakdown time (YMT) and estimated as;

$$\log_{10} \text{YMT} = 2.0872 - 0.04257T$$

where the YMT is in days and the temperature T is in degrees Celsius. The exponential growth rate (EGR) applies once yolk membrane breakdown is complete. EGR is estimated as:

$$\sqrt{\text{EGR}} = -0.1434 + 0.02601T$$

where EGR is in logs/hour and T in degrees Celsius.

In the Health Canada exposure assessment, YMT and EGR equations were estimated using essentially the same data as analysed in the USDA model. Nevertheless, the following slightly different equations were estimated.

$$\log_{10} \text{YMT} = 2.07 - 0.04T$$
$$\sqrt{\text{EGR}} = -0.13 + 0.04T$$

where EGR is in generations per hour and T in degrees Celsius. To convert from generations per hour to logs per hour, generations per hour is multiplied by $\log(2)$.

In the Whiting and Buchanan (1997) exposure assessment, growth in shell eggs was modelled as part of the distribution for initial numbers of *S.* Enteritidis in eggs (see Section 4.2.2 – Production). In this model, growth only occurred following breaking, mixing, pasteurization and storage of eggs. Yolk membrane breakdown was not a consideration. The growth model used was from a published study (Gibson, Bratchell and Roberts, 1988). A series of equations based on a Gompertz function are estimated for EGR as follows: the Gompertz equation is

$$L(t) = A + Ce^{(-\exp[-B(t-M)])}$$

where $L(t)$ is the log count of bacteria at time t (in hours), A is the starting log count of bacteria at $t = 0$ hours, C is the maximum logs of growth achievable, M is time when maximum growth rate is achieved and B is the maximum log growth rate at time M. From the Gompertz function, the EGR is calculable as:

$$\text{EGR} = BC/e$$

where $\ln(B) = -23.5 + 1.496s + 0.487t + 4.29p - 0.0608s^2 - 0.00563t^2 - 0.293p^2$ and C is a constant, e is the base of the natural logarithm, s is the salt concentration (%), p is pH, and t is temperature in Celsius.

A comparison of the EGR predicted for the three exposure assessment models is shown in Figure 4.16. The USDA equation predicts the slowest growth rates across all temperatures shown. The Whiting and Buchanan (1997) equation predicts slightly slower rates than the

Health Canada equation at lower temperatures, but faster rates at higher temperatures. All three equations predict no growth below 7°C.

Comparisons between the Health Canada and US SE RA exposure assessments for EGR are not direct. In the USDA model, growth rate is dependent on the internal egg temperature, not ambient temperature, while the Health Canada model uses the ambient temperature. Generally, internal egg temperature is greater than ambient temperature. Therefore, for a given ambient temperature, the EGR for the USDA model is, on average, a function of a slightly higher internal egg temperature.

In the US SE RA, microbial growth dynamics for *S*. Enteritidis in eggs are dependent on the internal temperature in the egg. Yet, available data on storage and handling usually reflect the ambient temperature surrounding eggs. Furthermore, the ambient temperature influences internal egg temperature in a variable manner depending on how eggs are stored and packaged. Therefore, equations are needed to predict the change in internal egg temperature across time as the ambient temperature changes.

The internal temperature of eggs shortly after lay is approximately 99°F (37°C). The interactions of initial internal egg temperature, ambient temperature and egg packaging conditions are used to predict the future internal egg temperature via a simple non-steady-state heat transfer equation.

$$\text{Log}[(T - T_0) / (T_I - T_0)] = -kt$$

where T is the internal egg temperature in Fahrenheit at a specific time t (in hours), T_i is the initial internal egg temperature, and T_0 is the ambient air temperature.

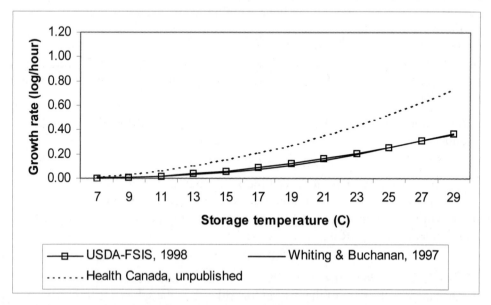

Figure 4.16. Comparison of *Salmonella* Enteritidis growth rates in eggs where yolk membrane permeability is complete, as predicted by three exposure assessments. Growth rate is shown as an increasing function of storage temperature.

The parameter k is a cooling constant (cooling rate per hour) that is estimated from available data using different packaging materials, methods and air flow in storage rooms (Table 4.8). The parameter values range from 0.008 for an egg in a box in the centre of a pallet, to 0.1 for eggs in a box, to 0.5 for individual eggs exposed to circulating air.

Table 4.8. Cooling constants estimated from available literature that describe rates of cooling for various storage situations

Situation	k (hours)	Reference
Pallet, cardboard and fibre flats, in-line	0.0075	Anderson, Jones and Curtis, 1992
Pallet, cardboard boxes	0.008	Czarik and Savage, 1992
Pallet, cardboard boxes, styrofoam	0.013	Czarik and Savage, 1992
Pallet, cardboard, off-line	0.035	Anderson, Jones and Curtis, 1992
Single cardboard case	0.052	Czarik and Savage, 1992
Flats, closed	0.07	Bell and Curley, 1966
Flats, folded shut	0.08–0.014	Bell and Curley, 1966
Pallet, plastic baskets, styrofoam	0.11	Czarik and Savage, 1992
Open stack	0.2–0.4	Bell and Curley, 1966
Fibre case, foam cartons with and without slots, moving air	0.24	Stadelman and Rhorer, 1987
Open stack, forced air	0.4–1.0	Bell and Curley, 1966
Cryogenic cooling	11	Curtis, Anderson and Jones, 1995

Among the population of shell eggs, k is a variability distribution for each storage period modelled. The values in Table 4.8 were estimated from the cooling characteristics of an egg in the centre of a pallet or box. Because these central eggs would be the warmest eggs in storage, these k values are thought to be conservative. Furthermore, these estimates varied between experimental replicates.

To reflect the natural egg-to-egg variability in the cooling rate, midpoint values were selected based on the expected storage conditions within a stage. The midpoint values were the modal values in a PERT(min, mode, max) distribution. Minimum and maximum values were assumed to be one-third lower or higher than the model. In the US SE RA model, uncertainty about these parameters was not explicitly considered.

In general, storage stages before egg processing reflect situations in which eggs are stored in boxes or flats and the modal k value is about 0.08. The variability in k is modelled as PERT(0.053 0.08 0.107) for storage before processing in the US SE RA model. During processing, eggs are exposed on all surfaces to the ambient air and k values tend to be higher (modal k about 0.5). During transportation, eggs are typically stored on pallets or in styrofoam containers and the modal k value is about 0.1. When eggs reach retail, wholesale and home storage conditions, they are usually in fibre cases or foam cartons, with some air movement around them. In the US SE RA model, it was assumed that k was non-variant for these stages and equalled 0.24.

Time and temperature distributions feed into the microbial growth equations via Monte Carlo simulation to determine a distribution for total logs of growth within contaminated eggs. Operationally, growth modelling considers the temperatures and periods each egg experiences in each of the stages listed in Table 4.7. The YMT is evaluated for a given temperature and compared with the time in each stage. If the YMT is exceeded, then the amount of growth is predicted at the current temperature.

Methods

To model *S.* Enteritidis growth during the distribution and storage of eggs, the inputs described above are needed. For each of the stages listed in Table 4.7, growth of *S.* Enteritidis in an egg is based on the period and temperature in the stage. Growth is accumulated across stages, so that the total number of organisms in each egg modelled through all stages represents the cumulative effect of the environmental conditions experienced by that egg between lay and consumption. Therefore, modelling of growth within each stage is a function of:

X = the number of organisms inside the egg at the start of the stage (logs per egg),

Y = the amount of YMT expended for that egg at the start of the stage (%),

T = the ambient temperature in that stage (Celsius), and

t = the time the egg spends in that stage (hours).

In the Health Canada exposure assessment, the following Excel® (Microsoft, Redmond, Washington) spreadsheet functions are executed to model growth in a stage:

1. YMT for the stage is calculated as: 24 (hours/day) * $10^{(2.07-0.04T)}$ = YMT (in hours)

2. The fraction of YMT used in the stage is calculated as:
   Output of 1 (above) $\div t$ = % of YMT used in stage.

3. The % of YMT used in the stage is added to the amount remaining at the start of the stage:
   % of YMT used in stage + Y = Cumulative % YMT used.

4. A statement of logic determines if the YMT has been expended so that growth may take place:
   IF(Cumulative % YMT used > 1, 1, 0).

5. If the logic of 4 (above) is false (=0), then no growth takes place in the stage and the next stage in the model is considered. If the logic of 4(above) is true (=1), then growth is modelled.

6. If growth is modelled, then the growth rate is calculated as:
   $(-0.13 + 0.04T)^2$ * $\log(2)$ = EGR(logs/hr).

7. To determine the time available for growth, a statement of logic is used:
   IF(Y > 1, 1, 0).

8. If the logic of 7 (above) is true (=1), then the YMT was already expended in the preceding stage(s). In this case, the time available for growth is the time spent in

the stage, t. Therefore, growth in the stage is modelled as:

EGR(logs/hr) * t = logs of growth in stage

Then logs per egg at the end of the stage equals logs of growth in stage plus X.

If the logic of 7 (above) is false (=0), then the time available for growth is some fraction of the total time spent in the stage. One method for calculating this is:

[(Cumulative % YMT used) – 1] * YMT (in hours) = time for growth

Then, logs of growth in the stage is calculated as:

EGR(logs/hr) * time for growth = logs of growth in stage, and this is added to X to calculate the logs per egg at the end of the stage.

An alternative method for calculating time for growth is best explained using an example. Assume that t equals 20 hours and YMT for the stage is calculated to be 80 hours. If Y equals 85% (i.e. 85% of the YMT was expended in previous stages) and the %YMT used in the stage equals 25% (i.e. 20 ÷ 80), then the Cumulative %YMT used in the stage equals 110%. Therefore, there was an excess of 10% of YMT used in the stage. This stage accounts for 25% of the cumulative YMT for the egg, and 10% of this time was not needed before growth could occur. So, 40% (10% ÷ 25%) of the time in the stage the YMT exceeded 100% and growth could occur. To determine the time for growth in this case, 40% is multiplied by t (time in stage). If t equals 20 hours, then the time for growth equals 8 hours. This contrasts with the previous method, where the time for growth is a function of YMT (in hours). Using the previous method, however, results in the same estimated time for growth (i.e. (110% – 100%) * 80 = 8 hours).

In the US SE RA, growth is modelled as described above, except that internal temperature is calculated as a function of ambient temperature in the stage and the growth equations are slightly different. Modelling internal egg temperature complicates this model. Given the internal egg temperature at the beginning of a stage (T_i) the ambient temperature in the stage (T_0), and the time in the stage (t), the average internal egg temperature (T) is predicted as:

$$T = \exp(-kt / 2) * (T_i - T_0) + T_0$$

where k is a cooling constant reflecting the storage practices in the stage. The average internal egg temperature is selected using the midpoint of the cooling curve. This average internal temperature then predicts the YMT for the stage. Subsequent calculations are similar to those described above.

Figure 4.17 shows the predicted logs of growth for the Health Canada and US SE RA exposure assessments. Most contaminated eggs have no growth between lay and consumption. The Health Canada model predicts that 96% of contaminated eggs do not grow *S.* Enteritidis bacteria. The US SE RA model predicts 90% of contaminated eggs do not grow *S.* Enteritidis bacteria. The expected values of the distributions in Figure 4.17 are 0.09 and 0.53 logs of growth for the Health Canada and US SE RA models, respectively.

Figure 4.18 illustrates the predicted level of growth for just those contaminated eggs in which growth occurs. In the US SE RA model, most of these eggs experience very low levels of growth, but a substantial fraction also experience the maximum possible growth of 10 logs. The Health Canada model predicts moderate growth (i.e. 1 or 2 logs) more frequently than the US SE RA model. The higher frequency of contaminated eggs with the maximum possible growth in the latter model results in a larger expected value.

The differences in growth between the USDA and Health Canada models are substantial. These differences result from differences in the time and temperature distributions modelled. These differences affect the time until yolk membrane breakdown, as well as the growth following yolk membrane breakdown.

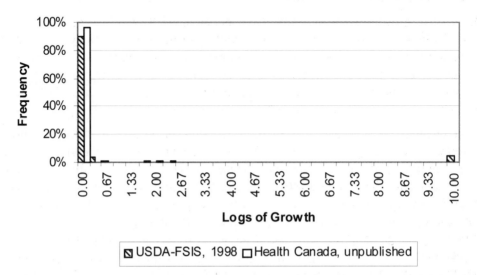

Figure 4.17. Comparison of predicted logs of growth in all *Salmonella* Enteritidis-contaminated eggs for two exposure assessment models.

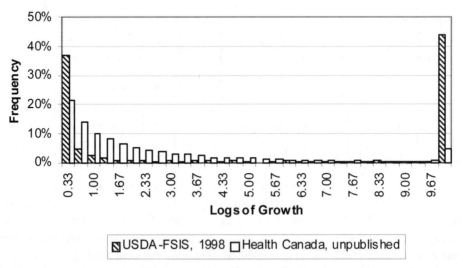

Figure 4.18. Distributions of logs of growth for those contaminated eggs in which growth occurs. In this case, frequency represents the proportion of those contaminated in which growth occurred.

The yolk membrane breakdown concept in eggs represents a complete threshold to multiplication of *S.* Enteritidis in eggs.   For growth to occur, an egg must typically experience elevated temperatures over a sustained period.   For example, either model predicts that yolk membrane breakdown requires about 18 days if the eggs are stored at room temperature (20°C). This time increases to about 46 days if the storage temperature is 10°C (50°F). The average time between lay and consumption in either model is less than 25 days. A contaminated egg in which growth occurs represents a situation where extraordinarily high temperatures or times existed. Consequently, it is reasonable for no – or very little – growth to occur if yolk membrane breakdown is not achieved, or for considerable growth to occur if yolk membrane breakdown is complete.

There are 16 growth pathways explicitly modelled in US SE RA. These pathways include branches for home or institutional egg users, pooling or not pooling of eggs, and cooking or not cooking of egg meals. These pathways are explained in Section 4.2.5 – Preparation and Consumption. In general, the amount of growth is primarily a function of whether the eggs are consumed in homes or institutions. In contrast, growth is modelled for all eggs in the Health Canada exposure assessment, without capturing growth for any specific path.

Sixteen average internal egg temperatures are summarized for the US SE RA model and compared with the average of all eggs in the Health Canada model (Figure 4.19). Pathways US1-US8 are those paths where eggs are marketed to home consumers. The average internal egg temperatures for these paths are nearly uniform across these different pathways, and only slightly different from the average ambient temperature of eggs in the Health Canada model. In contrast, pathways US9-US16 are those paths where eggs are sold to food service institutions. Average internal temperatures are also very similar among these pathways, but are higher than the average temperature in the Health Canada model. The variability in average temperature per egg is also consistently large for all of the USDA pathways.

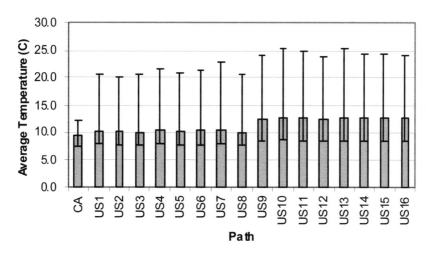

Figure 4.19. Comparison of average temperature between lay and consumption for all eggs in the Health Canada exposure assessment model (CA) and average internal egg temperature for 16 pathways in the US SE RA (US) exposure assessment model. Pathways US1-US8 model eggs consumed at home, and pathways US9-US16 model eggs consumed at food service institutions.

A different pattern in the 16 US SE RA pathways is noted for the time between lay and consumption (Figure 4.20). The home pathways (US1-US8) are associated with longer times than the food service institutional pathways (US9-US16). The food service pathways are shorter in time than, but have nearly the same variability as, the average time for the Health Canada model.

The inverse relationship between time and temperature noted for the 16 pathways in Figures 4.19 and 4.20 is expected. Because cooling of the internal contents of eggs is time dependent in the US SE RA model, longer times, on average, allow the internal temperature of eggs to fall farther than shorter times.

Because consumer storage represents the longest period during which eggs are held (on average), the storage temperature in this stage is responsible for much of the difference in growth between the USDA and Health Canada models. Figure 4.21 shows differences in the frequency of consumer storage temperatures above 7°C between these models. In both models, 7°C was the most likely temperature. Health Canada modelled consumer storage that included both home and institutional user behaviours. The maximum storage temperature in the Health Canada model is 25°C, but most temperatures are less than 11°C. US SE RA modelled storage in the home and at institutions. In that model, the same storage temperature distribution applied to both home and institution users, but home storage was longer than institutional storage. Ten percent of all eggs were modelled as experiencing temperatures above 7°C and the maximum temperature modelled was 32°C. Temperatures between 7°C and 32°C were equally frequent in the US SE RA model.

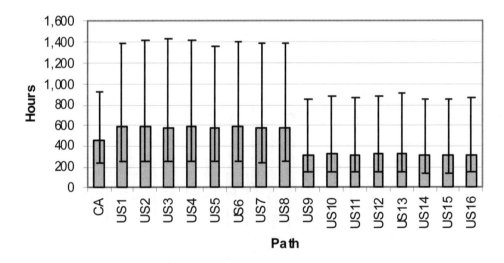

Figure 4.20. Comparison of average total time between lay and consumption for all eggs in Health Canada exposure assessment model (CA) and for the 16 pathways in the US SE RA. Pathways US1-US8 model eggs consumed at home, and pathways US9-US16 model eggs consumed at food service institutions.

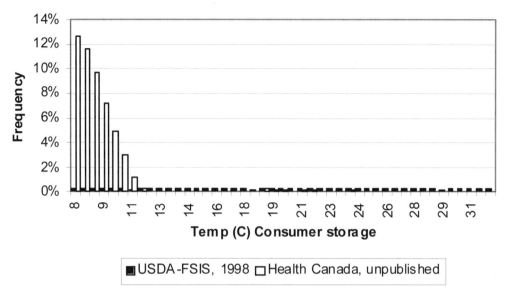

Figure 4.21. Frequency distributions storage used in USDA and Health Canada models for temperature during consumer. Distributions shown here only describe the frequency of eggs experiencing temperatures above 7°C (i.e. refrigeration temperature). The Health Canada distribution is skewed towards lower temperatures, but the FSIS-USDA distribution is uniformly distributed between 8°C and 32°C.

Because consumer storage temperature is important when predicting the total logs of growth, the results of a survey on United States of America refrigeration practices (Audits International, 1999) were examined. This survey found that for several perishable products examined in home refrigerators, 8% were above 7°C, with a maximum observed temperature of 21°C. Neither the USDA nor Health Canada models' distributions precisely reflect the Audits International findings. The US SE RA assumes 10% of temperatures above 7°C, but allows for temperatures up to 32°C. The Health Canada model assumes 51% of eggs above 7°C and the maximum temperature is 25°C. Nevertheless, the Audits International survey did not specifically address eggs. Eggs are possibly more prone to abuse by retailers and consumers than the products included in the survey. It is therefore difficult to determine which of the two models more correctly reflects consumer storage behaviour. Furthermore, it is likely that consumer behaviour in Canada and the United States of America is different.

To demonstrate the importance of consumer storage temperatures in predicting *S.* Enteritidis growth in eggs, the two models' predictions were compared when similar inputs were specified (Figure 4.22). In this case, the USDA model was modified by using Health Canada's consumer storage temperature distribution as the input for both homes and institutions. The resulting distributions are very similar (Figure 4.22). The modified USDA model predictions are slightly different because that model uses internal egg temperature and the modified input distributions are ambient temperature. On average, the USDA model predicts that internal egg temperature is greater than ambient temperature. Besides consumer storage, the other step where eggs are stored for longer times is retail storage. Retail storage temperature is more truncated in the Health Canada model than the USDA model. Nevertheless, Figure 4.22 shows that the underlying mathematics of modelling *S.* Enteritidis growth in eggs are similar in the two assessment models. Furthermore, the contrast between Figures 4.18 and 4.22 demonstrate the dramatic effect that differences in input temperature distributions can have on predicted *S.* Enteritidis growth. Therefore, evidence concerning these distributions is critical to improving model accuracy.

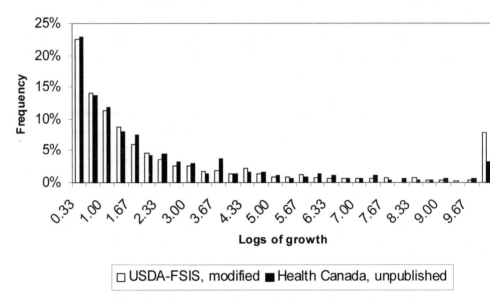

Figure 4.22. Comparison of distributions for logs of growth in contaminated eggs when growth occurs. In this case, the USDA-FSIS model is modified such that the consumer storage temperature distribution is the same as that used in the Health Canada model.

Summary

Time and temperature inputs are important in modelling the distribution and storage of *S.* Enteritidis-contaminated eggs, but there is a lack of reliable data to describe these distributions. Therefore, time and temperature data specifically addressing eggs is needed.

In neither exposure assessment was variability explicitly separated from uncertainty. For instance, eggs are exposed to a variety of temperatures during each stage listed in Table 4.7. The Health Canada exposure assessment models this as if we knew – with certainty – that most eggs in wholesale storage are stored at 4°C, with some eggs stored at temperatures as low as 2°C or as high as 7°C, but with no eggs exceeding those extremes. A PERT distribution was used here to model a symmetrical frequency distribution with a mode of 4°C. In contrast, the US SE RA assumes that consumer storage at 32°C is just as likely as consumer storage at 8°C. In this case, a uniform distribution was used that extended from 8°C to 32°C. A better approach to the ones used previously is to model a number of different frequency distributions that describe time and temperature of storage and handling. This requires running a series of simulations in which each simulation uses one frequency distribution for all iterations in that simulation. The results of different simulations describe the uncertainty in the model's predictions. Such an approach has been termed second-order modelling and its intent is to separate variability from uncertainty in modelling results.

The differences in average times and temperatures for the home and institutional pathways of the US SE RA highlight the need to explicitly model complete pathways to consumption. In the Health Canada exposure assessment, all end users are modelled as drawing from the same growth distribution, but the USDA analysis shows that there can be different times and temperatures for eggs between these end users. These differences can result in differences in growth of *S.* Enteritidis between eggs consumed at homes and in institutions. It is therefore recommended that *S.* Enteritidis exposure assessments model growth and preparation-consumption pathways jointly, rather than as independent predictions.

Predictive microbiology should be a common component of any exposure assessment of *S.* Enteritidis in eggs. Because environmental conditions differ on an international level, time and temperature distributions may be different between analyses. However, microbial behaviour within eggs is expected to be consistent regardless of location. The equations estimated in the Health Canada and US SE RA assessments for lag period duration and growth rate were reasonably similar. These equations were estimated from relevant evidence concerning *S.* Enteritidis in eggs. The Whiting and Buchanan (1997) growth curve was actually estimated from experimental data using *Salmonella* in broth. Therefore, this equation is not preferred to either of the others. Lacking additional evidence of *S.* Enteritidis growth dynamics, its is recommended that either the USDA or Heath Canada equations be used in future exposure assessments.

The results and conclusions of these microbial growth models are dependent on conventional assumptions regarding mechanisms of egg contamination. These mechanisms imply that *S.* Enteritidis contamination in eggs is initially restricted to the albumen and that such contamination enters eggs during their formation inside the hen's reproductive tissues. Additionally, the growth kinetics estimated for these models are assumed to be representative of all *S.* Enteritidis strains. Should these assumptions not hold (e.g. *S.* Enteritidis contamination might occasionally occur within the yolk at the time of lay), then the growth kinetics might differ from those presented.

4.2.4 Egg products processing

Processing of eggs into egg products involves the commercial breaking of shell eggs and subsequent processing of their contents for a variety of uses. Egg products are used in the commercial food industry as ingredients in a myriad of products. Institutional users of bulk eggs frequently prepare pasteurized egg products. These products are also sold at retail for home use. In addition, some egg products have non-food uses, such as in the cosmetic and pharmaceutical industries.

In the United States of America, the egg products industry is large and complex. It processes nearly one-third of all eggs. There are numerous product lines and a variety of treatments applied in different processing plants. Generally, there are three intermediate products: liquid whole egg, liquid albumen and liquid yolk. USDA's Food Safety and Inspection Service (FSIS) regulations exist to ensure that egg products are pasteurized or otherwise treated to reduce the risk of foodborne disease (9 CFR 590).

The inclusion of egg products processing in an exposure assessment of *S.* Enteritidis in eggs is controversial. Since the emergence of *S.* Enteritidis as an important food safety pathogen, little evidence has linked pasteurized egg products with human illness caused by *S.* Enteritidis. Because the products are pasteurized, it is generally assumed that they are safe. Although a large outbreak in the United States of America indirectly implicated raw liquid egg products, this was a case of cross-contamination and not a result of consuming pasteurized egg products (Hennessy et al., 1996). Nevertheless, pasteurization is not necessarily completely effective. Furthermore, implicating egg products as the source of *S.* Enteritidis in outbreak investigations would be difficult. As egg products are usually mixed with other ingredients, implicating an egg product as the source of an outbreak would typically require ruling out the other potential sources in a mixed food.

Because egg products are not thought to be a risk for *S.* Enteritidis illness in humans, little research exists outlining how, and to what extent, raw liquid egg is contaminated before pasteurization. The two exposure assessments that have modelled egg products processing (Whiting and Buchanan, 1997; USDA-FSIS, 1998) have concentrated on internally contaminated eggs as the source of *S.* Enteritidis in bulk volumes of liquid egg prior to pasteurization. Nevertheless, the US SE RA did discuss the implications of alternative sources of the *S.* Enteritidis in raw liquid egg.

Inputs to an exposure assessment of *S.* Enteritidis in egg products should include the concentration of *S.* Enteritidis in raw liquid product and the effectiveness of pasteurization. An output of this model would describe the distribution for number of *S.* Enteritidis bacteria surviving the pasteurization process. The prevalence of contaminated containers sold for preparation and consumption is another output, as is the prevalence and concentration of contaminated servings generated from these containers.

Contamination sources

Data

The Whiting and Buchanan (1997) and USDA exposure assessments both considered the likelihood of contamination of raw liquid egg products to be a function of the prevalence of infected flocks in the United States of America and the frequency of contaminated eggs produced by infected flocks. Contamination levels in raw liquid egg were predicted based on the number of bacteria within contaminated eggs. Average values for these inputs, discussed in Section 4.2.2, are summarized in Table 4.9. Whiting and Buchanan (1997) actually modelled two scenarios for flock prevalence (10% and 45%), but Table 4.9 summarizes flock prevalence in this case as the midpoint of these two prevalences. This midpoint was chosen to simplify the analysis. It should also be noted that the most likely contamination level in Whiting and Buchanan's (1997) model was 30 *S.* Enteritidis bacteria per contaminated egg. Nevertheless, heavy contamination levels are also possible in this model, albeit at low frequencies, and the average number of bacteria per egg is much larger.

Table 4.9. Inputs used to model likelihood of contamination and contamination levels of raw liquid egg products in two exposure assessment models

Input	Whiting and Buchanan, 1997	USDA-FSIS, 1998
Prevalence of infected flocks	27.50%	37.50%
Egg contamination frequency in infected flocks	2.00%	0.02%
Numbers of S. Enteritidis per contaminated egg	987	152

Two additional sources of data concerning raw liquid egg were considered in the USDA model. First, a study by Garibaldi, Lineweaver and Ijichi (1969) reported the most probable number of *Salmonella* bacteria per millilitre measured in samples of raw liquid egg (Table 4.10). While these data pertain to *Salmonella* in general and not *S.* Enteritidis specifically, it provides another perspective on contamination levels in liquid egg. The expected value of the distribution reported in Table 4.10 is 0.8 (MPN) *Salmonella* per millilitre of raw liquid egg prior to pasteurization.

Table 4.10. Reported numbers of *Salmonella* in raw liquid egg prior to pasteurization

Number of samples	MPN *Salmonella*	Frequency
187	0	65.2%
86	0.5	30.0%
10	2.25	3.5%
1	5.3	0.3%
2	24	0.7%
1	110	0.3%

SOURCE: Garibaldi, Lineweaver and Ijichi, 1969

Another source of data concerning the occurrence of *S.* Enteritidis in liquid egg is surveys completed by USDA in 1991 and 1995 (Hogue et al., 1997) (Table 4.11). These national surveys cultured 10-ml samples of liquid egg products over a period of one year each. Nevertheless, these surveys did not enumerate the bacteria per sample. About 50% (982 out of 1940) of liquid egg bulk tanks sampled were positive for *Salmonella* (any serotype) in these surveys. *S.* Enteritidis isolations were less frequent, but still substantial given the limited sample volume (10 ml).

Table 4.11. Results of national surveys of liquid whole egg prior to pasteurization in the United States of America

Year of survey	Samples collected	Salmonella sp.-positive	S. Enteritidis-positive
1991	1003	53%	13%
1995	937	48%	19%

SOURCE: Hogue et al., 1997

The Garibaldi, Lineweaver and Ijichi (1969) and Hogue et al. (1997) evidence were considered in the USDA modelling of contamination of raw egg product contamination from sources not limited to internally contaminated eggs. The implications of this analysis were not fully explored, however, because they were beyond the scope of that exposure assessment, which focused on the risk associated with internally contaminated eggs.

Methods

In the Whiting and Buchanan (1997) exposure assessment, the levels of *S.* Enteritidis per millilitre of liquid egg prior to pasteurization are modelled as the product of flock prevalence, egg contamination frequency in infected flocks, and levels of *S.* Enteritidis per millilitre of contaminated egg. Monte Carlo simulation calculates this product.

The US SE RA simulates the filling of a hypothetical 10 000-lb (~4 500 litre) bulk tank. This model also uses Monte Carlo simulation methods. Each randomly selected flock contributes its one-day production to a bulk tank and the total number of flocks contributing eggs to a bulk tank is determined iteratively until 10 000 lb are accumulated. For example, if each flock produces 20 000 eggs per day, this roughly equates to 2000 lbs (ca 900 kg) of liquid egg product per flock. Therefore a 10 000-lb bulk tank consists of eggs from five flocks in this example. Flock prevalence determines the probability of a flock being *S.* Enteritidis infected. Egg contamination frequency determines the number of contaminated eggs produced in a day of production by an infected flock. Similarly, the levels of *S.* Enteritidis per contaminated egg determines the total load of these bacteria in a 10 000-lb bulk tank when summed across all the flocks contributing to that bulk tank.

The resultant contamination levels predicted by the exposure assessments are shown in Figure 4.23. The Whiting and Buchanan (1997) distribution predicts generally greater levels of contamination per millilitre of liquid whole egg before pasteurization. Much of this difference results from the greater egg contamination frequencies modelled in that analysis. The mean log concentration is -2.1 for the Whiting and Buchanan (1997) distribution and -2.6 for the US SE RA distribution. If Whiting and Buchanan's high and low flock prevalence scenarios were considered separately, the resulting distribution would be shifted to the right and left, respectively, of the distribution shown in Figure 4.23.

The US SE RA included a separate analysis modelling *S.* Enteritidis contamination of liquid egg from all sources (i.e. not strictly limited to internal egg contamination), with the data from Garibaldi, Lineweaver and Ijichi (1969) adapted to model the concentration of *S.* Enteritidis per millilitre in a 10 000-lb (~4500 litre) bulk tank. The modelling approach is illustrated in Figure 4.24. As shown in cell B48, the negative results of Garibaldi and co-workers are assumed to equal a concentration of 1 in 100 000 ml. This concentration was arbitrarily determined. The @Risk cumulative function is used to model the average concentration in a single bulk tank (in cell B57) and the Poisson distribution predicts the number of organisms in the bulk tank (where Conversions!B13 is the number of litres in a bulk tank).

As discussed in the US SE RA report, use of the Garibaldi, Lineweaver and Ijichi (1969) data is problematic. It does not directly address *S.* Enteritidis contamination levels, nor is it necessarily temporally relevant. Nevertheless, it is the only observational data available on concentration of *Salmonella* in liquid egg.

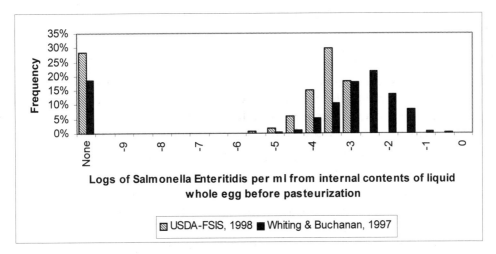

Figure 4.23. Comparison of predicted contamination distributions in liquid whole egg prior to pasteurization for two published exposure assessments. Contamination source is exclusively modelled as originating from the internal contents of shell eggs.

	A	B	C	D	
44	**Input – *Salmonella* Enteritidis in a bulk tank (all sources)**				
45		**Number of salmonellae in commercially broken eggs before pasteurization**			
46					
47		Number of samples	MPN Salmonella	Cumulative number of samples	Cumulative probability
48		187	0.00001	187	0.652
49		86	0.5	273	0.951
50		10	2.25	283	0.986
51		1	5.3	284	0.990
52		2	24	286	0.997
53		1	110	287	1.000
54		Garibaldi, Lineweaver and Ijichi, *Poultry Science*, p.1097, 1969			
55					
56					
57	Expected SE per ml of liquid egg	=RiskCumul(0.000001,150,C48:C53,E48:E53)			
58	Expected no. of SE in a bulk tank				
59		=RiskPoisson(Conversions!B13*1000*B57)			

Figure 4.24. Spreadsheet used to model *Salmonella* Enteritidis contamination of whole liquid egg prior to pasteurization when the source of *S.* Enteritidis is not limited to just internally contaminated eggs.

An alternative approach to estimating the underlying average concentration of S. Enteritidis bacteria in liquid egg is to use the USDA survey evidence to impute concentration (Table 4.11).   These surveys estimate prevalence, but one can infer concentration using a few assumptions. First, assume that the samples of liquid egg generally came from the same population of bulk tanks. In other words, one might assume that all the bulk tanks were basically similar and were filled with eggs from flocks that were roughly similar.   Furthermore, assume that there were no substantive differences between 1991 and 1995 surveys regarding the underlying prevalence and concentration of *Salmonella* and S. Enteritidis in the United States of America egg industry, and assume that if a sample contained S. Enteritidis, it was found positive (i.e. perfect test methods).

With these basic assumptions, one can model the prevalence of *Salmonella*-positive samples as:

$$@RISKBETA(982 + 1, 1940 - 982 + 1)$$

Furthermore, one can model the underlying average concentration as a Poisson process:

$$@RISKPOISSON(V * \lambda)$$

where V is the volume of sample collected (10 ml) and λ is the average concentration of *Salmonella* per millilitre in bulk tanks.

If the prevalence from the Beta distribution is known, then one knows the probability of a positive sample. Furthermore, if the average concentration is λ, then the probability that a sample contains one or more organisms is $1 - e^{(-V\lambda)}$ from the Poisson probability function (i.e. this is also the probability that the sample is positive).   Therefore, the following relationship where is derived, all but one of the elements are known:

$$@RISKBETA(982 + 1, 1940 - 982 + 1) = 1 - e^{(-V\lambda)}$$

Solving for λ results in the following function:

$$\lambda = -\ln(1 - @RISKBETA(982 + 1, 1940 - 982 + 1)) / V$$

Simulating this equation results in a distribution for λ with a mean of 0.07 *Salmonella* per millilitre in liquid egg bulk tanks. Given the large number of samples upon which this estimate is based, there is little associated uncertainty about the mean.

To compare this estimate with the Garibaldi, Lineweaver and Ijichi (1969) data in Table 4.10, consider V = 1 ml. A Poisson distribution with a mean of 0.07 predicts that 93% of 1-ml samples will contain less than 1 organism, and 99.99% will contain less than 2.25 organisms per millilitre.   Garibaldi and co-workers' results show that 95% of samples contained less than 1 organism per millilitre, and 98.6% contained less than 2.25 organisms per millilitre. While not a perfect match, the imputed distribution from the USDA data is similar except for the very large concentrations detected infrequently by Garibaldi and colleagues.

Stipulating the above analysis, the estimate of S. Enteritidis contamination can be refined. In this case, by use of the S. Enteritidis prevalence data from Table 4.11, and the solution for λ is 0.018. The probability of less than one S. Enteritidis organism in a 1-ml sample equals 98% for this mean.

Melding the estimated average *S.* Enteritidis concentration with the Garibaldi, Lineweaver and Ijichi (1969) data can be done by assuming that 98% of the time the concentration per millilitre of raw liquid egg follows a Poisson distribution, and the remaining 2% of the time it follows a cumulative distribution based on the non-zero data in Table 4.10. Furthermore, the Garibaldi data suggest that the variability in concentration per millilitre may be exponentially distributed. If one assumes that λ is the mean concentration among 98% of the bulk tanks, then the exponential distribution can be used to model variability in the mean concentration from one bulk tank to another.

Figure 4.25 illustrates how the direct use of the Garibaldi, Lineweaver and Ijichi (1969) data compares with the method described above. Modelling the Garibaldi data using a cumulative distribution results in bimodal pattern with many observations at or below 1 in 100 000 ml, but another peak at just under one per millilitre. Using the alternative approach based on the data in Table 4.11 results in concentrations centred on 1 per 30 ml. The expected values of the USDA and alternative method distributions are 0.8 and 0.4 *S.* Enteritidis organisms per millilitre, respectively.

Either of the approaches used to model *S.* Enteritidis contamination from all sources seems plausible, but the differences in the distributions, although not dramatic, suggest the need for better data than is currently available.

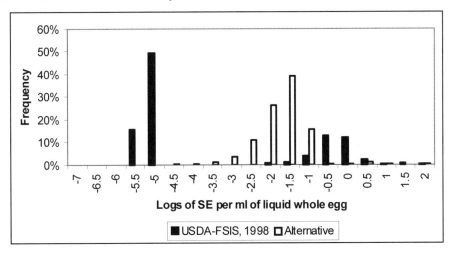

Figure 4.25. Comparison of distributions predicted by the US SE RA model and alternative methods using Beta and Poisson distributions, for contamination levels per millilitre in liquid whole egg. All sources of *Salmonella* Enteritidis (SE) are considered in these predictions. The US SE RA distribution is based on data from Garibaldi, Lineweaver and Ijichi (1969).

Pasteurization

Data

In the Whiting and Buchanan (1997) exposure assessment, thermal inactivation was based on evidence of 17 *S.* Enteritidis strains in liquid whole eggs (Shah, Bradshaw and Peeler, 1991). In the USDA model, additional studies by Humphrey et al. (1990) were included with the Shah, Bradshaw and Peeler (1991) data.

The US SE RA separately evaluated the performance of pasteurization on liquid albumen. This evaluation considered two publications (Schuman and Sheldon, 1997; Palumbo et al., 1996) and one set of unpublished data (G. Froning, University of Nebraska) that examined different pH levels of the albumen. These studies showed that the pH of albumen was critical to the effectiveness of pasteurization of this product.

The US SE RA also evaluated pasteurization of egg yolk. There were three studies included in this evaluation (Humphrey et al., 1990; Palumbo et al., 1995; Schuman et al., 1997).

Methods

For comparison of methods used in the two exposure assessments, this discussion is limited to the effectiveness of pasteurization as applied to liquid whole egg. The Whiting and Buchanan (1997) exposure assessment only considered this product. In that model, a regression equation was estimated for log (D), where D is the time in *seconds* to achieve a one log reduction in *S.* Enteritidis organisms. The equation estimated was:

$$\text{Log(D)} = 19.104 - 0.2954T$$

where T is in degrees Celsius. A standard deviation for Log D in this equation was approximated as 0.25.

The USDA model estimated the following regression equation:

$$\text{Log(D)} = 13.027 - 0.2244T$$

where D is *minutes* to achieve a 1 log reduction. The standard deviation for Log D was estimated as 0.16.

To model the variability in pasteurization effectiveness for liquid whole egg, values of D were estimated for specific time and temperature combinations. The USDA-FSIS standards require the application of 60°C for 3.5 minutes for liquid whole egg. For this temperature, log(D) equals 1.4 seconds for the Whiting and Buchanan (1997) exposure assessment and - 0.44 minutes for the US SE RA. In log space, the log of the log reduction for 3.5 minutes is calculated as log(3.5) − log(D). To determine the mean log reduction, the antilog is calculated as $10^{(\log(3.5)-\log(D))}$. Nevertheless, there is variability in this value and this variability is modelled using the lognormal distribution (i.e. $10^{(@\text{RISKLOGNORM}((\log(3.5)-\log(D)),\ \text{s.d})}$, where s.d. is the standard deviation estimated for each of the regression equations.

Figure 4.26 shows the predicted distributions for thermal inactivation of *S.* Enteritidis in whole liquid egg. The USDA distribution implies slightly greater reductions than the Whiting and Buchanan (1997) distribution, but both distributions imply substantial variability. The apparent improved effectiveness of the USDA curve reflects the incorporation of additional evidence not considered in the Whiting and Buchanan (1997) assessment. This additional evidence also slightly reduced the standard deviation of the estimated regression.

To model the concentration of *S.* Enteritidis remaining in a volume of liquid whole egg after pasteurization, both exposure assessments used Monte Carlo techniques. Given an initial log concentration in raw liquid whole egg (A), and the log reduction caused by pasteurization (B), the logs remaining equals A – B.

The distribution for level of contamination after pasteurization was simulated for both exposure assessments (Figure 4.27). The distributions in Figure 4.27 were estimated using initial contamination distributions representing *S.* Enteritidis that originated from internally contaminated eggs (i.e. Figure 4.23). Given that a typical 10 000-lb lot of liquid whole egg consists of about 4.4 million millilitres, or 6.6 logs, this graph suggest that *S.* Enteritidis bacteria surviving pasteurization is infrequent. For example, <0.1% of bulk tanks have concentrations greater than 1 in 10 million millilitres in the US SE RA. About 2% of bulk tanks have concentrations greater than 1 in 1 million millilitres in the Whiting and Buchanan (1997) assessment despite the greater incoming concentrations and less effective pasteurization estimated in that model.

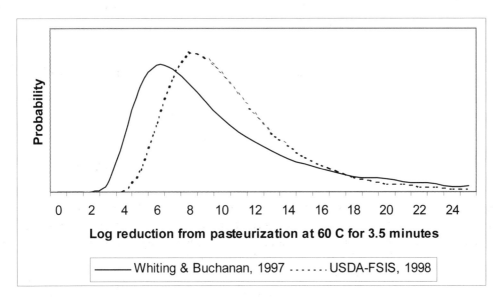

Figure 4.26. Comparison of predicted effectiveness of a specific pasteurization protocol applied to liquid whole egg for two exposure assessments.

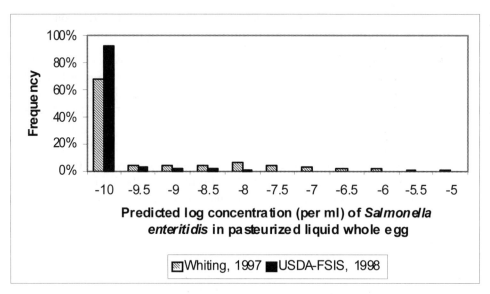

Figure 4.27. Comparison of residual concentration of *Salmonella* Enteritidis per millilitre of liquid whole egg following pasteurization for two completed exposure assessments. Contamination of liquid whole egg prior to pasteurization was modelled as only originating from internally contaminated eggs. Because of the extremely small likelihood of contamination remaining at low concentrations, the distributions were truncated at –10 logs.

If one were to assume that all liquid whole egg product was collected and pasteurized in 10 000-lb units, then it is possible to estimate the frequency of bulk tanks in which one or more *S.* Enteritidis survive. Furthermore, this analysis can be conducted assuming the incoming concentration is based on all sources of *S.* Enteritidis (Figure 4.25). In this case, the number of *S.* Enteritidis in a bulk tank prior to pasteurization, X, is modelled as a Poisson($V\lambda$) distributed variable where V is 4.4 million millilitres and λ is the concentration of *S.* Enteritidis per millilitre based on the distributions in Figure 4.25.

Given X, the number of *S.* Enteritidis in a bulk tank, and P, the log reduction resulting from pasteurization, one can model the number of bacteria remaining after pasteurization as:

$$@RISKPOISSON(10^{(\log(X)-P)}).$$

This algorithm is applied using the two distributions in Figure 4.25, as well as the USDA pasteurization effectiveness distribution (Figure 4.28). Regardless of the initial contamination concentration assumed from Figure 4.25, more than 95% of the bulk tanks have no bacteria remaining after pasteurization. Of those with some residual *S.* Enteritidis after pasteurization, the most likely number is less than 10 organisms per 4.4 million millilitres of liquid whole egg. If we assume that a typical serving size is 100 ml of liquid whole egg, and the concentration in a particular bulk tank is 10 organisms per 4.4 million millilitres, then the likelihood of a serving containing one *S.* Enteritidis bacteria is 0.02%. The probability of a serving containing more than one organism is exceedingly small. Given that at least some liquid whole egg products will be further heat treated (i.e. cooked) before consumption, the risk is even less than suggested by this analysis.

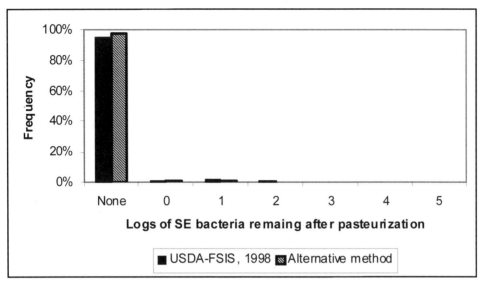

Figure 4.28. Predicted numbers of *Salmonella* Enteritidis (SE) organisms remaining after pasteurization of 10 000-lb (~4500 litre) bulk tanks. Incoming concentrations are modelled using the distributions in Figure 4.25 for the US SE RA model and an alternative estimate. In both cases, it is presumed that *S.* Enteritidis originates from all sources, including but not limited to internally contaminated eggs.

Summary

The methodologies used by Whiting and Buchanan and US SE RA are similar. The differences in results between these models are mostly caused by differences in modelled concentrations before pasteurization. These differences originate in the respective production models as described Section 4.2.2 – Production. The methods described here are reasonable given the limited epidemiological information linking *S.* Enteritidis in egg products to adverse human health events. Furthermore, data concerning concentrations of *S.* Enteritidis prior to pasteurization are lacking. Therefore, much uncertainty attends the modelling of initial contamination. To predict the effectiveness of regulatory standards concerning egg products, there is a need for additional data concerning the concentration of *S.* Enteritidis in raw liquid egg before pasteurization. For validation purposes, it would also be useful to collect data on the concentration of *S.* Enteritidis post-pasteurization.

A dramatic result of these analyses is the implied variability in pasteurization effectiveness. This finding is supported by subsequent analysis (van Gerwen, 2000). The standard errors of the estimated regression equations are assumed to represent variability in log(D) at all temperatures. The assumption seems warranted given the stipulations of linear regression analysis. Nevertheless, the standard error term arguable represents variability only. The unknown effect of measurement error might contribute to the calculated standard error of a regression analysis. However, quantifying the effect of measurement error requires more information than is usually available from published studies.

Lacking evidence of the degree of measurement error, it seems reasonable to assume that the standard error term of a linear regression analysis represents variability in the system. To incorporate uncertainty into an exposure assessment, the linear regression itself can be re-estimated using bootstrapping or jackknifing techniques to explore the effect of different or fewer observations. Such techniques then allow the explicit separation of variability and uncertainty in this analysis.

Implicit in these models is the assumption that pasteurized egg products are contaminated because pasteurization fails to eliminate all the *S.* Enteritidis organisms. No consideration is given to possible re-contamination of egg products following pasteurization. If egg products are not hygienically handled after pasteurization, this limitation may be important.

This discussion of egg products processing has also assumed that commercially broken eggs will be pasteurized, but there may be situations where liquid egg is not pasteurized. In such cases, the estimation procedures discussed for raw liquid egg would serve as the starting point for an analysis of the risk that these products pose to consumers. While the risk seems high in the examples described here, other situations may predict different results.

4.2.5 Preparation and consumption

The preparation and consumption stage is concerned with the end-users of eggs, the manner in which these end-users store and prepare their eggs, and the effectiveness of practices these end-users apply to destroy *S.* Enteritidis bacteria in prepared meals. Inputs include pooling practices, serving sizes, pathway probabilities, and cooking effectiveness.

This stage considers eggs following their production, distribution and storage. Therefore the number of bacteria within contaminated eggs and the lag period remaining for these eggs are fixed at the beginning of the Preparation and Consumption stage.

The output of this stage is a distribution of the doses of *S.* Enteritidis bacteria in servings. This distribution may be refined to reflect the frequency of servings that contain various levels of *S.* Enteritidis bacteria for specific end users (e.g. homes or institutions), or specific meal types (e.g. pooled or non-pooled egg dishes), or specific cooking practices (e.g. raw versus cooked meals).

Given the multiple pathways within the Preparation and Consumption stage, and the dependency of *S.* Enteritidis amplification and destruction on the pathway modelled, this part of a risk assessment model is likely to be the most complicated. Complexity is expected because each pathway must be modelled separately, and multiple iterations are necessary per pathway to accurately represent the variability of growth.

Egg pooling and serving size

Pooling refers to the practice of breaking eggs into containers and using the combined eggs to make multiple servings of egg dishes or for use in multiple recipes. Pooling is usually done to save time and control portion size. Pooling does not mean simply combining eggs. As an example of pooling, several dozen eggs could be broken into a large bowl and mixed before a restaurant opens for breakfast. Then as orders for scrambled eggs are taken, portions are ladled from the bowl and cooked. In contrast, mixing a dozen eggs into a cake batter would not constitute pooling because the cake could not be made with less than a dozen eggs. Pooling is essentially exposing consumers to more eggs than they ordered.

As a result of pooling, *S.* Enteritidis bacteria from a single egg are immediately spread to all eggs in the pool, and the bacteria are given immediate opportunity to grow without needing to wait for a breakdown in yolk membrane integrity.

The likelihood that eggs are pooled probably differs between home use and institutional use, as is the number of eggs that constitute a pool. Following pooling, there is possible storage before cooking. In addition, the likelihood that eggs are undercooked probably differs between eggs cooked at home versus those cooked at institutions.

When eggs are consumed as single eggs, there is a 1:1 correspondence between contaminated eggs and servings. Using eggs as ingredients results in a greater than unitary correspondence between contaminated eggs and servings.

Data

Data are needed to estimate the fraction of all eggs consumed in the home versus institutional settings. Similarly, data are needed that describe the fraction of eggs consumed in pooled dishes, the fraction of meals consisting only of eggs or where eggs are used as ingredients, and the fraction of meals undercooked. These probabilities can be considered fixed in the model; they do not vary but they are uncertain. Additional data are needed to describe how eggs are handled after they are pooled in homes and institutions. These inputs will have both variability and uncertainty associated with them.

Unfortunately, for both the Health Canada and US SE RA exposure assessments, little data were available to estimate these inputs. Therefore distributions were defined based on the opinions of the analysts, with comments from reviewers of the models.

The assumed probability of pooling and number of eggs per pool for the two risk assessments are summarized in Table 4.12. Generally, the @RISKPERT(minimum, most likely, maximum) distribution was used. This distribution is an alternative to the more traditional triangular distribution. The PERT distribution has a smooth shape that assigns less probability weight to the distribution tails than does the triangular distribution. Nevertheless, the value of both these distributions is that the user can define the most likely value and absolute minimum and maximum values, based on opinion. The uniform distribution is another alternative that is less informed. That distribution only requires the user to define minimum and maximum values.

Table 4.12. Pooling inputs used in two exposure assessments.

Input	Location	Health Canada, 2000	USDA-FSIS, 1998
Probability of a pooling	Home	Pert(25%,30%,35%)	Pert(0%,2%,10%)
	Institution	Pert(25%,35%,45%)	Pert(2%,5%,20%)
Pool size (servings per pool)	Home	Pert(1.5,2.5,3)	Pert(2,4,12)
	Institution	Pert(25,50,75)	Uniform(6,48)

When eggs are used as ingredients in the Health Canada exposure assessment, the number of servings generated is the same as the number of servings when eggs are pooled. In contrast, the US SE RA analysed a computerized recipe program to determine the number of servings when eggs are used as ingredients in the home. The distribution ranged from 2 to 10 servings, with a mean of 6 servings per egg. In the case of eggs used as ingredients in institutions, the distribution used for pooled servings was doubled.

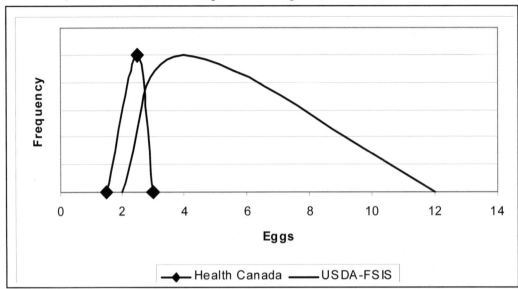

Figure 4.29. Modelled size of egg pools in homes for Health Canada and US SE RA exposure assessments. Distribution assumptions are as shown in Table 4.12.

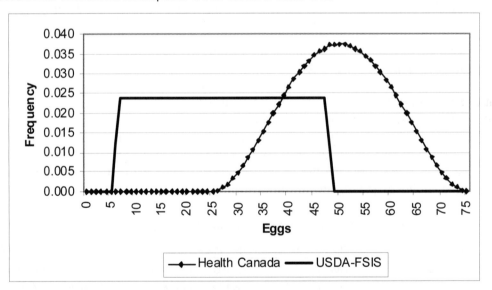

Figure 4.30. Modelled size of egg pools in institutions for Health Canada and US SE RA exposure assessments. Distribution assumptions are as shown in Table 4.12.

Methods

Both the Health Canada and US SE RA exposure assessments consider pooling of eggs. Nevertheless, there is considerable disparity in how pooling is modelled and in the subsequent results. In institutions, the Health Canada assessment models larger pools than does the US SE RA. In the home, the pools are larger for the US SE RA than the Health

Canada assessment. Figures 4.29 and 4.30 show the results of modelling the pool sizes for homes and institutions for the two exposure assessments.

Furthermore, the Health Canada exposure assessment assumes a single destination for pooled eggs – scrambling. Scrambling eggs is highly effective at destroying bacteria in the Health Canada model. The US SE RA allows pooled eggs to be served as egg meals or incorporated as ingredients in recipes. The distinction is important because of the post-pooling storage that is explicitly modelled in the US SE RA. This post-pooling storage gives eggs an immediate opportunity to grow without needing to wait for a breakdown in yolk membrane integrity. In the Health Canada model, pooling has no effect on the number of *S.* Enteritidis bacteria in a serving. Nevertheless, pooling does increase the likelihood of illness from a single egg because there are more exposures to the bacteria. In the US SE RA model, the number of *S.* Enteritidis bacteria in a serving is decreased as the pool size increases, but the model also simulates post-pooling growth and assumes that bacteria are able to grow immediately after pooling.

The attributable risk in the modelled output of the two risk assessments differs significantly. Though pooled eggs account for 17.7% of all servings in the Health Canada exposure assessment, only 6% of the risk of *S.* Enteritidis illness comes from pooled eggs. In contrast, pooled eggs in the US SE RA account for 13.1% of all servings while contributing to 26.8% of the illnesses (Table 4.13).

Table 4.13. Percentage of illnesses attributed to pooled eggs and the proportion of pooled eggs

	Health Canada	**US SE RA**
Percentage of illnesses from pooled eggs	6.0%	26.8%
Percentage of pooled eggs	17.7%	13.1%

Pathway probabilities

The Preparation and Consumption stage considers the effect of end user location. For example, eggs consumed in the home are likely to be handled differently from eggs consumed in restaurants or other food service institutions. It seems likely that eggs stored in the home are exposed to different storage times and temperatures compared with those stored in institutions.

Ideally, one would have access to data generated from studies that sought to chronicle the life of a contaminated egg subsequent to its production. If possible, such studies would report the number of bacteria at the beginning of the egg's life, then describe the effect of time and temperature as the egg moved from the farm to consumption. Having done this for thousands of eggs, we would theoretically have a very good understanding of how eggs are handled during marketing and preparation. Unfortunately, no such data exist. Therefore, understanding the effect of preparation and consumption on *S.* Enteritidis in eggs requires considering the limited evidence and dividing the problem into elements small enough to make expert opinion meaningful.

Pathway probabilities ultimately are used to determine the fraction of all egg-containing servings consumed by each of the endpoints defined by the risk analyst. The sum of the endpoint probabilities should equal 100% to signify that all servings are accounted for in the model. Nevertheless, the endpoints do not fully describe the risk of *S.* Enteritidis in eggs. These endpoints serve only to categorize the general pathways that eggs may travel. The consequence of the servings consumed at a particular pathway endpoint is a distribution of number of contaminated servings at different dose levels.

A very simple model might assume that the distribution for number of *S.* Enteritidis per serving is the same regardless of what it consisted of or where or how the serving was prepared or consumed. In this simple example, the initial contamination level in the egg and the growth dynamics within the egg, as well as the effectiveness of cooking, is always the same for all the pathways. The conclusion of such an analysis would be that the most risk is associated with the most probable pathway, but such a conclusion is trite. Microbial growth dynamics and cooking effectiveness are completely independent of the pathways in this example. Essentially, all that has been done is to apportion consumers into categories and the largest category is where most illnesses occur.

A more complicated, but more rewarding, approach to building an exposure assessment model is to construct pathways and their inputs such that the endpoint is dependent on the pathway. To varying degrees, this was done in the Health Canada and US SE RA exposure assessments.

Data

Opinion – expert and otherwise – plays an important part in defining the shape and content of distributions when data are lacking. Often, expert opinion is based on data that have not yet been sufficiently analysed. In such cases, the exposure assessment helps to document this data.

Absence of data increases uncertainty. Such uncertainty should be reflected in more dispersed inputs and outputs. Uncertainty distributions that are too narrow incorrectly imbue the model output with more confidence than is warranted. Furthermore, the narrow uncertainty associated with the output serves as a disincentive to collect additional information.

In general, little data are available for calculating path probabilities in the preparation and consumption stage. In the Health Canada exposure assessment, a survey of Canadians was used as evidence concerning the probability that single eggs were fried or scrambled. Otherwise, most probabilities were estimated based on opinion.

In the US SE RA, evidence on the probability that pooled eggs are consumed as single egg meals and are undercooked came from a 1996-97 Food Consumption and Preparation Diary Survey. This survey showed that 27% of all egg dishes were consumed as undercooked meals. Another survey estimated that each person consumed undercooked eggs 19 times per year (Lin, Morales and Ralston, 1997). The FDA Food Safety Survey was also cited as evidence for the probability that a pooled egg is used as an ingredient in the home and is not cooked (Klontz et al., 1995). The Lin, Morales and Ralston (1997) study also showed that the average frequency was 0.4 raw eggs consumed per consumer per year.

Although both risk assessments described their uncertainty in path probabilities as distributions (usually PERT distributions), the average probabilities assumed by each model are summarized in Tables 4.14 and 4.15.

Table 4.14. Summary of average pathway probabilities assumed in Health Canada exposure assessment.

	Location	Meal type	Health Canada
Fraction of eggs used	Home	Ingredients	45%
	Institution	Single egg	55%
	Home	Single egg	55%
	Institution	Ingredients	45%
Fraction of meals served raw	Home	Ingredients	2%
	Institution	Ingredients	2%
Fraction of meals lightly cooked	Home	Ingredients	30%
	Institution	Ingredients	30%
Fraction of meals well cooked	Home	Ingredients	68%
	Institution	Ingredients	68%
Fraction of meals fried	Home	Single egg	45%
	Institution	Single egg	60%
Fraction of meals boiled	Home	Single egg	25%
	Institution	Single egg	1%
Fraction of meals scrambled	Home	Single egg	30%
	Institution	Single egg	39%

Table 4.15. Summary of average pathway probabilities assumed in US SE RA.

	Location	Meal type	US SE RA	
			Pooled	Not pooled
Fraction of eggs used	Home		3%	97%
	Institution		7%	93%
	Home	Ingredients	30%	30%
	Institution	Single egg	50%	70%
	Home	Single egg	70%	70%
	Institution	Ingredients	50%	30%
Fraction of meals served raw	Home	Ingredients	2%	2%
	Institution	Ingredients	15%	15%
Fraction of meals well cooked	Home	Ingredients	98%	98%
	Institution	Ingredients	85%	85%
Fraction of meals thoroughly cooked	Home	Single egg	67%	67%
	Institution	Single egg	67%	67%
Fraction of meals lightly cooked	Home	Single egg	33%	33%
	Institution	Single egg	33%	33%

Methods

The Health Canada and US SE RA exposure assessments model eggs through distinct pathways. The Whiting and Buchanan (1997) risk assessment considers a single product (mayonnaise) consumed in the home. Therefore only the methods used in the Health Canada and US SE RA models are here compared.

The Health Canada exposure assessment models twelve combinations of location (home ([H] or food service facility [F]), use (egg meal [M] or an ingredient in a recipe [R]), and type of cooking (boiled [B], scrambled [S] or fried [F] for egg meals; raw [R], lightly cooked [L] or well cooked [W] for recipes). Each of these paths is replicated for three types of growth: none, some or maximum growth. A total of 36 distinct pathways are thus modelled (Figure 4.31).

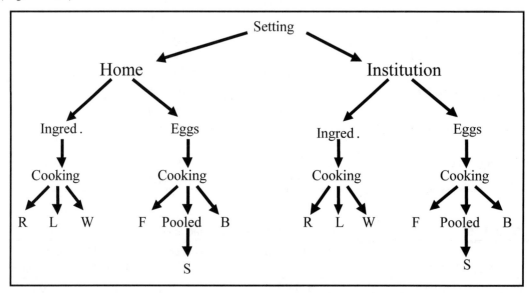

Figure 4.31. Schematic diagram of pathways modelled in the Health Canada exposure assessment. R is raw, L is lightly cooked, W is well cooked, F is fried, S is scrambled and B is boiled.

The Health Canada exposure assessment considers growth to be independent of pathway. Thus, regardless of the eventual location or use of the egg, there is a 0.962 probability that no growth will occur, 0.036 probability that some growth will occur and 0.002 probability that maximal growth will occur. Table 4.16 lists the twelve pathways and the associated endpoint probability for each.

The US SE RA (Figure 4.32) models sixteen combinations of location (home [H] or institution [I]), pooling (pooled [P] or not pooled [N]), use (single egg meal [E] or

Table 4.16. Summary of average endpoint path probabilities in the Health Canada exposure assessment. Codes are explained in the text.

Name of path	Probability of path
FMF	0.0825
FMS	0.0536
FMB	0.0014
FRR	0.0023
FRL	0.0338
FRW	0.0765
HMF	0.1856
HMS	0.1238
HMB	0.1031
HRR	0.0068
HRL	0.1013
HRW	0.2295

ingredient in a recipe [I]), and type of cooking (thorough [T] or lightly cooked [L] for egg meals, cooked [C] or raw [R] for ingredients). This model considers growth to be dependent on pathway. In other words handling of eggs in homes may differ from handling of eggs in institutions. Modelling this dependency avoids situations where the model depicts results of particular time and temperature inputs that should not occur in a particular setting. This can be done by collecting output from each of the pathways and then integrating the results by weighting them by the pathway probabilities. Table 4.17 lists the sixteen pathways and the associated endpoint probabilities for each. As can be seen in the table, one path accounts for nearly 34% of all eggs (HNET) while another for only 0.01% (HPIR).

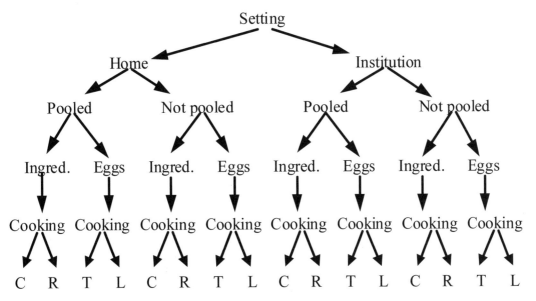

Figure 4.32. Schematic diagram of pathways modelled in the US SE RA. C is cooked and R is raw. T is thoroughly cooked and L is lightly cooked.

From the distribution and storage chapter, it can be recalled that the probability of eggs being consumed in the home is about 75%. That information and the information in Table 4.15 can be used to illustrate how final path probabilities in Table 4.17 can be calculated. For example, the HNET pathway represents the fraction of eggs that are consumed in the home in non-pooled single-egg meals that are thoroughly cooked. The fraction of all eggs consumed that travel the HNET pathway can be calculated as the product of the following terms: the probability that eggs are consumed in the home (75%), the probability that home eggs are not pooled (97%), the probability that pooled

Table 4.17. Summary of average endpoint path probabilities in US SE RA Codes are explained in the text.

Name of path	Probability of path
HPET	0.0104
HPEL	0.0054
HPIC	0.0066
HPIR	0.0001
HNET	0.3374
HNEL	0.1738
HNIC	0.2144
HNIR	0.0047
IPET	0.0057
IPEL	0.0029
IPIC	0.0073
IPIR	0.0013
INET	0.1061
INEL	0.0547
INIC	0.0586
INIR	0.0103

eggs in the home are consumed as single egg meals (70%), and the probability that these meals are thoroughly cooked (67%). The product of these probabilities equals 34%. This is approximately the same as that shown in Table 4.17 for the HNET pathway. Differences are partly due to rounding error and the fact that the probabilities are actually skewed distributions for which the mean of the product does not precisely equal the product of the mean.

Cooking

Data

Data are available for d-values and z-values for *S.* Enteritidis in eggs. Unfortunately, these values are not helpful unless information on cooking times and temperatures is also available. Inputs to both exposure assessments are thus based on results of direct measurements of log reduction for different types of cooking when applied to single egg meals.

Some data pertaining to expected log reduction when eggs are undercooked was cited in the US SE RA (Humphrey et al., 1989b). This evidence provided estimates of the effectiveness of boiling, frying or scrambling eggs at suboptimal temperatures.

Methods

The Health Canada and the US SE RA exposure assessments use almost identical inputs to model the log reduction predicted from various types of cooking (Table 4.18).

Table 4.18. Comparison of distributions used to model the log reduction from cooking different single-egg servings in the Health Canada and US SE RA exposure assessments.

Variable	Distribution type	Health Canada, 2000			USDA-FSIS, 1998		
		Min.	ML	Max.	Min.	ML	Max.
Log reduction – fried eggs	Pert	1	4	7	0	4	7
Log reduction – scrambled	Pert	4	6	7	0	6	7
Log reduction – boiled	Pert	0.5	1	7	0	1	7

NOTES: Min. = minimum; ML = most likely; Max. = maximum.

When eggs are used as ingredients in recipes, however, the Health Canada and the US SE RA exposure assessments differ markedly in how they model the log reductions. Figure 4.33 shows that the Health Canada exposure assessment uses a bimodal distribution with peaks around 3 and 10 logs, while the US SE RA has an equal likelihood of any log reduction from 0 to 8 logs.

Table 4.19 summarizes the pathways and events within pathways including cooking for the two exposure assessments. Although average values are shown in this table, it is important to realize that specific values can vary from one egg to the next. The table is organized to associate similar pathways defined in the two exposure assessments. Hence the pooled egg pathways, "_P_ _", in the US SE RA are associated with the scrambled cooking pathways, "_ _S", of the Health Canada assessment. Those USDA pathways modelling eggs consumed in either homes or institutions that were prepared as single-egg meals and not thoroughly cooked did not neatly fit within one of the Health Canada pathways. These pathways actually would fit in all three of the cooking types (i.e. HMF, HMB and HMS) modelled in the Health Canada assessment. For simplicity, these pathways are simply separated in Table 4.19. Other associations were similarly made to simplify this presentation. Nevertheless, direct comparisons for path probabilities are problematic based on these associations.

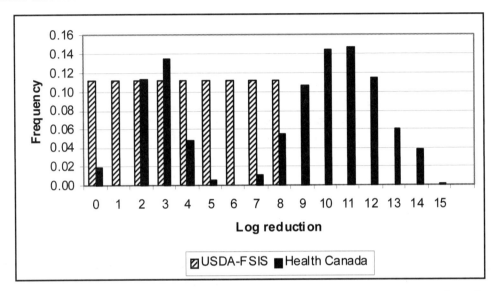

Figure 4.33. Comparison of predicted effectiveness of cooking meals containing eggs as ingredients for two risk assessments.

Table 4.19 also illustrates some of the similarities between these two analyses. Average initial contamination levels are the same for all pathways within each exposure assessment, and these are similar between the assessments. Average logs of growth are the same for all pathways in the Health Canada assessment (i.e. 0.14 logs per egg), but vary by location (e.g. home or institution) and pooling practices (e.g. pooled or not pooled) in the US SE RA. The logs of growth for institutional users of non-pooled eggs is similar for both assessments (i.e. 0.14 versus 0.24 logs per egg). For all other pathways, the USDA model predicts more growth. In particular, the USDA pooled pathways average one log of growth more than the Health Canada pathways. Such results reflect the explicit modelling of growth after pooling in the USDA model. Cooking effectiveness is also similar between both models. The only dramatic difference is the average log reduction for well cooked meals containing eggs as ingredients (i.e. 10.2 log reduction in the Health Canada assessment).

A quantitative farm-to-table model of *S.* Enteritidis will contain the components shown in Table 4.19. Using a Monte Carlo approach, the initial logs of bacteria are added to the logs of growth to determine the pre-cooking exposure. Log reduction from cooking is then subtracted to determine the exposure dose remaining. These calculations are completed for each of the pathways. Because the inputs (i.e. initial logs, logs of growth and logs reduction) are random variables, Monte Carlo simulation does the calculations iteratively to determine a final distribution for each pathway. Pathway probabilities and number of servings then serve to weight each distribution. In this manner, these multiple distributions can be integrated into a single exposure distribution.

Table 4.19. Summary of average values predicted by two exposure assessments for all pathways modelled.

Pathway codes[1]		Path probabilities		Log initial concentration		Logs of growth		Logs cooking reduction	
HC[2]	USDA[3]	HC[2]	USDA[3]	HC[2]	USDA[3]	HC[2]	USDA[3]	HC[2]	USDA[3]
Home settings									
N.A.	HNEL	N.A.	0.1738	N.A.	2.2	N.A.	0.52	N.A.	3.8
HMF	HNET	0.1856	0.3374	1.9	2.2	0.14	0.52	4.0	7.0
HMB		0.1031		1.9		0.14		1.9	
HMS	HPET	0.1238	0.0104	1.9	2.2	0.14	1.41	5.8	7.0
	HPEL		0.0054		2.2		1.41		3.8
HRR	HPIR	0.0068	0.0001	1.9	2.2	0.14	1.41	0.0	0.0
	HNIR		0.0047		2.2		0.52		0.0
HRL	HPIC	0.1013	0.0066	1.9	2.2	0.14	1.41	2.3	4.0
HRW	HNIC	0.2295	0.2144	1.9	2.2	0.14	0.52	10.2	4.0
Institutional settings									
N.A.	INEL	N.A.	0.0547	N.A.	2.2	N.A.	0.24	N.A.	3.8
IMF	INET	0.0825	0.1061	1.9	2.2	0.14	0.24	4.0	7.0
IMB		0.0014		1.9		0.14		1.9	
IMS	IPET	0.0536	0.0057	1.9	2.2	0.14	1.48	5.8	7.0
	IPEL		0.0029		2.2		1.48		3.8
IRR	IPIR	0.0023	0.0013	1.9	2.2	0.14	1.48	0.0	0.0
	INIR		0.0103		2.2		0.24		0.0
IRL	IPIC	0.0338	0.0073	1.9	2.2	0.14	1.48	2.3	4.0
IRW	INIC	0.0765	0.0586	1.9	2.2	0.14	0.24	10.2	4.0

NOTES: (1) Pathway codes are explained in the text. (2) HC = Health Canada Exposure Assessment (Health Canada, 2000). (3) USDA = US SE RA (USDA-FSIS, 1998). (4) N.A. = not applicable

4.2.6 Summary

Absence of data should increase the uncertainty in an exposure assessment. If one can imagine that replacing a triangular distribution based on expert opinion with an empirical distribution based on limited test data would increase uncertainty, then the original distribution must be too narrow.

Careful attention should be directed to those areas in exposure assessments in which the product changes form or the units change. Pooling eggs into a container creates a product distinctly different from shell eggs. This product is able to support immediate bacterial growth and its storage must be modelled as a unique event.

Not all available data are necessarily useful to an exposure assessment. Detailed information on certain processes often can not be used without more information. In the case of *S.* Enteritidis, knowing the d-values does not help in construction of a model unless cooking times and temperatures either are known or can be modelled. Thus, the high degree of uncertainty and variability in cooking effectiveness inputs noted in this comparison of models emphasizes the need for more research on these inputs.

Given the dearth of published evidence on relevant egg consumption and preparation practices among populations of end users, the preparation and consumption component of an exposure assessment is the most difficult to accurately model. Unfortunately, even with perfect information, this component is very complicated. Multiple pathways reflecting multiple end users, products, practices and cooking effectiveness levels guarantee that assessing the preparation and consumption component is fraught with difficulties. Nevertheless, the advances inherent in both the Health Canada and USDA models provide reasonable starting points for subsequent analyses.

Neither model includes the possibility of re-contamination of egg meals following cooking. These models also do not account for possible cross-contamination of other foods from *S.* Enteritidis-contaminated eggs. These limitations might be addressed in future models.

4.3 EXPOSURE ASSESSMENT MODEL, MODEL PARAMETERS AND ASSUMPTIONS

4.3.1 Introduction

The previous section compared and contrasted previous exposure assessments of *S.* Enteritidis in eggs. This section intends to describe a simple exposure assessment for the purpose of completing a hypothetical risk-characterization exercise. Results from the exposure assessment model described here were combined with the dose-response model described in Section 3.4 to yield the risk-characterization results for *S.* Enteritidis in eggs in Section 5.

The exposure model developed here combines and modifies the US SE RA and Health Canada exposure models described in the preceding section. Generally, where input types were similar, the Health Canada parameters were used. If an input type was missing in one model, then the other model's parameters were used (e.g. the Health Canada model did not consider cooling constants, therefore these were specified by the US SE RA model). The exposure model structure was generally based on the US SE RA model.

It should be noted that this model is necessarily indicative of North American management practices, but it is not intended to reflect any specific country's risk. The effects of different parameter settings have been evaluated within the context of this model (e.g. different flock prevalence levels, different storage times and temperature profiles). Such differences are intended to reflect a wide array of situations, some of which might be indicative of particular countries or regions. Nevertheless, there are undoubtedly exceptions in specific regions that render this a poor model for assessing risk. Despite such limitations, it is hoped that the general framework and analysis completed here is of some value to a country or region as they begin conducting their own risk assessment of *S.* Enteritidis in eggs.

4.3.2 Model overview

The general structure of the *S.* Enteritidis in shell eggs risk assessment is outlined in Figures 4.34a, b & c. The exposure assessment model consists of production, shell egg processing and distribution, and preparation modules.

The production module of the exposure assessment (Figure4.34.a) is concerned with predicting the fraction of contaminated eggs among the population of all eggs produced per unit time. This fraction is determined by considering the flock prevalence, the within-flock prevalence, and the fraction of eggs laid by infected hens that are contaminated with *S.* Enteritidis.

The shell egg processing and distribution module of the exposure assessment (Figure 4.34b) is concerned with predicting the amount of growth of *S.* Enteritidis in contaminated eggs due to storage and handling of eggs between the farm and retail or institutional storage. Growth within each step of this module is a function of the storage time, temperature and environment. Environment is reflected in the cooling constants (k) for each step. In contrast to the production module, which estimates a population fraction of contaminated eggs, this module simulates individual contaminated eggs.

The preparation module (Figure 4.34c) is concerned with the effects of egg storage, egg meal preparation (e.g. serving sizes, mixing of eggs together), and the effectiveness of cooking in reducing the amount of *S.* Enteritidis in contaminated eggs. As in the previous module, growth of *S.* Enteritidis during steps in this module is a function of storage time, temperature and the value of k. Furthermore, pooling practices influence the number of servings per contaminated egg, and product type and serving size influence the amount of *S.* Enteritidis per serving after cooking. This module also simulates individual contaminated eggs.

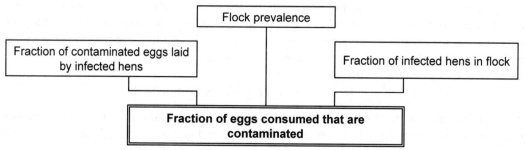

Figure 4.34a. Schematic diagram of production module.

4.3.3 Production

The production model is a simplification of the US SE RA and Health Canada models, and models in the following manner the likelihood that an egg is contaminated.

If a flock is infected, the fraction of *S.* Enteritidis-contaminated eggs among all eggs a flock produces (FEggs_Flock) depends on the fraction of hens that are *S.* Enteritidis-infected in that flock (FHen_Flock) and the fraction of eggs an infected hen lays that are *S.* Enteritidis-contaminated (FEggs_Hen). This is described as:

$$FEggs\_Flock = FHen\_Flock \times FEggs\_Hen \qquad \text{Equation 4.1}$$

Among the population of all infected flocks in a region or country, the fraction of hens infected per flock varies. In other words, it is not true that every flock in a region or country contains exactly the same proportion of infected hens. Therefore, the input FHen_Flock should be represented by a variability distribution.

By definition, flock prevalence (Prev) describes the proportion of flocks for which FHen_Flock is >0%. If we know the flock prevalence, then we know that the fraction of flocks in which 0% of hens are infected (FHen_Flock = 0%) is 1-Prev. For example, if 60% of flocks are infected (Prev = 60%), then 40% of flocks are not infected and FHen_Flock is 0% for these flocks.

As a convention, one can represent a variable input in bold. If referring to a specific value from the variability distribution, the input will not be in bold. Therefore, **FHen_Flock** refers to a distribution and FHen_Flock refers to a particular value from that distribution that occurs with frequency f(FHen_Flock).

Mathematically, this convention means:

$$\textbf{FHen\_Flock} = \{FHen\_Flock_i, f(FHen\_Flock_i)\} \qquad \text{Equation 4.2}$$

where {} describes the set of all possible values of FHen_Flock$_i$
and $\sum_i f(FHen\_Flock_i) = 1.0$.

FEggs_Flock describes the flock-to-flock variability in egg contamination frequency for all infected flocks in a region or country. The expected value of this distribution is:

$$EV\,[\textbf{FEggs\_Flock}] = Prev \times \left(\sum_{i>0\%} FHen\_Flock_i \times f(FHen\_Flock_i) \right) \times FEggs\_Hen$$

$$\text{Equation 4.3}$$

Equation 4.3 calculates the fraction of *S.* Enteritidis-contaminated eggs among all eggs produced in a region. Consequently, 1-EV[FEggs_Flock] equals the fraction of all eggs that contain zero *S.* Enteritidis (i.e. not contaminated) at the time of lay. Eggs that are not contaminated are not modelled further in this exposure assessment.

A flock is defined as a group of hens of similar ages that are housed and managed together. A farm may contain more than one flock if, for example, two poultry structures (e.g. buildings) exist on the farm and there is little commingling of the birds between the structures. In such a case, one flock on the farm might be affected with *S.* Enteritidis while the other is not.

Flock prevalence (Prev) is assumed to be a scalar value in this model, but three levels are evaluated: 5%, 25% and 50% (Table 4.20). Such a convention can be interpreted either as examining the effect of uncertainty about flock prevalence, or as examining the influence of different country or region situations.

Flock prevalence is the proportion of all flocks in a country or region that are infected. A flock is considered infected if *S.* Enteritidis exists in the flock or its environment. At any given time, there is a fixed proportion of flocks that are *S.* Enteritidis infected, but flock prevalence might theoretically vary according to season. For example, some flocks might only be infected during the summer. This could happen if the flocks became infected because of exposure to *S.* Enteritidis from, for example, migratory waterfowl that gained access to the flock. In this case, flock prevalence would theoretically increase in the summer, and be lower during the rest of the year. However, the available evidence suggests that most flocks are infected early in their production cycle and their likelihood of infection is independent of season. Furthermore, unless a flock manager specifically takes steps to rid the flock of *S.* Enteritidis it is unlikely that an infected flock will become non-infected during its lifetime. Therefore, the assumption that flock prevalence is constant across seasons seems reasonable.

Flock prevalence in a country might also vary from year to year. For example, it seems likely that *S.* Enteritidis flock prevalence in the United States of America was very low before the 1980s. Subsequently, *S.* Enteritidis became established in a substantial proportion of the United States of America commercial flocks and flock prevalence increased (although the lack of surveillance prior to recognition of the problem prohibits quantitative estimates). Survey evidence suggests that flock prevalence stabilized somewhat in the 1990s (Hogue et al., 1997). For the purposes of this risk characterization, it is assumed that a country is dealing with a *S.* Enteritidis problem that has stabilized and that control programmes are expected to commence in the near term. Nevertheless, if a country is in the early stages of an epidemic, it might be important to consider the future risk of illness for its population as the epidemic worsens and eventually stabilizes at higher endemic levels.

Table 4.20. Description of assumed production model inputs and parameters.

Production model inputs	Distribution	Parameters	
Prev (Prevalence of infected flocks)	Uncertain scalar	5%, 25% or 50%	
FHen_Flock (Percentage of infected hens within infected flocks)	Variable (Lognormal)	Mean: 1.89%	S.D: 6.96%
FEggs_Hen (Prevalence of contaminated eggs from infected hens)	Uncertain scalar (Beta distributed)	Alpha: 12	Beta: 1109

Within-flock prevalence (FHen_Flock) is the proportion of infected hens within infected flocks. Because evidence suggests that this proportion is not constant among infected flocks or even in the same infected flock across time, within-flock prevalence is a random variable in the model. A probability distribution was estimated for within-flock prevalence by statistical fitting to data cited in the USDA and Health Canada assessments (Hogue et al., 1997; Poppe et al., 1991). It was assumed that the Hogue et al. survey detected 76% of infected flocks. Therefore, the data were adjusted to indicate that 24% of infected flocks had within-flock prevalence levels less than this survey's lowest observed prevalence (i.e. 0.33%). A statistical fitting software (BestFit®; Palisade Corp., Newfield NY) determined that a lognormal distribution best fitted the data ($\chi^2 = 0.66$, P>0.90) (Table 4.20).

It was assumed that the fraction of contaminated eggs an infected hen lays (FEggs_Hen) is a scalar value. It is biologically plausible that this input varies during the period the hen is infected. Furthermore, the value is also likely to be influenced by the infecting strain of *S.* Enteritidis, the strain of hen and environmental and managerial factors. In the US SE RA, egg contamination frequency was modulated based on the class of flock (e.g. high or low prevalence, moulted or not moulted). In that risk assessment, within-flock prevalence was not explicitly modelled. Instead, empirical evidence concerning the proportion of contaminated eggs produced by infected flocks was used. In contrast, the exposure assessment model developed here for FAO/WHO explicitly includes the variability in within-flock prevalence, but assumes the frequency at which infected hens lay contaminated eggs is constant. Therefore fluctuations in egg contamination frequency between infected flocks – resulting from differences in *S.* Enteritidis strain, hen strain or environmental and managerial factors – are assumed to be reflected by the within-flock prevalence variability distribution (FHen_Flock). For example, when FHen_Flock is high, the egg contamination frequency from that type of flock is correspondingly high (and vice versa).

FEggs_Hen is derived from data cited in the Health Canada risk assessment, where 11 of 1119 eggs were found to be *S.* Enteritidis contaminated from naturally infected hens (Humphrey et al., 1989a) (Table 4.20).

The frequency at which infected hens lay contaminated eggs was compared with the US SE RA model inputs and outputs. The US SE RA model predicts that an average of 1 in 20 000 eggs produced in the United States of America is contaminated with *S.* Enteritidis. The average flock prevalence for that model was 37%. From this information and the within-flock prevalence described above, the frequency at which infected hens lay contaminated eggs was calculated using Equation 3. The answer, 0.7%, was the 11th percentile of a Beta distribution from the Humphrey et al. (1989a) data. Although those data may reflect a more virulent strain of *S.* Enteritidis than occurs in the United States of America, the US SE RA model results are reasonably consistent with the Humphrey et al. (1989a) results.

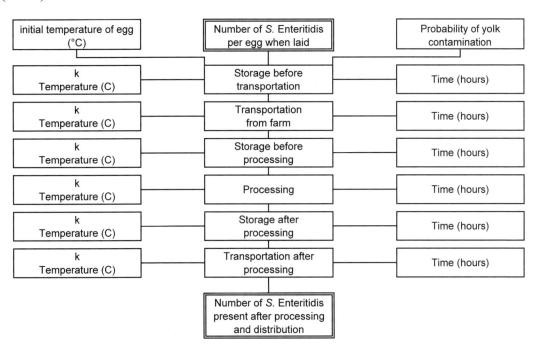

Figure 4.34b. Schematic diagram of shell egg processing and distribution module. k is cooling constant.

4.3.4 Shell egg processing and distribution

This part of the model combines the US SE RA and Health Canada inputs and structure, with most storage times and ambient temperatures based on the latter model. The US SE RA cooling constants (denoted as k) are used to model the transition of internal egg temperature given ambient temperature and time of storage, with input settings as shown in Table 4.21.

The values in Tables 4.21 and 4.22 arguably represent conditions in North America. The PERT distributions representing egg-to-egg variability in storage time, temperature and k values reflect the North American climate and local management practices. Other countries will have different ambient temperatures and times of storage. To examine the effect of the assumed variability distributions for time and temperature, the baseline parameter values in these default distributions were arbitrarily adjusted up and down by 10% and the adjusted distributions denoted as "elevated" and "reduced" time-temperature scenarios, respectively. These adjustments can be interpreted as effects of uncertainty about the true distributions or as different scenarios applicable to different countries or regions. In all simulations, the lowest temperature is truncated at 4.4°C to avoid excessive refrigeration or freezing of eggs.

Table 4.21. Shell egg processing inputs[1] used in the baseline scenario of the risk characterization exercise. These inputs are based on those used in the US SE RA and Health Canada models.

Inputs	Distribution
Number of *S. Enteritidis* per egg when laid	=ROUND(RiskTexpon(152,1,400) 0)
Initial temperature of egg (°C)	=37
Probability of yolk contamination	=RiskBeta(1,33)
Storage temperature before transportation (°C)	=IF(RiskBinomial(1,RiskUniform(0.9 0.95)), RiskPert(10,13,14),RiskPert(18,25,40))
Value of k	=RiskPert(0.0528 0.0800 0.1072)
Storage time before transportation (hours)	=RiskUniform(0,IF(RiskBinomial(1,RiskPert(0.6 0.7 0.8)), RiskUniform(56,84),RiskUniform(84,168)))
Temperature during transportation (°C)	=Storage temperature before transportation
Value of k	=RiskPert(0.0528 0.0800 0.1072)
Time for transportation (hours)	=RiskPert(0.5,2,8)
Storage temp. before processing (°C)	=RiskPert(11,13,14)
Value of k	=RiskPert(0.0528 0.0800 0.1072)
Storage time before processing (hours)	=RiskUniform(1,24)
Temperature addition at processing	=RiskNormal(5.6,0.56)
Temperature at processing (°C)	=RiskPert(15,20,25)
Value of k	=RiskPert(0.3300 0.5000 0.6700)
Time for processing (hours)	=RiskPert(0.1 0.2 0.5)
Storage temperature after processing (°C)	=RiskPert(11,13,14)
Value of k	=RiskPert(0.0053 0.0080 0.0107)
Storage time after processing (hours)	=RiskPert(12,48,168)
Transportation temperature (°C)	=RiskPert(7,10,32)
Value of k	=RiskPert(0.0660 0.1000 0.1340)
Transportation time (hours)	=RiskUniform(1, 6)

NOTES: PERT distribution has parameters RiskPert(minimum, most likely, maximum). Uniform distribution has parameters RiskUniform(minimum, maximum). Truncated exponential distribution has parameters RiskTexpon(mean, minimum, maximum). Beta distribution has parameters RiskBeta(number positive +1, number negative + 1). Binomial distribution has parameters RiskBinomial(number of samples, probability of positive).

Table 4.22. Shell egg storage distributions used in the baseline scenario of the risk characterization exercise. These inputs are based on those used in the US SE RA and Health Canada models.

Inputs	Distribution
Retail Storage	
Retail storage time (hours)	=RiskTlognorm(7, 10, 1, 30)*24
Retail storage temperature (°C)	=RiskPert(4.4,7,12)
k value	=0.24
Home storage	
Home storage time (hours)	=RiskUniform(0,RiskTlognorm(14,10,1,60)*24)
Home storage temperature (°C)	=IF(RiskBinomial(1,RiskPert(0.001 0.005 0.02)),
	RiskPert(15,20,25),RiskPert(4.4,7,12))
k value	=0.24
Institutional storage	
Institutional storage time (hours)	=RiskUniform(12, 147)
Institutional storage temperature (°C)	=RiskPert(4.4,4.4,7)
k value	=0.24
Home pooling	
Time post pooling (hours)	=RiskCumul(0,48,4 0.8)
Temperature post pooling (°C)	=RiskCumul(4.4,32,7 0.8)
Pool size	=ROUND(RiskPert(2,4,12) 0)
Institutional pooling	
Time post pooling (hrs)	=RiskCumul(0,48,4 0.8)
Temperature post pooling (°C)	=RiskCumul(4.4,32,7 0.7)
Pool size	=ROUND(RiskUniform(6,48) 0)
Ingredient use	
Home serving size	=RiskDiscrete({2,4,6,8,9,10},{0.0233 0.1938, 0.6047,
	0.1473 0.0078 0.0233})
Institutional serving size	=ROUND(RiskUniform(6,48) 0)

NOTES: PERT distribution has parameters RiskPert(minimum, most likely, maximum). Uniform distribution has parameters RiskUniform(minimum, maximum). Binomial distribution has parameters RiskBinomial(number of samples, probability of positive). Truncated lognormal distribution has parameters RiskTlognorm(mean, stand deviation, minimum, maximum). Cumulative distribution has parameters RiskCumul(minimum, maximum, range of values, cumulative probabilities of each value in range). Discrete distribution has parameters RiskDiscrete(range of values, probability weight of each value in range).

One feature not examined in the earlier models is the possibility that some eggs might be contaminated in the yolk at lay. The US SE RA and Health Canada models assumed that all internally contaminated eggs were only contaminated in the albumen of the egg at the time of lay. More recent evidence makes it difficult to ignore the possibility of some yolk-contaminated eggs (Gast and Holt, 2000a, b).

Based on our understanding of the growth of *S.* Enteritidis in eggs, yolk-contaminated eggs theoretically would support immediate amplification of numbers of *S.* Enteritidis in eggs after lay. In contrast, albumen-contaminated eggs experience a lag phase during storage and processing until there is sufficient breakdown in the yolk membrane to allow access of *S.* Enteritidis organisms to critical yolk nutrients.

If all contaminated eggs were contaminated in the yolk, then one would expect enumeration of *S.* Enteritidis per egg to demonstrate very large numbers after just a few days of storage (even at room temperature). For example, according to some growth equations, one would expect 5.5 logs of growth after just one day at room temperature, yet the evidence from a limited number of naturally contaminated eggs suggests that none of 32 eggs stored up to 21 days at room temperature had levels of *S.* Enteritidis consistent with being yolk contaminated (Humphrey et al., 1991). Using this evidence in a beta distribution, it was calculated that 2.9% of contaminated eggs are contaminated in the yolk, on average. Nevertheless, this estimate is undoubtedly biased upwards. It assumes that before considering the data from the 32 eggs, we were uniformly uncertain about the prevalence of yolk-contaminated eggs. In other words, before consideration of the data, it was believed that this prevalence could be 0% with the same confidence that it could be 100%. Of course, the prior belief was truly more in favour of very low prevalences, but here it has been decided to ignore the effect of such prior beliefs. For a specific application, however, it is expected that a more informed input be used to estimate the prevalence of yolk-contaminated eggs in a particular region or country.

As shown in Figure 4.34b, this module is a series of steps during which *S.* Enteritidis can increase within a contaminated egg. The model used is described below.

Let G_i be the amount of growth during step *i*. Think of G_i as a multiplier of the organisms that were in a contaminated egg before step *i*. If there was no growth in the egg during step *i*, then $G_i = 1.0$ (or 0 logs of growth). If there was one log of growth during step *i*, then $G_i = 10$.

Mathematically, G_i can be represented as $G_i = g(T_i, t_i)$. In other words, growth in a step (e.g. storage before processing) is some function, *g*, of the temperature distribution, T, and the storage time distribution, t, for that step. The functional relationship is complex and involves the influence of storage time and temperature on yolk membrane breakdown time and the exponential growth rate (EGR) for *S.* Enteritidis in eggs. The algorithms for modelling these dependencies were discussed earlier.

The output of this module is a variability distribution for the number of *S.* Enteritidis in contaminated eggs. Let SE_egg be this variability distribution. Then

$$SE\_egg = InitSE \times G_1 \times G_2 \times ... \times G_6 \qquad\qquad \text{Equation 4.4}$$

where InitSE is the variability distribution for the initial number of *S.* Enteritidis in contaminated eggs at the time of lay, and G_1 through G_6 are the growth predicted to occur during the six steps of this module (i.e. from storage before transportation through to transportation after processing).

Using Monte Carlo methods, SE_egg can be estimated. This describes the variability in number of *S.* Enteritidis in contaminated eggs after shell egg processing and distribution.

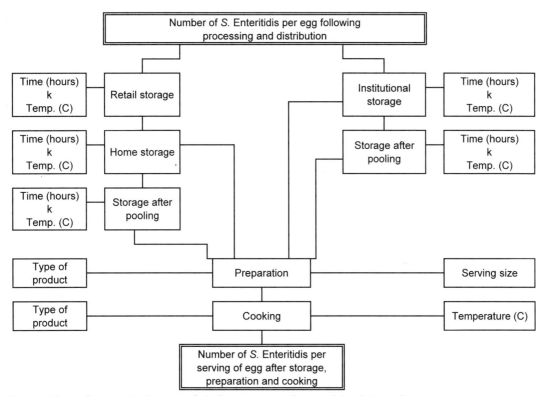

Figure 4.34c. Schematic diagram of shell egg preparation module. k is cooling constant.

4.3.5 Egg products processing

The egg products processing stage is concerned with predicting the *S.* Enteritidis contamination in bulk liquid egg products before and after pasteurization. The US SE RA model was used to simulate the numbers of *S.* Enteritidis organisms remaining after pasteurization of 10 000-lb containers of whole liquid egg. Only *S.* Enteritidis contributed via internally contaminated eggs are considered in this risk characterization. As noted in Section 4.2.4, any modelling of *Salmonella* contamination from sources other than internally contaminated eggs is based on scant quantitative data. Furthermore, the available qualitative data on the occurrence of species or strains other than *S.* Enteritidis in bulk liquid egg does not explain the sources or transfer mechanisms involved.

The US SE RA did not originally consider the potential for growth of *S.* Enteritidis inside eggs prior to being sent for breaking and pasteurization. To model this growth, this report used the shell egg processing and distribution stage, but limited the total amount of time between lay and breaking. Consistent with United States of America data (Ebel, Hogue and Schlosser, 1999), it was assumed that 69%, 30% and 1% of eggs pasteurized were nest run, restricted and graded, respectively. Nest run eggs are not washed or graded before being sent to pasteurization. Restricted eggs are those washed and graded eggs that are found inappropriate for sale as shell eggs. These eggs include eggs with cracked shells, thin shells, eggs with internal blood spots, or eggs that are leaking their contents. Graded eggs are eggs that are deemed suitable for sale as shell eggs, but for some reason are re-routed to egg products plants (e.g. eggs returned from retail stores). Each of these types of eggs may be stored a variable amount of time before they are broken and pasteurized. Nest run, restricted and graded eggs are stored an average of 2, 5 and 11 days in the model.

4.3.6 Preparation and consumption

Generally, the inputs used to model preparation and consumption practices for eggs were those of the US SE RA model (Tables 4.22 and 4.23). Pathway probabilities were assumed constant for this analysis (Table 4.24).

At the beginning of the preparation module, contaminated eggs have some number of *S.* Enteritidis that is described as SE_egg. As shown in Figure 4.34c, the preparation module simulates each contaminated egg as it traverses one of several pathways to eventual consumption. A contaminated egg might go to retail (and eventually home) or institutional users. It might be pooled with other eggs or not be pooled. Growth can occur during any of the storage steps that a contaminated egg experiences. Growth is modelled as described for the storage and distribution module.

A contaminated egg might be served as an egg-based meal or as an ingredient. Therefore, the effect of cooking depends on which path it follows. The number of servings to which that egg contributes also depends on its pathway.

An output of the preparation module is a variability distribution for the number of *S.* Enteritidis per serving for each of the possible pathways, SE_serving$_j$ (where the subscript *j* refers to a specific pathway). For example, Monte Carlo methods are used to estimate SE_serving for the pathway in which eggs are consumed in the home, the eggs are pooled, the eggs are consumed in egg-containing dishes, and the eggs are thoroughly cooked. This is a variability distribution because the input to the preparation module, SE_egg, is a variability distribution, as are factors such as storage time and temperature, k values, and cooking effectiveness within the preparation module.

The penultimate output of the preparation module is the classic risk triplet, which describes the exposure risk, ER, for the population of egg consumers. This can be represented as

$$ER = \{path_j, f(path_j), SE\_serving_j\}$$

where the symbols {} represent the complete set of paths, *path$_j$* identifies a specific path, *f(path$_j$)* is the likelihood of that pathway among all possible pathways, and SE_serving$_j$ is the consequence of that path (i.e. the exposures resulting from contaminated eggs).

If one integrates across all the possible doses (i.e. SE_serving) to calculate their likelihoods within all the pathways, you derive the exposure distribution for the population (Expos). This is the ultimate output of the exposure assessment and is represented as

$$Expos = \{dose_i, f(dose_i)\}$$

which is a variability distribution for the dose of *S.* Enteritidis per serving ingested by the consuming population. This distribution is combined with the dose-response function to calculate the likelihood of illness, on a per serving basis, from *S.* Enteritidis in eggs. This integration occurs in the risk characterization exercise in Section 5.

Table 4.23. Shell egg cooking and post-cooking handling distributions used in the baseline scenario of the risk characterization exercise. These inputs are based on those used in the US SE RA and Health Canada models.

Inputs	Distribution
Home cooking – pooled	
Fully cooked eggs (log reduction)	=RiskUniform(6,8)
Eggs cooked as ingredients (log reduction)	=RiskUniform(0,8)
Boiled (log reduction)	=RiskPert(0,1,7)
Fried (log reduction)	=RiskPert(0,4,7)
Scrambled (log reduction)	=RiskPert(0,6,7)
Home cooking – non-pooled	
Fully cooked eggs (log reduction)	=RiskUniform(6,8)
Eggs cooked as ingredients (log reduction)	=RiskUniform(0,8)
Boiled (log reduction)	=RiskPert(0,1,7)
Fried (log reduction)	=RiskPert(0,4,7)
Scrambled (log reduction)	=RiskPert(0,6,7)
Institutional cooking – pooled	
Fully cooked eggs (log reduction)	=RiskUniform(6,8)
Eggs cooked as ingredients (log reduction)	=RiskUniform(0,8)
Boiled (log reduction)	=RiskPert(0,1,7)
Fried (log reduction)	=RiskPert(0,4,7)
Scrambled (log reduction)	=RiskPert(0,6,7)
Institutional cooking – non-pooled	
Fully cooked eggs (log reduction)	=RiskUniform(6,8)
Eggs cooked as ingredients (log reduction)	=RiskUniform(0,8)
Boiled (log reduction)	=RiskPert(0,1,7)
Fried (log reduction)	=RiskPert(0,4,7)
Scrambled (log reduction)	=RiskPert(0,6,7)
Post cooking storage	
Home egg handling	
Time (hours)	=RiskExpon(0.25)
Temp (°C)	=RiskUniform(4.4,32)
Home ingredient handling	
Time (hours)	=RiskExpon(1)
Temp (°C)	=RiskUniform(4.4,32)
Institutional egg handling	
Time (hours)	=RiskExpon(1)
Temp (°C)	=RiskUniform(4.4,32)
Institutional ingredient handling	
Time (hours)	=RiskExpon(1)
Temp (°C)	=RiskUniform(4.4,32)

NOTES: Uniform distribution has parameters RiskUniform(minimum, maximum). PERT distribution has parameters RiskPert(minimum, most likely, maximum). RiskExpon distribution has a single parameter RiskExpon.

Table 4.24. Probabilities used in the baseline scenario of the risk characterization exercise. These inputs are based on those used in the US SE RA and Health Canada models.

Inputs	Probability
Egg goes to institutional consumer	=0.25
Home pooling	=0.02
Institutional pooling	=0.05
Home-pooled egg used as egg	=0.90
Home-pooled egg used as egg – undercooked	=0.33
Home-non-pooled egg used as egg	=0.90
Home-non-pooled egg used as egg – undercooked	=0.33
Institutional-pooled egg used as egg	=0.70
Institutional-pooled egg used as egg – undercooked	=0.33
Institutional-non-pooled egg used as egg	=0.90
Institutional-non-pooled egg used as egg – undercooked	=0.33
Home-pooled egg used as ingredient – not cooked	=0.02
Home-non-pooled egg used as ingredient – not cooked	=0.02
Institutional-pooled egg used as ingredient – not cooked	=0.30
Institutional-non-pooled egg used as ingredient – not cooked	=0.30
Cooking by boiling	=0.22
Cooking by frying	=0.49
Cooking by scrambling	=0.29

4.4 REFERENCES CITED IN CHAPTER 4

9 CFR 590. Code of Federal Regulations [of the United States of America]; Title 9 – Animals and animal products; Chapter III – FSIS, USDA; Part 590 – Inspection of eggs and egg products (Egg Products Inspection Act). See:
http://www.access.gpo.gov/nara/cfr/waisidx_02/9cfr590_02.html

Anderson, K.D., Jones, F.T., & Curtis, P.A. 1992. Legislation ignores technology. *Egg Industry*, Sept./Oct: 11–13.

Audige, L., & Beckett, S. 1999. A quantitative assessment of the validity of animal-health surveys using stochastic modelling. *Preventive Veterinary Medicine*, **38**: 159–176.

Audits International. [1999]. Information on United States cold temperature evaluation. Submitted to FAO/WHO in response to 2000 call for data.

Bell, D.D., & Curley, R.G. 1966. Egg cooling rates affected by containers. *California Agriculture*, June: 2–3.

Cowling, D.W., Gardner, I.A., & Johnson, W.O. 1999. Comparison of methods for estimation of individual-level prevalence based on pooled samples. *Preventive Veterinary Medicine*, **39**(3): 211–225.

Curtis, P.A., Anderson, K.D., & Jones, F.T. 1995. Cryogenic gas for rapid cooling of commercially processed shell eggs before packaging. *Journal of Food Protection*, **58**: 389–394.

Czarick, M., & Savage, S. 1992. Egg cooling characteristics in commercial egg coolers. *Journal of Applied Poultry Research*, **1**: 389–394.

Ebel, E.D., Hogue, A.T., & Schlosser, W.D. 1999. Prevalence of *Salmonella enterica* serovar enteritidis in unpasteurized liquid eggs and aged laying hens at slaughter: Implications on epidemiology and control of the disease. p.341–352, *in:* A.M. Saeed, R.K. Gast and M.E. Potter (eds). Salmonella enterica *serovar enteritidis in humans and animals: Epidemiology, pathogenesis, and control.* Ames, Iowa IA: Iowa State University Press.

Garibaldi, J.A., Lineweaver, H., & Ijichi, K. 1969. Number of *Salmonellae* in commercially broken eggs before pasteurization. *Poultry Science,* **48**: 1096–1101.

Gast, R.K. 1993. Research note: evaluation of direct plating for detecting *Salmonella* Enteritidis in pools of egg contents. *Poultry Science,* **72**: 1611–1614.

Gast, R.K. 1994. Understanding *Salmonella* Enteritidis in laying chickens: the contributions of experimental infections. *Poultry Science,* **21**: 107–116.

Gast, R.K., & Beard, C.W. 1992. Detection and enumeration of *Salmonella* enteritidis in fresh and stored eggs laid by experimentally infected hens. *Journal of Food Protection,* **55**: 152–156.

Gast, R.K., & Holt, P.S. 2000a. Deposition of phage type 4 and 13a *Salmonella* Enteritidis strains in the yolk and albumen of eggs laid by experimentally infected hens. *Avian Diseases,* **44**: 706–710.

Gast, R.K., & Holt, P.S. 2000b. Influence of the level and location of contamination on the multiplication of *Salmonella* Enteritidis at different storage temperatures in experimentally inoculated eggs. *Poultry Sciences,* **79**: 559–563.

Gerner-Smidt, P., & Wegener, H.C. 1999. *Salmonella enterica* serovar Enteritidis in Denmark. *In:* A.M. Saeed, R.K. Gast and M.E. Potter (eds). Salmonella enterica *serovar enteritidis in humans and animals: Epidemiology, pathogenesis, and control.* Ames, Iowa IA: Iowa State University Press.

Gibson, A.M., Bratchell, N., & Roberts, T.A. 1988. Predicting microbial growth: growth responses of *Salmonella* in a laboratory medium as affected by pH, sodium chloride and storage temperature. *International Journal of Food Microbiology,* **6**(2): 155–178.

Hammack, T.S., Sherod, P.S., Bruce, V.R., June, G.A., Satchell, F.B., & Andrews, W.H. 1993. Growth of *Salmonella* enteritidis in grade A eggs during prolonged storage. *Poultry Science,* **72**(2): 373–377.

Hennessy, T.W., Hedberg, C.W., Laurence, S., White, K.E., Besser-Wiek, J.M., Moen, M.E., Feldman, J., Coleman, W.W., Edmonson, L.M., MacDonald, K.L., Osterholm, M.T., & the Investigation Team. 1996. A national outbreak of *Salmonella* Enteritidis infections from ice cream. *New England Journal of Medicine,* **334**: 1281–1286.

Henzler, D.J., Kradel, D.C., & Sischo, W.M. 1998. Management and environmental risk factors for *Salmonella* enteritidis contamination of eggs. *American Journal of Veterinary Research,* **59**(7): 824–829.

Henzler, D.J., Ebel, E., Sanders, J., Kradel, D., & Mason, J. 1994. *Salmonella* Enteritidis in eggs from commercial chicken layer flocks implicated in human outbreaks. *Avian Diseases,* **38**(1): 37–43.

Hogue, A.T., Ebel, E.D., Thomas, L.A., Schlosser, W.D., Bufano, N., & Ferris, K. 1997. Surveys of *Salmonella* Enteritidis in unpasteurized liquid egg and spent hens at slaughter. *Journal of Food Protection,* **60**(10): 1194–1200.

Holt, P.S. 1995. Horizontal transmission of *Salmonella* Enteritidis in moulted and unmoulted laying chickens. *Avian Diseases*, **39**: 239–249.

Holt, P.S. 1998. Airborne horizontal transmission of *Salmonella* Enteritidis in molted laying chickens. *Avian Diseases*, **42**(1): 45–52.

Holt, P.S., & Porter, R.E. Jr. 1992. Effect of induced moulting on the course of infection and transmission of *Salmonella* Enteritidis in white leghorn hens of different ages. *Poultry Science*, **71**: 1842–1848.

Holt, P.S., & Porter, R.E. Jr. 1993. Effect of induced moulting on the recurrence of a previous *Salmonella* Enteritidis Infection. *Poultry Science*, **72**: 2069–2078.

Holt, P.S., Buhr, R.J., Cunningham, D.L., & Porter, R.E. 1994. Effect of two different moulting procedures on a *Salmonella* Enteritidis infection. *Poultry Science*, **73**: 1267–1275.

Humphrey, T.J. 1993. Growth of *Salmonella* enteritidis in egg contents. p.29–36, *in: Proceeding from the 5th European Symposium on the Quality of Eggs and Egg Products*. Tours, France, 4–8 October 1993.

Humphrey, T.J. 1999. Contamination of eggs and poultry meat with *Salmonella* enterica serovar Enteritidis. p.183-192, *In:* A.M. Saeed, R.K. Gast and M.E. Potter (eds). Salmonella enterica *serovar enteritidis in humans and animals: Epidemiology, pathogenesis, and control*. Ames, Iowa IA: Iowa State University Press.

Humphrey, T.J., Baskerville, A., Mawer, S., Rowe, B., & Hopper, S. 1989a. *Salmonella* Enteritidis phage type 4 from the contents of intact eggs: a study involving naturally infected hens. *Epidemiology and Infection*, **103**: 415–423.

Humphrey, T.J., Greenwood, M., Gilbert, R.J., Chapman, P.A., & Rowe, B. 1989b. The survival of Salmonellas in shell eggs cooked under simulated domestic conditions. *Epidemiology and Infection*, **103**: 35–45.

Humphrey, T.J., Chapman, P.A., Rowe, B., & Gilbert, R.J., 1990. A comparative study of the heat resistance of salmonellas in homogenized whole egg, egg yolk, or albumen. *Epidemiology and Infection*, **104**: 237–241.

Humphrey, T.J., Chart, H., Baskerville, A., & Rowe, B. 1991. The influence of age on the response of SPF hens to infection with *Salmonella* Enteritidis PT4. *Epidemiology and Infection*, **106**: 33–43.

Kilsby, D.C., & Pugh, M.E. 1981. The relevance of the distribution of microorganisms within batches of food to the control of microbiological hazards from foods. *Journal of Applied Bacteriology*, **51**: 345–354.

Kinde, H., and 10 co-authors. 1996. *Salmonella* Enteritidis phage type 4 infection in a commercial layer flock in southern California: Bacteriologic and epidemiological findings. *Avian Diseases*, **40**(3): 665–671.

Klontz, K.C., Timbo, B., Fein, S., & Levy, A. 1995. Prevalence of selected food consumption and preparation behaviour associated with increased risks of food-borne disease. *Journal of Food Protection*, **58**: 927–930.

Lin, C.-T.J., Morales, R.A., & Ralston, K. 1997. Raw and undercooked eggs: The dangers of Salmonellosis. *Food Review,* **20**(1): 27–32.

Martin, S.W., Meek, A.H., & Willeberg, P. 1987. *Veterinary epidemiology: Principles and methods.* Ames, Iowa IA: Iowa State University Press.

Morales, R.A., & McDowell, R.M. 1999. Economic consequences of *Salmonella enterica* serovar Enteritidis infection in humans and the United States egg industry. *In:* A.M. Saeed, R.K. Gast and M.E. Potter (eds). Salmonella enterica *serovar enteritidis in humans and animals: Epidemiology, pathogenesis, and control.* Ames, Iowa IA: Iowa State University Press.

Morris, G.K. 1990. *Salmonella enteritidis* and eggs: Assessment of risk. *Dairy, Food, and Environmental Sanitation,* **10**(5): 279–281.

Palumbo, M.S., Beers, S.M., Bhaduri, S., & Palumbo, S.A. 1995. Thermal resistance of *Salmonella* spp. and *Listeria monocytogenes* in liquid egg yolk and egg yolk products. *Journal of Food Protection,* **58**(9): 960–966.

Palumbo, M.S., Beers, S.M., Bhaduri, S., & Palumbo, S.A. 1996. Thermal resistance of *Listeria monocytogenes* and *Salmonella* spp. in liquid egg white. *Journal of Food Protection,* **59**: 1182–1186.

Poppe, C., Irwin, R., Messier, S., Finley, G., & Oggel, J. 1991. The prevalence of *Salmonella enteritidis* and other *Salmonella* spp. among Canadian registered commercial chicken broiler flocks. *Epidemiology and Infection,* **107**: 201–211.

Poppe, C., Johnston, R.P., Forsberg, C.M., & Irwin, R.J. 1992. *Salmonella* Enteritidis and other *Salmonella* in laying hens and eggs from flocks with *Salmonella* in their environment. *Canadian Journal of Veterinary Research,* **56**: 226–232.

Schlosser, W.D., Henzler, D.J., Mason, J., Hurd, S., Trock, S., Sischo, W.M., Kradel, D.C., & Hogue, A.T. 1995. *"Salmonella enteritidis Pilot Project" Progress Report.* Washington, DC: United States Government Printing Office.

Schlosser, W.D., Henzler, D.J., Mason, J., Kradel, D., Shipman, L., Trock, S., Hurd, S.H., Hogue, A.T., Sischo, W., & Ebel, E.D. 1999. The *Salmonella* enterica serovar Enteritidis Pilot Project. p.353-365, *In:* A.M. Saeed, R.K. Gast and M.E. Potter (eds). Salmonella enterica *serovar enteritidis in humans and animals: Epidemiology, pathogenesis, and control.* Ames, Iowa IA: Iowa State University Press.

Schoeni, J.L., Glass, K.A., McDermott, J.L., & Wong, A.C.L. 1995. Growth and penetration of *Salmonella enteritidis, Salmonella heidelberg* and *Salmonella typhimurium* in eggs. *International Journal of Food Microbiology,* **24**(2): 385–396.

Schuman, J.D., & Sheldon, B.W. 1997. Thermal resistance of *Salmonella* spp. and *Listeria monocytogenes* in liquid egg yolk and egg white. *Journal of Food Protection,* **60**: 634–638.

Schuman, J.D., Sheldon, B.W., Vandepopuliere, J.M., & Ball, H.R., Jr. 1997. Immersion heat treatments for inactivation of *Salmonella enteritidis* with intact eggs. *Journal of Applied Microbiology,* **83**: 438–444.

Shah, D.B., Bradshaw, J.G., & Peeler, J.T. 1991. Thermal resistance of egg-associated epidemic strains of *Salmonella* Enteritidis. *Journal of Food Science,* **56**: 391–393.

Stadelman, W.J., & Rhorer, A.R. 1987. Egg quality: Which is best – in-line or off-line production? *Egg Industry,* **93**: 8–10.

Sunagawa, H., Ikeda, T., Takeshi, K., Takada, T., Tsukamoto, K., Fujii, M. Kurokawa, M., Watabe, K., Yamane, Y., & Ohta, H. 1997. A survey of *Salmonella* Enteritidis in spent hens and its relation to farming style in Hokkaido, Japan. *International Journal of Food Microbiology*, **38**(2-3): 95–102.

Todd, E.C.C. 1996. Risk assessment of use of cracked eggs in Canada. *International Journal of Food Microbiology*, **30**: 125–143.

Van Gerwen, S.J.C. 2000. Microbiological risk assessment of food. A stepwise quantitative risk assessment as a tool in the production of microbiologically safe food. Wageningen University, The Netherlands [PhD Thesis].

USDA-FSIS. 1998. *Salmonella* Enteritidis Risk Assessment. Shell Eggs and Egg Products. Final Report. Prepared for FSIS by the *Salmonella* Enteritidis Risk Assessment Team. 268 pp. Available on Internet as PDF document at: www.fsis.usda.gov/ophs/risk/contents.htm.

Vose, D. 1996. *Quantitative Risk Analysis – A Guide to Monte Carlo Simulation Modelling*. Chichester, England: Wiley.

Whiting, R.C., & Buchanan, R.L. 1997. Development of a quantitative risk assessment model for *Salmonella* Enteritidis in pasteurized liquid eggs. *International Journal of Food Microbiology*, **36**(2-3): 111–125.

Zwietering, M.H., de Wit, J.C., Cuppers, H.G., & van't Riet, K. 1994. Modelling of bacterial growth with shifts in temperature. *Applied Environmental Microbiology*, **60**: 204–213.

5. RISK CHARACTERIZATION OF *SALMONELLA* ENTERITIDIS IN EGGS

5.1 SUMMARY

In risk characterization for *S.* Enteritidis in eggs, the output of exposure assessment was combined with hazard characterization, and the probability that an egg serving results in human illness was demonstrated. Changes in predictive risk upon changes in the flock prevalence and time-temperature scenarios are investigated. Key uncertainties that might have certain influence on the result are also shown. In addition, effects of risk management options are quantitatively compared and evaluated. It should be noted that the risk assessment of *S.* Enteritidis in eggs was intentionally conducted so as not to be representative of any specific country or region. The probability of illness and the compared effects of possible management options therefore only reflect the data environment used in this assessment.

5.2 RISK ESTIMATION FOR *S.* ENTERITIDIS IN EGGS

5.2.1 Model overview

The general structure of the *S.* Enteritidis in eggs risk assessment is outlined in Figure 5.1. The exposure assessment model consists of three stages: production; shell egg processing and distribution; and preparation and consumption; and combined with egg products processing if appropriate. This information is combined with the dose-response model from the *Salmonella* hazard characterization to estimate human illnesses resulting from exposures predicted by the exposure assessment to provide the risk characterization. The parameters used for the beta-Poisson dose-response function were described in Hazard Characterization (Table 3.16 in Section 3.5.2). One simulation of the entire model consists of 30 000 iterations, sufficient to generate reasonably consistent results between simulations.

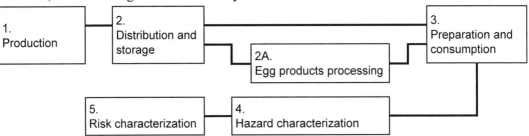

Figure 5.1. Schematic diagram showing the stages of the risk assessment of *Salmonella* Enteritidis in eggs

5.2.2 Results

The final output of the shell egg model is the probability that an egg serving results in human illness. This probability is determined as the weighted average of all egg servings (both contaminated and not contaminated) in a population. Clearly, the risk per serving is variable when we consider individual egg servings (e.g. a serving containing 100 organisms is much more likely to result in illness than a serving containing just 1 organism), but the meaningful measure is the population likelihood of illness. This risk per serving can be interpreted as the likelihood of illness given that a person consumes a randomly selected serving.

Three values for flock prevalence (5%, 25% and 50%) were considered. As explained earlier, three scenarios for egg storage time and temperature were also considered (reduced, baseline and elevated). The combination of these uncertain inputs generates nine different outputs from the model.

The lowest risk of illness is predicted when flock prevalence is 5% and storage times and temperatures are reduced (Table 5.1). In this scenario, the calculated risk is 2 illnesses in 10 million servings (0.00002%). The highest risk is predicted when flock prevalence is 50% and storage times and temperatures are elevated. In this case, the calculated risk is 4.5 illnesses in each million servings (0.00045%).

Table 5.1. Predicted probabilities of illness per egg serving based on different flock prevalence settings and different egg storage time and temperature scenarios.

Flock prevalence	Time-temperature scenarios		
	Reduced	Baseline	Elevated
5%	0.00002%	0.00002%	0.00004%
25%	0.00009%	0.00012%	0.00022%
50%	0.00017%	0.00024%	0.00045%

Changes in risk are approximately proportional to changes in the flock prevalence. For example, 5% flock prevalence is one-fifth of 25%. Correspondingly, the risk of illness for scenarios with 5% flock prevalence is one-fifth that of scenarios with 25% flock prevalence. Similarly, doubling flock prevalence from 25% to 50% also doubles the risk of illness if all other inputs are constant.

Under the baseline conditions using data set for this model, for any constant flock prevalence, the risk decreases by almost 25-30% from the baseline time-temperature scenario to the reduced time-temperature scenario. This risk increases by almost 90% between the baseline and elevated time-temperature scenarios. Although the degree of change in risk would be altered from baseline conditions, these simulations show, for example, that changing storage times and temperatures from farm to table results in disproportionately large effects on risk of illness.

The final output of the egg products model is a distribution of the numbers of *S.* Enteritidis remaining in 10 000-lb (~4500 litre) containers of liquid whole egg following pasteurization. The *S.* Enteritidis considered in this output are only those contributed by internally contaminated eggs. This output serves as a proxy for human health risk until the model is extended to consider distribution, storage, preparation – including additional processing – and consumption of egg products. Figure 5.2 shows the output for the 25% flock prevalence, baseline scenario. About 97% of the pasteurized lots are estimated to be *S.* Enteritidis-free, and the average level is about 200 *S.* Enteritidis remaining in each lot.

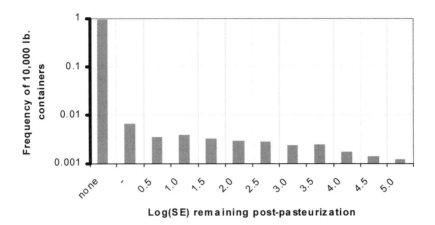

Figure 5.2. Predicted distribution of *Salmonella* Enteritidis (SE) contributed by internally contaminated eggs remaining in 10 000-lb (~4500 litre) containers of liquid whole egg after pasteurization. This distribution is predicted based on an assumed 25% flock prevalence, and the baseline egg storage times and temperatures in the model. Note that the y-axis is in $\log_{10}$ scale.

5.2.3 Uncertainty

Key uncertainties considered in this analysis relate to within-flock prevalence, frequency of egg contamination from infected hens, frequency of contaminated eggs laid in which the yolk is contaminated, and dose-response parameters.

Within-flock prevalence (FHen_Flock) is a distribution fitted to available data (Table 4.20 and Figure 5.3). Uncertainty regarding the mean of this distribution is estimated by re-sampling from the estimated lognormal distribution with a sample size equivalent to the original data and re-calculating the mean of the simulated data (i.e. bootstrap methods). For simplicity, it was assumed that the standard deviation of this lognormal distribution was constant and equal to 6.96% (Table 4.20). Uncertainty in this curve was calculated by assuming that the uncertainty about the mean was normally distributed. The standard deviation of the mean calculated from 1000 bootstrap replicates was 0.38%. The 5th and 95th confidence bounds are shown in Figure 5.3.

Frequency of egg contamination from infected hens is assumed constant in the model, but its uncertainty is modelled using a beta distribution with inputs from Humphrey et al.(1989). The frequency of yolk-contaminated eggs is constant in the model, but its uncertainty is modelled using a beta distribution reflecting the outcome of Humphrey et al. (1991). Uncertainty regarding the dose-response parameters is modelled as described in the hazard characterization section.

Uncertainty about the probability of illness per serving is shown to increase as the assumed flock prevalence increases (Figure 5.4). For any given flock prevalence, the uncertainty distribution has a constant coefficient of variation (i.e. standard deviation/average). Therefore, as the average probability of illness increases, its uncertainty increases proportionately.

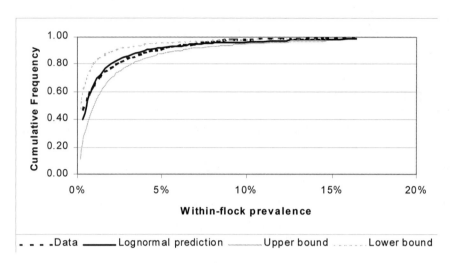

Figure 5.3. Cumulative frequency distributions for within-flock prevalence (FHen_Flock). The curve predicted by available data from infected flocks is shown relative to the best fitting lognormal distribution curve. Upper and lower bound curves are predicted using the 95th and 5th confidence intervals of the mean of the best fitting lognormal distribution.

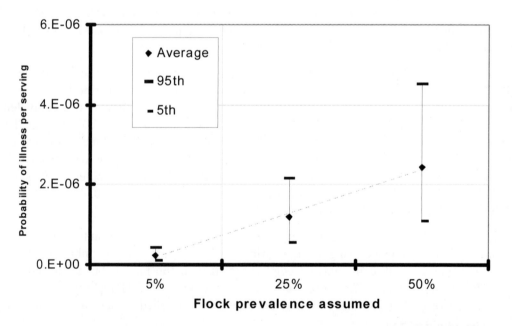

Figure 5.4. Uncertainty in probability of illness for different flock prevalence inputs assuming the baseline egg storage times and temperatures. Error bars represent the 90% confidence intervals for calculated uncertainty distributions.

Uncertainty not considered in this analysis relates to flock prevalence, predictive microbiology equations, time and temperature of storage, and pathway probabilities. Nevertheless, by changing the input values for flock prevalence, storage time and storage temperature, some evidence is provided regarding the effect of these inputs on risk (i.e. the sensitivity of the predicted risk per serving to these model inputs).

5.2.4 Discussion

The range in risk of illness predicted by this model extends from at least 2 illnesses per 10 million shell egg servings to 45 illnesses per 10 million servings. The scenarios considered represent a diversity of situations that approximate some countries or regions in the world. Nevertheless, no specific country is intentionally reflected in this model's inputs or outputs.

The effect of different flock prevalence levels on per serving risk is straightforward to calculate from this model. Nevertheless, the impact of changing egg storage times and temperatures is not trivial. These effects must be simulated to estimate the result. The model shows that change of 10% (either increase or decrease) in storage times and temperatures result in greater than a 10% change in the predicted risk per serving.

The uncertainty of probability of illness per serving was proportional to the average probability in each scenario considered. That finding suggests that we should be able to simulate scenarios and directly calculate uncertainty based on the average risk predicted by this model.

5.3 RISK MANAGEMENT OPTIONS FOR *S.* ENTERITIDIS IN EGGS

5.3.1 Estimation of the risk of illness from S. Enteritidis in eggs in the general population at different prevalence and concentration levels of contamination

The model was used to estimate the relative effects of different prevalence and concentration levels of *S.* Enteritidis in contaminated eggs. Prevalence can either be the proportion of flocks containing one or more infected hens (i.e. flock prevalence) or the proportion of infected hens within infected flocks (i.e. within-flock prevalence). The risk associated with different flock prevalence levels was illustrated in Table 5.1. That analysis illustrated that risk was generally proportional to the flock prevalence level. Reducing the proportion of infected flocks is therefore associated with a proportional decline in the likelihood of illness per serving among the population of all servings. One can also examine the risk of illness per serving for different within-flock prevalence levels, as well as for different starting concentrations of *S.* Enteritidis per egg.

To model the effect of within-flock prevalence on risk, the 1st, 50th and 99th percentile values of the within-flock prevalence distribution (0.1%, 0.5% and 22.3%, respectively) were simulated (Figure 5.5). The point of this analysis is to isolate the effect of within-flock prevalence on likelihood of illness by considering within-flock prevalence to be non-variant, but examining three different levels. This analysis also provides insight as to the effect of assuming different *average* within-flock prevalence levels on probability of illness. For these simulations, flock prevalence was assumed to be 25%. In the baseline time-temperature scenario, risk per serving was 6×10^{-8}, 3×10^{-7} and 1×10^{-5} for within-flock prevalence levels of 0.1%, 0.5% and 22.3%, respectively. The results show that risk of illness per serving changes in direct proportion to changes in within-flock prevalence. This effect occurs regardless of the time-temperature scenario considered. Consequently, the risk per serving if all infected flocks had within-flock prevalence levels of 10% (i.e. 10 of every 100 hens are infected) is 100 times the risk per serving when the within-flock prevalence is fixed at 0.1% (i.e. 1 in every 1000 hens is infected). In terms of control, these results suggest that reducing the proportion of infected hens in flocks provides a direct means of reducing illnesses from contaminated eggs.

Different initial levels of *S.* Enteritidis in eggs at the time of lay were modelled by assuming that all contaminated eggs started with 1, 10 or 100 organisms (Figure 5.6). The baseline egg storage time and temperature scenario was assumed, but flock prevalence was varied. For a flock prevalence of 5%, risk per serving was about 2 per 10 million regardless of whether the initial number of *S.* Enteritidis per egg was 1, 10 or 100. For flock prevalence levels of 25% and 50%, a more detectable change in risk per serving occurs between eggs initially contaminated with 1, 10 or 100 *S.* Enteritidis. For example, at 25% flock prevalence, the risk per serving increases from 8 per 10 million to 10 per 10 million as the number of *S.* Enteritidis in eggs at lay increases from 1 to 100. Nevertheless, for one-log changes in the initial numbers of *S.* Enteritidis, the resulting change in probability of illness is much less than one log.

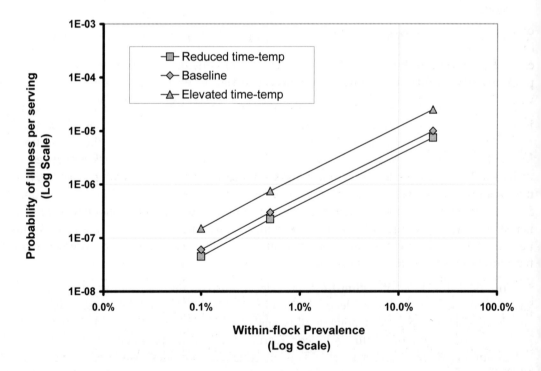

Figure 5.5. Predicted probability of illness, assuming that within-flock prevalence is either 0.1%, 0.5% or 22.3% (1st, 50th, or 99th percentiles of the lognormal distribution used in the model, respectively). Three egg storage time and temperature scenarios are considered. Flock prevalence is assumed to be 25%.

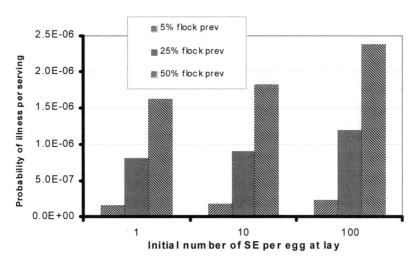

Figure 5.6. Predicted probability of illness per serving, assuming that the number of *Salmonella* Enteritidis (SE) per contaminated egg at lay is 1, 10 or 100. Three flock-prevalence levels are considered. Egg storage times and temperatures are assumed to be the baseline settings.

The dose-response function used in this risk characterization predicts that the probability of illness given an average dose of 1, 10 or 100 organisms is 0.2%, 2.2% or 13%, respectively. If all contaminated eggs were consumed raw immediately after lay, one would expect these probabilities to be appropriate to predict illnesses. The production module predicts that contaminated eggs are produced at a frequency of about 5×10^{-5} (~1 in 20 000) when flock prevalence is 25%. If all contaminated eggs contained just one organism, with no growth or decline before consumption, the predicted risk per serving should be $5 \times 10^{-5} \times 0.002$, or 10^{-7}. Similarly, the risk per serving if all eggs were contaminated with 10 and 100 organisms would be 10^{-6} and ~7×10^{-6}, respectively.

Figure 5.7 compares these predicted risks – when no growth or cooking is assumed – to the predictions shown in Figure 5.6 for 25% flock prevalence. When just a single *S*. Enteritidis organism is in contaminated eggs, Figure 5.6 implies that allowing growth inside eggs elevates the risk. Yet when contaminated eggs contain 10 or 100 organisms, Figure 5.6 implies that cooking of egg meals substantially reduces the risk. The explanation for these findings is that, regardless of the initial contamination, the combined effect of growth and cooking is to stabilize the risk per serving to nearly one per million; whereas if growth and cooking are not modelled, the risk per serving only depends on a dose-response function that is increasing at an increasing rate across the dose range considered. Therefore, it can be concluded from Figures 5.5 and 5.6 that the model's output is relatively less sensitive to initial numbers of *S*. Enteritidis than other inputs that influence growth and cooking.

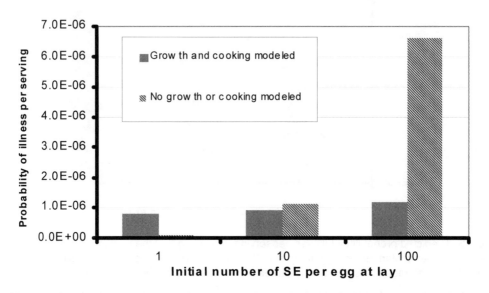

Figure 5.7. Predicted risk of illness when the exposure assessment model includes effects of growth and cooking compared with cases when no growth or cooking is modelled, for situations where the initial number of S. Enteritidis in contaminated eggs at lay is 1, 10 or 100. Flock prevalence is assumed to be 25% and baseline egg storage times and temperatures are assumed when growth and cooking are modelled.

5.3.2 Estimation of the change in risk likely to occur from reducing the prevalence of infected flocks and destroying breeding or laying flocks, and estimation of the change in risk likely to occur from reducing the prevalence of S. Enteritidis-positive eggs through testing of flocks and diversion of their eggs to pasteurization, and including the effect of pasteurization

As shown previously, risk of illness per serving decreases as the percent of infected flocks (i.e. flock prevalence) decreases. Table 5.2 illustrates the influence of flock prevalence on risk of illness per serving. Because the model includes uncertain inputs, risk per serving is also uncertain and this table summarizes uncertainty as the mean, 5th, and 95th percentile values (rounded to the nearest significant digit) of the predicted distribution.

Table 5.2. Predicted uncertainty in risk of illness per egg serving for different flock prevalence levels.

Flock prevalence	Mean	5th	95th
0.01%	0.00000005%	0.00000002%	0.00000009%
0.10%	0.0000005%	0.0000002%	0.0000009%
5.00%	0.00002%	0.00001%	0.00004%
25.00%	0.0001%	0.0001%	0.0002%
50.00%	0.0002%	0.0001%	0.0005%

We can use the results in Table 5.2 to predict the reduction in risk for a country or region that decides to control infected flocks. For example, consider a country with 5% of its flocks containing one or more infected hens. If such a country were to institute a programme with 98% effectiveness in reducing flock prevalence, then successful implementation of the programme would result in a flock prevalence of about 0.1%. The model predicts, in this case, that the mean risk of illness per egg serving would decrease from 200 per thousand million to 5 per thousand million. Pre-harvest interventions, such as those used in Sweden and other countries, might result in flock prevalence levels of 0.1% or lower.

Although the model predicts that probability of illness per serving is proportional to flock prevalence, the question remains: how can we reduce prevalence of infected flocks? To accomplish this seemingly requires either preventing uninfected flocks from becoming infected, or treating infected flocks to render them uninfected.

Treatment of breeding flocks to render them uninfected has been used in The Netherlands (Edel, 1994). Antibiotic treatment of the flock followed by competitive exclusion culture administration might succeed in eliminating the organism from infected hens, but environmental reservoirs may still exist to re-infect hens once the effects of the antibiotic have worn off. Furthermore, application of this method to commercial flocks may not be feasible or economic.

Preventing uninfected flocks from becoming infected is where most attention is focused in control programmes. Uninfected flocks can become infected via vertical transmission (i.e. infected eggs before hatch result in exposure of a cohort via horizontal transmission following hatching), via feed contamination, or via environmental sources (i.e. carryover infection from previously infected flocks). Control programmes may attempt to eliminate these avenues of exposure by applying one or more actions.

1. Test breeding flocks to detect *S.* Enteritidis infection, followed by destruction of the flock, if infected, to prevent it from infecting commercial flocks consisting of its future offspring.

2. Require heat treatment of feed before its sale (thereby eliminating *S.* Enteritidis and other pathogens).

3. Following depopulation of an infected flock, intense cleaning and disinfecting of poultry environments known to be contaminated. Such an approach must also eliminate potential reservoirs (e.g. rodents).

Most control programmes use all three interventions to preclude *S.* Enteritidis-infected flocks. The control programme in Sweden consists of such an approach (Engvall and Anderson, 1999). The Pennsylvania Egg Quality Assurance Program in the United States of America also used such an approach (Schlosser et al., 1999). However, discerning the efficacy of each intervention is difficult. Ideally, one would like to know what percent of newly infected flocks result from vertical transmission, feed contamination or previously contaminated environments.

Giessen, Ament and Notermans (1994) present a model for determining the relative contribution of risk of infection from vertical, feed-borne (or other outside environmental sources) and carryover environmental contamination. Comparing the model with data collected in The Netherlands, it appears that carryover infection was the dominant contributor to infection risk. Such a conclusion is based on the shape of a cumulative frequency curve for flock infection, which suggests that most flocks are infected soon after placement in commercial facilities. There is also evidence that the prevalence of infected breeder flocks is very low in The Netherlands.

Data from the United States of America *Salmonella* Enteritidis Pilot Project (Schlosser et al., 1999) suggest a fairly constant prevalence of positive samples collected in flocks by age, and that infection did not necessarily increase over time. Nevertheless, these data do not describe the age when infection was introduced. Roughly, 60% of the poultry flocks tested in this project were *S.* Enteritidis-positive. Additional evidence presented shows that 6 of 79 pullet flocks (8%) tested were *S.* Enteritidis-positive. These data suggest that the risk of infection from vertical transmission might be about 8%. Furthermore, there is some suspicion that feed contamination is an important source of *S.* Enteritidis for United States of America poultry flocks.

The data from The Netherlands and the United States of America suggest that the carryover route may account for >80% of the risk of flock infection in countries where *S.* Enteritidis is endemic. If true, then complete control of breeder flocks might only be expected to achieve ≤20% reduction in the prevalence of *S.* Enteritidis-infected flocks in such countries.

Results of an aggressive monitoring programme for breeder flocks in The Netherlands between 1989 and 1992 have been reported (Edel, 1994). For egg-sector breeding flocks, there is some suggestion that prevalence of infected flocks was reduced by about 50% per year. Effectiveness was less dramatic for meat-sector breeding flocks. This programme involved regular faecal testing of all breeder flocks, as well as regular testing of hatchery samples from day-old chicks. Positive flocks were depopulated until mid-1992, when treatment with enrofloxacin and a competitive exclusion culture was allowed as an alternative to the expense of prematurely depopulating a breeding flock. If a programme with 50% effectiveness in reducing prevalence of infected flocks each year were implemented for 3 years, one might predict that prevalence would fall to 12% (0.5^3) of the prevalence at programme start.

To reduce the risk of carryover infection for commercial flocks, it is thought that aggressive cleaning and disinfection must be completed after an infected flock is depopulated and before another flock is placed to begin a new production cycle. Cleaning and disinfection must also include an effective long-term rodent-control programme. Analysis of efforts in Pennsylvania to reduce the prevalence of infected commercial flocks suggests a decline from 38% to 13% during three years of programme operation (White et al., 1997). This programme routinely screened flocks for evidence of *S.* Enteritidis and required thorough cleaning, disinfection and rodent control once positive flocks had been depopulated. Another study in Pennsylvania (Schlosser et al., 1999) found 16 of 34 (47%) poultry environments that were initially *S.* Enteritidis-positive were negative for the pathogen following cleaning and disinfection.

Risk characterization of test and diversion programmes depends on the specific testing used in commercial flocks. For example, the Swedish programme collected three pooled samples, each consisting of 30 faecal droppings, during two or more examinations of egg production flocks during each production cycle (Engvall and Anderson, 1999). In The Netherlands, the breeder-flock monitoring programme testing protocol required the collection of 2 pools of 50 caecal droppings each every 4 to 9 weeks of production (Edel, 1994). The *Salmonella* Enteritidis Pilot Project's protocol required collection of swabs from each manure bank and egg belt in a hen house on three occasions in each production cycle (Schlosser et al., 1999).

Regardless of the size or type of sample collected, it would seem that a testing protocol that examines commercial flocks frequently and diverts eggs soon after detection should result in a meaningful reduction in the contaminated shell eggs marketed each year.

To examine the effect of test and diversion with the present model, two protocols were assumed, with either one or three tests administered to the entire population of egg production flocks. The single test would be administered at the beginning of egg production. Under the three-test regime, testing at the beginning of egg production would be followed by a second test four months later, and the third administered just before the flock is depopulated. Each single test consists of 90 faecal samples randomly collected from each flock. A flock is considered positive if one or more samples contained *S*. Enteritidis.

For the within-flock prevalence distribution used in this model, a single test of 90 faecal samples was likely to detect 44% of infected flocks. This was calculated using a discrete approximation to Equation 5.1, where a summation replaces the integral and discrete values of *p*, the within-flock prevalence. This equation assumes that an infected hen sheds sufficient *S*. Enteritidis in her faeces to be detected using standard laboratory methods.

Probability of flock testing positive = $\int_0^1 1-(1-p)^{90} f(p)dp$ Equation 5.1

If a flock was found positive on a test, its entire egg production was diverted to pasteurization. It was assumed that the egg products industry normally uses 30% of all egg production (consistent with the United States of America industry). Therefore eggs going to breaker plants from flocks other than those mandatorily diverted were adjusted to maintain an overall frequency of 30% (i.e. the percentage of eggs sent to breaker plants from test-negative infected flocks, and non-infected flocks, was reduced proportionally).

Test-positive flocks' premises were assumed to be cleaned and disinfected following flock depopulation. The effectiveness of cleaning and disinfection in preventing re-infection of the subsequent flock was assumed to be 50%. Furthermore, it was assumed that carryover infection was responsible for flocks becoming infected. Consequently, houses that were not effectively cleaned and disinfected resulted in infected flocks when they were repopulated.

Assuming a starting prevalence of 25% and the baseline egg storage time and temperature scenario, the effectiveness of the two testing protocols was estimated over a four-year period. Probability of illness per shell egg serving in each year was calculated for each protocol (Figure 5.8). Testing three times per year for four years reduced the risk of human illness from shell eggs by more than 90% (i.e. >1 log). Testing once a year for four years reduced risk by over 70%. At the end of the fourth year, the flock prevalences for the one-test and three-test protocols were 7% and 2%, respectively. Therefore, assuming the cost of testing three times per year to be three times greater than the cost of testing once a year (ignoring producer costs or market effects from diversion of eggs), then the flock prevalence results suggest a roughly proportional difference (i.e. 7%/2% $\approx$ 3) in the protocols. However, the reduction in risk per serving of the one-test protocol is greater than one-third of the three-test protocol. In other words, the one-test protocol achieves a 70% reduction while a testing protocol that is three times more costly achieves a 90% reduction (i.e. a 20% improvement). Such a result is not surprising when we consider that the single (or first) test at the beginning of the year most substantially affects risk. This is because flocks detected on the first test have their eggs diverted for the entire year, while flocks detected on a second test have their eggs diverted for just over half the year. Furthermore, flocks detected on the third test are tested so late in production that diversion of their eggs does not influence the population risk at all.

While egg diversion from positive flocks reduces the public health risk from shell eggs, it might be expected that there is some increased risk from egg products. Mandatory diversion causes more contaminated eggs to be sent to pasteurization. Nevertheless, the average quality of contaminated eggs is improved by diversion in this model.

It was assumed in the model that all diverted eggs were nest run (i.e. stored usually less than 2 days). Without mandatory diversion, 97% of lots were *S.* Enteritidis-free post-pasteurization and the average number of surviving *S.* Enteritidis in a 10 000-lb (~4500 litre) bulk tank was 200 (assuming 25% flock prevalence and the baseline egg storage times and temperatures). If a single test is used to determine which flocks are diverted, there are still 97% of vats that are *S.* Enteritidis-free and they average 140 *S.* Enteritidis per lot. The decrease in the average number of *S.* Enteritidis per lot is due to the increased proportion of nest run eggs that are diverted. Nest run eggs are stored for a shorter period and consequently contribute fewer organisms. If two tests are used, then there are 97% of vats that are *S.* Enteritidis free, and the average is 130 per lot. If three tests are used, there is no additional effect on egg products beyond the second test because the third test occurs just as the flock is going out of production.

Although not a direct measure of public health risk, these results suggest that the risk from egg products decreases as flocks are detected and diverted. However, this effect is conditional on nest run eggs being substantially less contaminated than restricted or graded eggs. Alternative scenarios might result in some increase in risk from diversion.

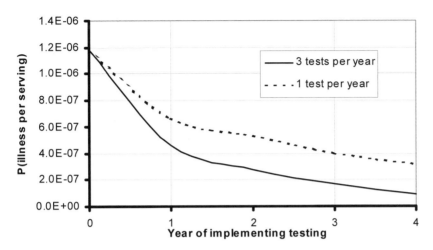

Figure 5.8. Predicted probability of illness per serving from shell eggs per year after implementing two testing protocols. It is assumed that all flocks in the region are tested each time and that initial flock prevalence is 25%. Baseline egg storage times and temperatures are used for the four years.

5.3.3 Estimation of the change in risk likely to occur from the use of competitive exclusion or vaccinating flocks against *S.* Enteritidis

The effects of competitive exclusion (CE) treatment are difficult to quantify from field evidence. Sweden and The Netherlands are examples of countries that include the use of CE in their *S.* Enteritidis control programmes. Nevertheless, such treatment is only one component of these programmes and its effect is not clearly separable from other components. CE has been studied in experimental settings for newly hatched chicks. The intent of CE inoculation in chicks is to quickly establish an indigenous intestinal flora to resist *S.* Enteritidis colonization. Efficacy of preventing infection appears to depend on the CE culture used, timing of exposure, dose of exposure, and possibly the addition of lactose (Corrier and Nisbet, 1999). Field evidence of CE efficacy in mature hens comes from the United Kingdom and from The Netherlands. In both countries, antibiotic treatment was applied to flocks known to be infected and the hens were subsequently inoculated with CE cultures. The intent of CE inoculation for hens was to quickly restore intestinal flora – that had been destroyed by the antibiotic treatment – to assist the hens in resisting future *S.* Enteritidis exposures. In the UK, 20 of 22 trials that combined antibiotic and CE treatments succeeded in preventing re-infection of flocks for a 3-month study period (Corrier and Nisbet, 1999). Infection status was determined from cloacal swab samples in treated flocks. In The Netherlands, combining antibiotic and CE treatments resulted in preventing 72% (n = 32) of flocks becoming re-infected. Two such combined treatments protected 93% of flocks from re-infection.

Vaccination for *S.* Enteritidis has been examined extensively in experimental settings, but less so in field trials. Experimentally, several types of vaccines have been evaluated: killed bacterins of various strains, live bacterins of attenuated strains, and surface antigen extracts of various strains. Injected killed bacterins are thought to have limited efficacy in preventing intestinal colonization of hens with *S.* Enteritidis, although such bacterins may, through stimulation of humoral antibody, reduce internal organ (including ovary) infection. Live bacterins – or surface antigen vaccines – may be more effective at modulating intestinal colonization by *S.* Enteritidis because these products may elicit the cell-mediated immune response needed to resist colonization. Nevertheless, most commercially available vaccines are currently of the killed variety.

Evidence concerning the effectiveness of *S.* Enteritidis bacterins in controlling infection has been reported for some Pennsylvania flocks (Schlosser et al., 1999). A total of 19 flocks from two farms used a bacterin to control their *S.* Enteritidis infection and sampling results were compared with 51 flocks that did not use a bacterin. Only a slight difference was noted in environmentally-positive samples collected in vaccinated (12%) and unvaccinated (16%) flocks. Yet, the overall prevalence of *S.* Enteritidis-positive eggs was 0.37 per 10 000 in vaccinated flocks and 1.5 per 10 000 in unvaccinated flocks. These results support the hypothesis that bacterins may not influence risk of colonization, but may reduce systemic invasion of *S.* Enteritidis, with resultant egg contamination. Nevertheless, this analysis did not control for confounding factors (e.g. rodent control, adequacy of cleaning and disinfection) that may have influenced the differences between vaccinated and unvaccinated flocks.

To evaluate the effect of vaccination against *S.* Enteritidis using the present model, it was assumed that flocks would need to be tested to determine their status prior to use of a vaccine. A single test or two tests four months apart, with 90 faecal samples per test, were assumed. The vaccine was assumed to be capable of reducing the frequency of contaminated eggs by approximately 75% (e.g. 0.37 per 10 000 for vaccinated flocks ÷ 1.5 per 10 000 for non-vaccinated flocks).

Assuming 25% flock prevalence and the baseline egg storage time and temperature scenario, the probability of illness per serving for a single test and vaccination protocol is about 70% of a non-vaccination protocol (Figure 5.9). Risk is reduced to 60% of the non-vaccination protocol if two tests are applied.

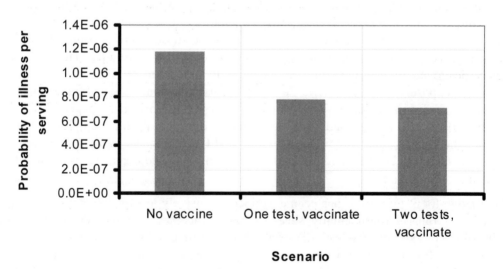

Figure 5.9. Comparison of predicted probability of illness per serving between three scenarios: when no vaccination is used; when one test is applied at the beginning of production and positive flocks are

all vaccinated; and when a second test is applied four months after the first test and additional test-positive flocks are vaccinated. Flock prevalence is assumed to be 25%, and the baseline egg storage time and temperature scenario is used.

Given the efficacy of bacterin use implied by the field evidence, one can assume that universal vaccination might reduce baseline risk to 25% of the risk resulting from a non-vaccinated population. However, the cost of vaccinating the entire population of laying hens could be high. The scenarios considered here assume that before a flock is vaccinated some testing is done to determine if that flock is infected. Nevertheless, the cost of testing all flocks must be weighed against the cost of vaccination. Also, more field research concerning the true efficacy of vaccination should be conducted before the cost of vaccination is borne by more than a few producers (i.e. if costs are to be paid by the public or shared across the entire industry).

5.3.4 Estimation of the change in risk likely to occur from minimizing the number of S. Enteritidis organisms in eggs through refrigeration of eggs after lay and during distribution, or requiring a specific shelf life for eggs stored at ambient temperatures

Interventions intended to minimize the dose of *S.* Enteritidis in contaminated eggs focus on preventing any growth of the pathogen after the egg is laid. Most evidence suggests that naturally contaminated eggs contain very few *S.* Enteritidis organisms at lay. If eggs are consumed soon after lay, or if eggs are kept refrigerated during storage, then the number of *S.* Enteritidis remains relatively unchanged prior to preparation of egg-containing meals.

Available predictive microbiology models suggest that eggs stored at 10°C will not grow *S.* Enteritidis for an average of 46 days. If most eggs are stored at <10°C and are consumed within 25 days, then interventions intended to improve egg handling will only influence the fraction of eggs that are time and temperature abused.

The effect of mandatory retail storage times and temperatures were evaluated using slightly different baseline assumptions (Table 5.3). These hypothetical settings might be typical in a country that does not have egg refrigeration requirements. The effects of time and temperature restrictions were evaluated assuming a flock prevalence of 25%.

Table 5.3. Hypothetical baseline input distributions for egg storage time and temperatures, assuming no egg storage requirements.

Inputs	Distributions
Storage temperature before transportation (°C)	=RiskPert(0,14,35)
Storage time before transportation (hours)	=RiskUniform(0,3)*24
Storage temperature after processing (°C)	=RiskPert(5,14,30)
Storage time after processing (hours)	=RiskUniform(1,5)*24
Retail storage temperature (°C)	=RiskPert(0,14,35)
Retail storage time (hours)	=RiskPert(1,9.5,21)*24

NOTES: PERT distribution has parameters RiskPert(minimum, most likely, maximum). Uniform distribution has parameters RiskUniform(minimum, maximum).

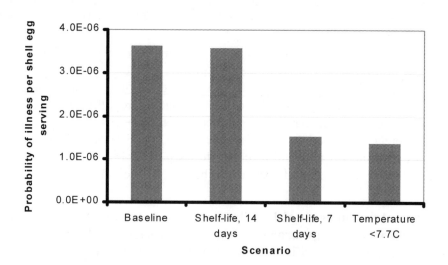

Figure 5.10. Probability of illness per serving of shell eggs given mandatory shelf lives of <14 or <7 days at retail, or mandatory retail storage temperature <7.7°C. Egg storage times and temperatures are modelled as for the baseline scenario, except for changes noted in Table 5.3. These changes to baseline egg storage times and temperatures were made to represent a country or region that does not routinely refrigerate eggs. Flock prevalence was assumed to be 25%.

Truncating retail storage time to a maximum of either 14 days or 7 days simulated a shelf-life restriction scenario. Truncating the retail storage temperature to less than 7.7°C simulated a refrigeration requirement. The results are summarized in Figure 5.10.

Restricting shelf life to less than 14 days reduced the predicted risk of illness per serving by a negligible amount (~1%). However, keeping retail storage temperature at no more than 7.7°C reduced risk of illness per serving by about 60%. If the shelf life was reduced to 7 days, risk per serving was also reduced by about 60%.

5.4 DISCUSSION

This model was purposely configured and parameterized to not reflect any specific country or region, although its results might be indicative of many country situations. A generic risk assessment such as this one provides a starting point for countries that have not developed their own risk assessment. It can serve to identify the data needed to conduct a country-specific risk assessment, as well as to provoke thinking concerning policy development and analysis.

Control of prevalence – either the proportion of flocks infected or the proportion of infected hens within flocks – has a direct effect in reducing probability of illness per serving. On the whole, egg storage times and temperatures can disproportionately influence the risk of illness per serving. Numbers of organisms initially in eggs at the time of lay seems less important.

Testing flocks, combined with diversion of eggs from positive flocks, is predicted to reduce public health risk substantially. In the scenarios considered here, diversion of eggs from test-positive flocks also reduced the apparent risk from egg products. Vaccination may reduce risk of illness by about 75%, but is typically less effective because producers would only vaccinate test-positive flocks.

As discussed in the Exposure Assessment for *S.* Enteritidis in Eggs (Chapter 4), biological inputs may be constant between models for different countries or regions, yet little else is likely to be similar. The predictive microbiological inputs, the distribution of within-flock prevalence, and the frequency at which infected hens lay contaminated eggs are examples of biological inputs that might be constant from one country to another (although not necessarily). The effects of uncertainty regarding these biological inputs to the model have been examined. Nevertheless, there are many aspects of uncertainty not fully considered (e.g. alternative statistical distributions were not evaluated for the predictive microbiology equations or within-flock prevalence distributions). Furthermore, many of the inputs are both highly uncertain and variable between countries. For example, times and temperatures of egg storage may vary considerably within and between countries, but it is difficult for any country to precisely know its distributions for storage times and temperatures.

This model introduces two new concepts not included in previous exposure assessments for *S.* Enteritidis in eggs. First, it considers the possibility of eggs being laid with *S.* Enteritidis already inside the yolk. Such eggs defy previous model descriptions of the time and temperature dependence of *S.* Enteritidis growth in eggs. Although predicted to be uncommon, yolk-contaminated eggs can support rapid growth of *S.* Enteritidis in much shorter times than eggs contaminated in the albumen.

Second, this model considers the role of *S.* Enteritidis growth in eggs destined for egg products. While most eggs are modelled as being shipped very quickly to egg products plants (i.e. nest run eggs), some eggs can experience moderate or high levels of growth before being broken and pasteurized.

Many of the results generated by this model are contingent on epidemiological assumptions:

- It is assumed that infected hens produce contaminated eggs at a constant frequency that is independent of host, bacterial strain or environmental factors.

- A homogeneous population of layer flocks is assumed (e.g. same size, same basic management and environment). This model also does not consider the effect of moulting practices on egg contamination frequency.

- It is assumed that within-flock prevalence is random and independent of hen age or other host, bacterial strain or environmental factors.

These may be reasonable default assumptions, but more research is needed to determine their appropriateness. Changing these assumptions could generate results that differ from the model, and the model can be adapted to consider such changes.

5.5 REFERENCES CITED IN CHAPTER 5

Corrier, D.E., & Nisbct, D.J. 1999. Competitive exclusion in the control of *Salmonella* enterica serovar Enteritidis infection in laying poultry. *In:* A.M. Saeed, R.K. Gast and M.E. Potter (eds). Salmonella enterica *serovar enteritidis in humans and animals: Epidemiology, pathogenesis, and control.* Ames, Iowa IA: Iowa State University Press.

Edel, W. 1994. *Salmonella* Enteritidis eradication programme in poultry breeder flocks in The Netherlands. *International Journal of Food Microbiology*, **21**: 171–178.

Engvall, A., & Anderson, Y. 1999. Control of *Salmonella enterica* serovar Enteritidis in Sweden. *In:* A.M. Saeed, R.K. Gast and M.E. Potter (eds). Salmonella enterica *serovar enteritidis in humans and animals: Epidemiology, pathogenesis, and control.* Ames, Iowa IA: Iowa State University Press.

Giessen, A.W. van de, Ament, A.J.H.A., & Notermans, S.H.W. 1994. Intervention strategies for *Salmonella* Enteritidis in poultry flocks: a basic approach. *International Journal of Food Microbiology*, **21**(1-2): 145–154.

Humphrey, T.J., Baskerville, A., Mawer, S., Rowe, B., & Hopper, S. 1989. *Salmonella* Enteritidis phage type 4 from the contents of intact eggs: a study involving naturally infected hens. *Epidemiology and Infection*, **103**: 415–423.

Humphrey, T.J. et al. 1991. Numbers of *Salmonella* Enteritidis in the contents of naturally contaminated hens' eggs. *Epidemiology and Infection*, **106**: 489–496.

Schlosser, W.D., Henzler, D.J., Mason, J., Kradel, D., Shipman, L., Trock, S., Hurd, S.H., Hogue, A.T., Sischo, W., & Ebel, E.D. 1999. The *Salmonella* enterica serovar Enteritidis Pilot Project. p.353-365, *In:* A.M. Saeed, R.K. Gast and M.E. Potter (eds). Salmonella enterica *serovar enteritidis in humans and animals: Epidemiology, pathogenesis, and control.* Ames, Iowa IA: Iowa State University Press.

White, P.L., Schlosser, W., Benson, C.E., Madox, C., & Hogue, A. 1997. Environmental survey by manure drag sampling for *Salmonella* Enteritidis in chicken layer houses. *Journal of Food Protection*, **60**: 1189–1193.

6. EXPOSURE ASSESSMENT OF *SALMONELLA* IN BROILER CHICKENS

6.1 SUMMARY

This section considers the development of an exposure assessment of *Salmonella* in broiler chickens. Initially, a general framework and the data requirements for such an assessment are considered. Data collected during the course of this work is then presented, and its usefulness for inclusion within an exposure assessment is discussed. Using appropriate data, an exposure assessment model is then developed. This model is general in nature, rather than being representative of any particular country or region. It is parameterized using two categories of data – country-specific data and general data – and these types are highlighted at the appropriate place in the model description. The output from the model is the probability of exposure by two routes: an undercooked serving of chicken, and cross-contamination resulting from preparation of that serving. For each of these routes, the number of organisms is also an output. These outputs are used to undertake a risk characterization, described in the next Chapter.

6.2 REVIEW OF LITERATURE, DATA AND EXISTING MODELS

6.2.1 Introduction

Purpose

This section describes the information available to develop a production-to-consumption exposure assessment of *Salmonella* in poultry, specifically broiler chicken. To date, no complete quantitative exposure assessments have been developed for this pathogen-commodity combination. This discussion considers the way in which such assessments could be developed, focusing on data requirements and possible methodologies. In addition, this report presents summaries of some of the available data and discusses the utility and limitations of existing data. This discussion is followed by a description of the exposure assessment model developed for the current FAO/WHO risk assessment of *Salmonella* in broiler chicken (Section 6.3). The assessment focuses on home preparation and consumption of the product.

Organization

A general model framework for conducting an exposure assessment for this pathogen-commodity combination is outlined. The framework covers the various stages on the production to consumption pathways that can be analysed as individual modules.

Each module identified is discussed in detail with respect to data requirements, possible modelling approaches and data availability. The discussions on data availability are followed by a presentation of data that has been collected for each module, together with an assessment of its use in conducting a full exposure assessment. Some of these data will be country specific, while the remainder will be general and can thus be used for the majority of countries. Collection and presentation of the data serves to illustrate the type of information that is currently available to individual member countries, and simultaneously demonstrates where information is lacking, and thus highlights critical data gaps.

The data summarized in the following sections have been collected from the literature, through the FAO/WHO calls for data, from discussions with *Salmonella* experts (microbiologists, veterinarians and epidemiologists) and other sources. Therefore the database is current up to the point of writing this report, but it is acknowledged that additional information may become available in the future.

Although no complete quantitative exposure assessments, from production to point of consumption, have been developed to date for *Salmonella* in poultry products, there are models that describe segments of poultry production and processing. These are also reviewed, together with a model for *Campylobacter* spp. in fresh broiler products.

6.2.2 Production-to-consumption pathways

Overall model pathway

A general aim of microbiological exposure assessment for any pathogen-commodity combination is to provide estimates of the extent of food contamination by the particular pathogen, in terms of both prevalence and numbers of organisms, together with information on commodity consumption patterns for the population of interest. Estimation of these outputs can involve consideration of a number of complex and interrelated processes that relate to all stages of the production-to-consumption pathway. Throughout this pathway, process-specific factors will influence both prevalence and numbers of organisms on the product, and hence final exposure. Such effects will be both inherently variable, due to, for example, differences in production and processing methods, and uncertain because some aspects lack appropriate information.

Given this complexity, it is often necessary to split the overall pathway into a number of distinct modules, each representing a particular stage from production to consumption (Lammerding and Fazil, 2000). Such an approach has previously been used for *S.* Enteritidis in eggs (USDA-FSIS, 1998), *Campylobacter jejuni* in fresh poultry (Fazil et al., unpublished; A.M. Fazil, personal communication) and *Escherichia coli* O157 in ground beef hamburgers (Cassin et al., 1998). The resulting exposure model is then integrated with a dose-response assessment to yield the risk characterization outcomes. This type of an approach has also been described as a Process Risk Model (Cassin et al., 1998).

A modular framework for an exposure assessment of *Salmonella* in fresh broilers is outlined in Figure 6.1. Outputs from one module are used as inputs to the subsequent module. In particular, the variables that are likely to flow from one module to the next are the prevalence of contaminated birds, carcasses or products (P) and the probable numbers of organisms per contaminated unit (N). Each module should describe, quantitatively, the changes in prevalence and numbers that occur within that step, attributable to specific factors, including, for example, the extent of cross-contamination, processing effects, the opportunity for temperature abuse, and the organism's ability to survive or grow under the conditions described.

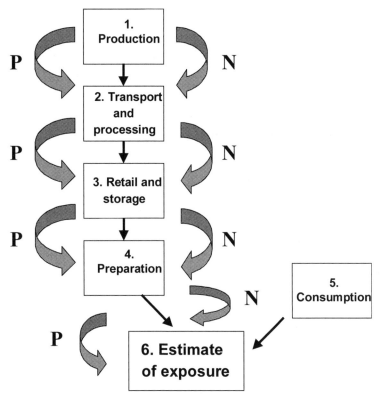

Figure 6.1. Modular pathway to describe the production-to-consumption pathway. Each step describes the changes to prevalence (P) and numbers of *Salmonella* (N) that occur within that specific module.

Individual modules of the overall pathway

The first module shown in Figure 6.1 relates to on-farm production of broilers. Here the aim is to estimate prevalence of *Salmonella*-positive birds (intestinal carriage of *Salmonella*) and the probable number of organisms per bird at the time of transportation for primary processing. This can involve taking into account various epidemiological and farm management factors that may influence these parameters.

Following farm production, the second module of the overall pathway considers transport and processing of broilers. This module models the effects of transport and the sequential processing steps on the prevalence and numbers of organisms. Important considerations are changes because of the type of transport facilities, processing methods and conditions, including changes in prevalence because of cross-contamination between negative and positive birds.

In the third module, the effects of retail distribution and storage in the home of the consumer are modelled. With respect to retail, both transportation and "on-shelf" storage are considered. Similarly, home storage includes transportation from retail source.

Preparation of the broiler chicken product is considered in the fourth module. Changes in prevalence and numbers of *Salmonella* present for the specific product purchased is determined by handling and cooking practices, and may include estimating impacts of cross-contamination. The outputs from this module – the estimated prevalence of contaminated products and number of organisms present in the food at time of consumption – are used in the calculation of exposure.

The amount of chicken consumed during a meal by various members of the population, and over a period, is quantified in the fifth module. This information is combined with the outputs from the previous module – i.e. the predicted likelihood that the pathogen will be in the food, and the predicted numbers of organisms present – to yield an estimate of the total number of *Salmonella* ingested. This information, together with the dose-response (i.e. the likelihood of illness associated with the number of *Salmonella* the consumer ingests), is then used to calculate the risk estimate in the risk characterization.

Data needs

Quantitative modelling of the individual exposure steps requires quantitative information. Data can be collected from a number of sources including, but not limited to:

- national surveillance data;
- epidemiological surveys;
- industrial surveys;
- research publications;
- unpublished research work; and
- government reports.

Often these data are publicly available, appearing, for example, in the published literature. However, other data, such as those collected through industry surveys, are often confidential and thus access becomes difficult. It is vital that confidence be built up between the risk managers, the assessors and those who can provide valuable data for risk assessment. Confidence building requires discussions and meetings (interactive risk communication) to discuss the type of data needed and what the data are being used for (the risk management activity). In addition, discussions provide insight into the data and how they were generated, with regards to sampling strategy, testing methods, etc. Such insight can be important for correct modelling, and thus the final results. Overall, good communications among all parties is essential.

In certain cases, adequate data may not be available. One way of dealing with this is to use expert opinion. Use of expert opinion introduces several considerations, such as how to choose experts, how to avoid biased judgement, how to elicit information and how to combine information from different experts. This area of study has been discussed by Kahneman, Slovic and Tversky (1982) and by Vose (2000).

In risk assessment, and particularly in the development of generic models (i.e. for application in general commodity production, processing, distribution and consumption management decision-making), data often come from many different sources. Two issues arise from this: first, what data to include within the model, and, second, how to combine such information. Determining what data to include involves consideration of applicability, such as whether the data are relevant for a particular country; whether the data are representative of the existing situation; and whether scientifically and statistically sound sampling and testing methods were used in the collection of the data. Furthermore, regardless of the data selection criteria, the rationale and process for selection must be transparent. The importance of transparency is also emphasized for combining data. Thus various methodologies could be used, such as weighting of information, but the assessor must clearly set out the methodology to ensure clarity and reproducibility.

Overall, data collection is probably the most resource-intensive part of modelling exposure and involves many issues that influence the quality of the risk assessment outcome.

Modelling approaches

The modelling approach used for individual stages of the overall pathway will necessarily depend on the data available to quantify input parameters and, in certain cases, the simplifying assumptions made until further data becomes available. Approaches are likely to differ from one exposure module to the next, depending on the parameters being described. Moreover, the risk management question will also determine the overall approach followed.

Static and dynamic approaches

Mathematical models can be described as either static or dynamic in nature. Dynamic models describe a process over time while static models consider the state of a process at one particular point in time. Dynamic models are generally constructed in terms of differential or difference equations that describe the rate of change of model variables over time. This approach has been used for several years to describe the spread of infectious diseases in both humans and animals (see Anderson and May, 1991). In contrast, static models consider the probability of an event happening at a certain time, such as the probability of infection from consumption of a chicken product, or over a period of time, such as the probability of introduction of infection in a year.

To date, most full quantitative risk assessments have been driven by static risk management questions and thus the output estimates of risk can usually also be termed static. However, many of the sub-modules of the assessment may involve dynamic modelling to some extent. In particular, in a microbial exposure assessment, the retail and storage step may involve dynamic modelling of the growth of the organism under conditions of temperature abuse (for an overview of bacterial growth modelling, see McMeekin et al., 1993; Baranyi and Roberts, 1995). Some modules of the pathway may require a combination of static and dynamic modelling; thus, preparation may involve a description of both growth (dynamic component) and cross-contamination (static component).

Uncertainty and variability

Modelling of each stage will have to account for the inherent variability of the specific process. The level of variability may be country or region specific, although it may be possible to generalize. Variability will arise due to causes that include seasonal effects, different procedures followed by different producers, differences in primary processing facilities, characteristics of the distribution chain, and consumption patterns. Variability cannot be reduced within a model because it describes the natural process.

In addition to variability, it will be necessary to model the uncertainty surrounding these processes. Such uncertainty will relate to the level of knowledge concerning a process and is usually reflective of the amount of available data.

Ideally, risk assessment models will explicitly separate uncertainty and variability; in essence, not separating means that one is neglected, and this can be a critical assumption with regard to further analysis. Various methods for such separation have been proposed (such as Vose, 2000), but, in reality, this often becomes complex. Ideally, factors that may be variable or uncertain, or both, should be identified and their influence on the risk assessment outcome described.

Deterministic versus stochastic models

Consideration of variability and uncertainty within exposure assessments leads to discussion of deterministic versus stochastic modelling. Deterministic models use point values (e.g. the mean of a data set) to describe inputs and thus to determine outputs. Stochastic models modify the data inputs to represent variability, uncertainty or both, using probability distributions. Probability distributions describe the relative weightings of each possible value and are characterized by a number of parameters that determine their shape, such as the mean and standard deviation or the most likely, minimum and maximum.

Consider the situation where prevalence of *Salmonella* infection in broilers is unknown in two countries and an expert has provided the following opinion.

	Minimum	Best estimate	Maximum
Country 1 (P1)	0.1	0.4	0.6
Country 2 (P2)	0.1	0.15	0.25

In this situation, in order to capture the expert's opinion, a triangular distribution could be used to describe the uncertainty about prevalence for each country (Figure 6.2):

$$P1 = \text{Triangular}(0.1, 0.4, 0.6) \qquad\qquad P2 = \text{Triangular}(0.1, 0.15, 0.25)$$

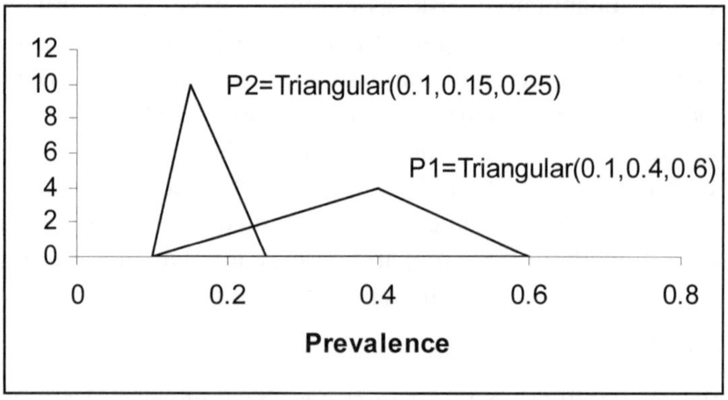

Figure 6.2. Probability distributions for P1 and P2

Stochastic models are most easily implemented on a computer using Monte-Carlo simulation. The technique of Monte-Carlo simulation involves repetition of the following events a large number of times (iterations):

1. Select a value for each input from its associated probability distribution (selection is determined by the shape of the distribution) to give one combination of input values.

2. Calculate the estimate of exposure for this combination of values.

3.   Store the calculated value.

The stored values are then combined to give a probability distribution for the estimate of exposure. There are numerous references in the literature (e.g. Haas, Rose and Gerba, 1999; Vose, 2000) explaining Monte Carlo techniques, and the uses of different probability distributions.

Consideration of the risk management question

Production-to-consumption exposure assessments require considerable time, data and other resources. The inherent uncertainty and variability associated with modelling individual exposure steps in a production-to-consumption exposure pathway increases its complexity. However, this type of an assessment provides the most information for risk managers when implementation of intervention strategies may be considered at any point of the food chain, and, perhaps more importantly, for identifying important information gaps. However, alternative approaches can also be useful, depending on the risk management information needs for decision-making, and the availability of adequate data. For example, the exposure assessment can begin at the point of retail sale of poultry products, using contamination data collected at that point. This approach, in effect, disregards the effects of individual factors occurring prior to retail sale that contributed to the microbiological status of the product. A similar approach has been taken to model exposure to chicken contaminated with fluoroquinolone-resistant *Campylobacter* (CVM, 2001). This methodology is useful when data are limited, or when the complexity of the process and associated uncertainties means that modelling becomes difficult and resource intensive, but it does not facilitate the investigation of specific control measures. In particular, the effects of mitigation at different stages throughout the exposure pathway cannot be quantified. Of course, in certain cases, the investigation of specific control strategies may not be required and thus the importance of the risk question is highlighted.

Defining the correct question is the most important part of any risk assessment. The risk question drives the model and hence the approach followed in any one module. As such, it must be stressed that this report does not present a prescribed formula for model development. Rather, general approaches are presented.

6.2.3 Primary production

The overall aim of the production module is to estimate, first, the prevalence of live broiler chickens contaminated with *Salmonella* at the time of leaving the farm for processing, and, second, the number of *Salmonella* per contaminated bird.

Sources of infection

Ideally, control of *Salmonella* within broiler flocks relies on knowledge of the source of infection. Possible sources include water, feed, litter, farm staff and the environment both inside and outside the broiler house (Mead, 1992). Furthermore, hatcheries are possible sources of infection, as is vertical transmission.

Many studies associated with the production of broilers have investigated factors that increase the prevalence of *Salmonella*. Rose et al. (1999) summarize the literature into five groups of risk factors:

- Inadequate level of hygiene, *Salmonella* contamination of the previous flock, with persistence inside the house.
- Contaminated day-old chicks and contaminated feed.
- The farm structure (>3 houses on the farm).
- Wet and cold seasonal conditions.
- Litter-beetle infestation of the house.

Several of the studies included within this summary focus on broiler-breeder farms rather than broiler chicken production farms. However, it may be assumed that the risk factors identified above are applicable to all poultry flocks. Of the above-listed factors, feed and hatcheries are regarded as principle sources of infection.

An ideal exposure assessment of *Salmonella* in broilers would include the calculation of the probability of infection from a number of possible sources. Such calculation could be based on, for example, the numbers of salmonellae a chicken is exposed to from each source and the subsequent consequences of exposure. Results from epidemiological studies could assist in this type of calculation. Given such a model, possible control strategies could be investigated in a quantitative manner.

In reality, data relating to the numbers of *Salmonella* organisms within feed, litter, etc., and the numbers to which a bird has been exposed, is extremely limited or simply unknown. Due to this limitation, previous microbial exposure assessments have started from the point of estimating the prevalence of contaminated, *Salmonella*-positive birds (Fazil et al., unnpublished; A.M. Fazil, personal communication; Hartnett et al., 2001). Although this approach inhibits the investigation of on-farm control strategies, it is currently the most likely approach that can be used for developing an exposure assessment of *Salmonella* in broilers.

Prevalence of Salmonella-*positive birds*

Prevalence in this document is defined to be the probability of a bird being infected with *Salmonella.* To estimate prevalence, data are required on positive (infected) birds at the point of leaving the farm for slaughter. Such data should be representative of the population of broilers and hence should cover a number of producers, flocks and seasons. Often, this type of information is not available (Hartnett et al., 2001), and, in this case, flock prevalence and within-flock prevalence can be estimated and used to generate an estimate of bird-level prevalence.

Flock prevalence

Flock prevalence is the proportion of flocks containing one or more infected – *Salmonella*-positive – birds. Flock prevalence is a national estimate, hence country-specific data are required. Estimation of flock prevalence requires consideration of the broiler production methods used. Differences in production practices occur not only between countries, but also within countries. For example, within the United Kingdom (and therefore probably in many other industrialized countries), many poultry companies may have their own feed mills, breeder flocks and hatcheries, thus differences between companies may exist. In addition, different breeds of birds may be used, both within a country and worldwide. Further, flock sizes, densities and the conditions under which a bird lives can also vary, such as free-range and organic birds versus mass-produced commercial birds. Many of these factors may influence the *Salmonella* status of a flock.

In addition to production methods, it is possible that climatic conditions may also influence flock prevalence. There is distinct seasonal effect in the outbreak of human *Salmonella* cases, which peak in the summer months. However, Angen et al. (1996) have showed a significant increase in prevalence of *Salmonella* in broiler chickens in Denmark during the autumn months of September–November, and Soerjadi-Liem and Cumming (1984) demonstrated a higher probability of *Salmonella* infection in Australian flocks during the cold and wet season. Climatic effects may in turn produce variation in flock prevalence between different geographical locations of a particular country.

Consequently, it is likely that flock prevalence may vary from region to region, from producer to producer, from season to season, and even from year to year. Testing all poultry before leaving the farm is impractical, and hence, data from sampling a portion of flocks are used to estimate the flock prevalence distribution, and should be defined by the associated uncertainty.

Within-flock prevalence

Within-flock prevalence refers to the proportion of birds within a single flock that are infected with *Salmonella*. Within-flock prevalence of *Salmonella* is very likely to vary from flock to flock for a number of reasons. Factors influencing such variability include the virulence of the *Salmonella* strain, levels of stress within the broiler house, and the occurrence of other avian diseases that may concurrently weaken resistance to *Salmonella*. As with flock prevalence, this variability should be represented within the exposure assessment model.

Ideally, the prevalence of *Salmonella* within flocks would be determined by sampling all broilers within all flocks just before leaving the farm for slaughter, but such comprehensive data collection is impractical. Therefore, as with flock prevalence, sample data could be utilized to obtain an estimate of the distribution for within-flock prevalence, together with a description of its associated uncertainty.

Note that intermittent shedding may affect the detection of *Salmonella* and thus birds and flocks testing negative by cloacal swabbing just prior to slaughter may nevertheless carry external contamination.

Number of Salmonella *in infected birds*

In addition to prevalence of *Salmonella*-positive broiler chickens, the number of organisms per positive bird is also a consideration, so that contamination in the processing environment can be modelled. Methods for determining the numbers of salmonellae in or on a bird can differ markedly, and a large degree of variability arises from different procedures. Results are reported in different units depending on the methodology, e.g. colony-forming units (CFU) or most probable number (MPN). In general, for risk assessments, CFU is the preferred unit of data, but MPN data can also be formulated such that they can be of use for estimation. In addition, the true number of organisms per bird is likely to vary from bird to bird. Consequently, there will be a large amount of variability in this estimate, and such variability may arise from a number of different sources.

Sampling information

For both prevalence and concentration, other information related to the collection of the data is also important. In particular, the test method used and its associated sensitivity and specificity must be considered. At the farm level, many different sample collection methods are used to determine the *Salmonella* status of individual broiler chickens or of the flock. For example, samples may be faeces, the caeca, cloacal swabs, and various environmental specimens. Other factors that influence results include the basis for the sampling strategy, the statistical validity of the sampling plan, information on farm management, the time of year of data collection, and the age of birds. Consequently, interpretation and combination of data can become difficult.

Summary of available data

Salmonella-*positive flocks and within-flock prevalence*

Tables 6.1 to 6.4 provide a summary of the flock and within-flock prevalence collected for this project. Initial observation of the data indicates that at the time of writing this report, information on *Salmonella* prevalence is missing for countries in a number of regions of the world. In particular, there is no or limited data for African, Asian and South American countries. Many countries within these regions provided some information for the 1995 Animal Health Yearbook (FAO-OIE-WHO, 1995), but information is restricted to details such as when the last case was reported and the level of occurrence. For other countries, no information appears to be have been reported.

For flock prevalence, in Table 6.1, much of the reported prevalence data include details of the numbers of flocks tested and the numbers of positive flocks. In cases where number of flocks tested and numbers positive are not provided, point estimates or ranges for flock prevalence are reported (e.g. studies by Mulder and Schlundt, in press; Hartung, 1999; White, Baker and James, 1997). In some cases (Tables 6.1 and 6.2), different sample materials are used to derive the flock prevalences, which introduces uncertainty. In addition, specificity and sensitivity of the various test protocols are rarely described. Few of the reports include information on how the results relate to the overall population of broiler chicken flocks, hence any variability due to, for example, differences between poultry companies (vertically integrated operations) is difficult to estimate. At the time of preparing this report, only one study (Soerardi-Liem and Cumming, 1984) had considered seasonality by sampling at different times of the year (Table 6.3).

Overall, it appears that flock prevalence is very variable between countries. However, it must be recognized that different sampling methods have been used in the different studies. In particular, in some reports environmental samples such as the litter, water and feed have been tested to determine positive flocks (for example, Lahellec et al., 1986; Jones et al., 1991a; Poppe et al., 1991). In contrast, other studies (such as Jacobs-Reitsma, Bolder and Mulder, 1994; Angen *et al.* 1996) involve direct testing of the broilers by examining the cloaca or caeca. Given these differences, comparison of country data must be undertaken with caution.

For within-flock prevalence, the data presented in Tables 6.3 and 6.4 indicate that there is very limited information relating to within-flock prevalence. In contrast to the flock prevalence data, several of the reported studies have considered variability among flocks, using the same sampling and testing protocols. For example, the data reported by Jacobs-Reitsma, Bolder and Mulder (1991) for the Netherlands show a large amount of variation in within-flock prevalence (a range of 0 to 80% for the caeca samples, and a range of 0 to 100% for the liver samples). Similarly, a wide range in values is reported from the Australian study by Soerjadi-Liem and Cumming (1984) (Table 6.3). As noted for some of the flock prevalence studies, the sample sizes reported in these surveys are small and thus there will be a large amount of uncertainty associated with any derived distributions for within-flock prevalence.

Number of organisms

At present, there are few data for numbers of *Salmonella* within infected broiler chickens (e.g. number per gram of faeces), or the numbers that may be present on feathers, skin, etc., of either birds that are infected, or birds that do not have intestinal carriage of the organisms but are surface contaminated. Most studies simply determine the presence or absence of salmonellae in the material tested. However, one study reported 100–1000 CFU of *Salmonella* per gram of gut content (Huis in 't Veld, Mulder and Snijders, 1994). Humbert (1992) reported that samples of *Salmonella*-positive faeces in the environment contain between 10^2 and 10^4 CFU salmonellae per gram. This small amount of information could be used to derive a distribution for the number of organisms, but there would be large associated uncertainty.

Data gaps

Overall, the following main data gaps have been identified for the primary production module.

- *Salmonella* prevalence information is available for some countries worldwide, but many of these studies give *limited details of study design*.

- Regions for which there is no or limited prevalence data include Africa, Asia and South America.

- No information relating to *sensitivity* or *specificity* of tests used is presented in the studies.

- There are very limited data relating to *numbers of organisms* per *Salmonella*-positive or contaminated bird.

Table 6.1. *Salmonella* flock prevalence data (see also Table 6.2).

Country (and year of sampling if stated)	Sample	No. of flocks tested	Percentage of positive flocks	Reference
Australia				
(April–Sept.) 1984	Caeca	7	86	Soerjadi-Liem &
(Oct.–March) 1984		13	46	Cumming, 1984
Austria				
1998	Cloaca	5 029	3.4	EC, 1998
1997		8 698	4.8	
1996		7 412	5.5	
Belgium				
1998	Faeces	122	36.1	EC, 1998
Denmark				
1998	Sock-samples	4 166	6.5	EC, 1998
1997		4 139	12.9	
1996		3 963	7.9	
1996-97	NS[1]	NS	5–10	Mulder and Schlundt, in
1995	NS	NS	25–30	press
1996	Caeca	7 108	16.8	Angen et al., 1996
Finland				
1998	Faeces	2 856	0.7	EC, 1998
1997		2 951	0.7	
1996		2 568	0.9	
France				
	NS	86	69.8	Rose *et al.*, 1999
	Walls, drinkers, litter, feed	180	53.3	Lahellec, Colin and Bennejean, 1986
Germany				
–	NS	58	12.0	Hartung, 1999
1998	NS	455	4.2	EC, 1998
1997	NS	691	5.8	
1996	NS	3 119	4.2	
Ireland	NS			
1998		1 732	20.7	EC, 1998
Italy				
1998	NS	1 093	3.1	EC, 1998
1997	NS	754	1.1	
Japan				
1995–96	Faeces	35	57.1	Murakami et al., 2001
Netherlands				
1998	NS	192	31.8	EC, 1998
1997	NS	63	25.4	
–	Caeca	181	27.0	Jacobs-Reitsma, Bolder and Mulder, 1994
–	NS	NS	Up to 25.0	MSF, 1990
–	Faeces (trucks, crates)	107	67.3	Goren et al., 1988
Norway				
–	NS	2 639	<0.01	ARZN, 1998
Sweden				
1998	Faeces	2 935	0.03	EC, 1998
1997		3 379	0.06	
1996		3 300	0.12	
UK				
–	Litter	3 073	18.5	Anon., 1999

Note: NS = not stated

Table 6.2. Flock prevalence and comparison of different sampling methods

Country	No. of flocks tested	Sample (no. of samples)	% Positive	Reference
Canada	294	Environment	76.9	Poppe et al., 1991
		Litter	75.9	
		Water	21.6[1]	
		Feed	13.4[2]	
Netherlands	141	Caeca	24.1	Goren et al., 1988
	92	Litter	19.6	
	49	Skin	12.0	
USA	267		4.5	
		Dead bird rinse (14)	14.3	Jones et al., 1991a
		Live bird rinse (14)	7.2	
		Faeces (155)	5.2	
		Environment (42)	2.4	
		Litter (14)	0	
		Water (14)	0	
		Feed (14)	0	

NOTES: (1) 63 of 292 flocks. (2) 39 of 290 flocks

Table 6.3. Seasonal flock and within-flock prevalence of *Salmonella* in Australian flocks based on caecal samples (Source: Soeradi-Liem and Cumming, 1984).

Season	No. of birds tested per flock	% positive birds
Autumn-winter (April–Sept.)	50	32
	50	36
	50	34
	50	92
	50	90
	50	40
	50	0
Spring-Summer (Oct.–March)	50	22
	50	12
	50	30
	50	10
	50	4
	50	22
	7 flocks, 50 birds each	0

Table 6.4. Within-flock prevalence and bird prevalence.

Country	No. of birds tested per flock (flocks sampled)	Caeca	Liver	Caeca 5–6 weeks (on-farm)	Skin and feathers 5–6 weeks	Caeca, 5–6 weeks (at processing)	Other	Source
Netherlands	3 399 (1)	14.3						[1]
Netherlands	10 (10)	20	10					[2]
		20	0					
		10	20					
		0	50					
		70	100					
		30	80					
		0	10					
		80	90					
		20	100					
USA	100 (3)	52[(1)]		15	9	2		[3]
		48[(1)]		17	5	4		
		66[(1)]		25	49	11		
Iraq	232 (NS)[(2)]						1.3[(3)]	[4]

NOTES: (1) Caecal samples at 3–4 weeks, on farm. (2) Not stated if from one or more flocks, therefore considered as individual bird prevalence. (3) Cloacal swabs.

SOURCES: [1] Goren et al., 1988. [2] Jacobs-Reitsma, Bolder and Mulder, 1991. [3] Corrier et al., 1995. [4] Hadad and Mohammed, 1986.

6.2.4 Transport and processing

The transport and processing module describes the processing of broiler chickens, from the point of leaving the farm to the time the finished product leaves the slaughterhouse. The outputs of this step should be an estimate of (i) the prevalence of *Salmonella*-contaminated product, and (ii) the numbers of organisms per contaminated product unit.

Transport and processing steps

Overview

There are many different sub-modules within this stage, some of which increase or decrease the level of *Salmonella* contamination. Figure 6.3, from Eley (1996), summarizes the main steps of the process. This discussion focuses on transport, stun-and-kill, scalding, de-feathering (plucking), evisceration and chilling, although the other operations are also briefly mentioned.

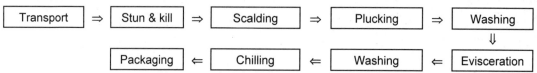

Figure 6.3. A flow chart describing transport and processing of raw poultry meat (from Eley,1996)

Many sources give detailed descriptions of the processing of poultry (e.g. Geornaras and von Holy, 1994; ACMSF, 1996). Each stage can potentially increase or decrease the prevalence of *Salmonella* in broilers, or increase or decrease the numbers of organisms on the exterior of the broiler chicken carcass, or a combination. Overall, it is probable that the stages will be similar in all regions of the world, although the changes in microbial load occurring at each step can differ, depending on the facilities, technologies and hygienic practices employed.

Transport

During transportation, birds are often stored in open crates that are placed on top of each other; thus, faeces can drop from an upper crate to a lower crate and cause cross-contamination. The stress of transport associated with factors such as vehicle conditions, length of journey, temperature and road conditions, will increase faecal excretion (and hence *Salmonella* excretion in *Salmonella*-positive birds) and therefore the possibility of cross-contamination is increased (ACMSF, 1996). There is an additional problem if the crates used are not thoroughly cleaned and disinfected between each collection of birds.

Stun and kill

Birds are stunned when their heads are submersed into water within which there is an electrical current. They are then killed by exsanguination. These procedures have not been identified as major cross-contamination steps. A second, more modern technique is using a mixture of gas, which is also unlikely to be a significant cross-contamination step.

Scalding

Scalding facilitates the removal of feathers. Birds are immersed in water, the temperature of which can depend on whether the bird is to be sold fresh or frozen. A scald tank with water that is too hot can cause discoloration of the skin, so broilers to be sold fresh are scalded at a lower temperature of 50–52°C (soft-scald), whereas birds to be sold frozen are scalded at higher temperatures, 56–58 C (hard-scald) (ACMSF, 1996). The temperatures have important implications of *Salmonella*. In particular, some *Salmonella* species may remain viable in the scald tanks for long periods (ICMSF, 1996). As a result, there is potential for cross-contamination.

The addition of chemicals to the scald tank water may reduce the potential for pathogen survival and hence cross-contamination. However, in certain areas of the world (e.g. Europe) regulations may not permit such practices due to the requirements to use only potable water and to demonstrate that no residues remain on the carcass.

There are a number of options for the mechanical system used for scalding, including spray systems, counter-current scald tanks and multi-stage scalding. More information from different areas of the world is required to assess the different systems used.

Plucking or de-feathering

During de-feathering, machinery mechanically removes the feathers from the birds using counter-rotating domes or discs that have rubber fingers mounted on them. De-feathering is regarded as a major site for contamination. In particular aerosol spread of microoganisms may occur as the feathers are removed (ACMSF, 1996). In addition, organisms can sometimes persist in machines due to inadequate cleaning.

Evisceration

Evisceration involves the removal of internal organs. Initially, the intestines remain attached so that they can be inspected. Due to this, the exterior of the bird may be contaminated if the intestines are damaged. Such damage can occur frequently since the machinery used for evisceration is not flexible with respect to the size of the bird. However, newer evisceration machines, which separate the carcass from the offal at the point where the offal becomes exposed, may overcome this problem.

Washing

Washing a carcass (in any form) should decrease the numbers of *Salmonella* residing on the exterior, although many studies have highlighted the attachment of *Salmonella* to the skin of broiler chickens during processing (e.g. Notermans and Kampelmacher, 1974, 1975). Depending on the method of washing, the prevalence of *Salmonella* may increase or decrease. For example, if washing takes place in an immersion tank, although *Salmonella* will be washed off those carcasses contaminated on their exterior, these organisms may then cross-contaminate an initially *Salmonella*-free carcass.

Chilling

The two most common methods of chilling are the immersion chiller and the air chiller. Different countries may use different chilling methods. For example, in the United States of America, immersion chilling is generally used, while in Europe immersion chilling can only be used for frozen poultry products. With immersion chilling, a counterflow current can be used such that a carcass is always moving towards cleaner water. Note that counterflow immersion chilling is a requirement of the EU, but it is not necessarily used in other parts of the world. Chlorine in the form of hypochlorite or chloride dioxide has been shown to reduce levels of cross-contamination within immersion chillers. Addition of chemicals to the chill tank is country dependent and, as with scalding, may depend on regulations. In the United States of America, in 1992, a decision was made to include chlorine in the chill tank (Waldroup *et al.*, 1992).

Portioning and packaging

Portioning and packaging of broiler chicken products can also potentially cause cross-contamination, but it is not considered to be significant. Briefly, a chicken can be portioned either by personnel from the processing plant or by machinery. The usual order of removal is neck skin, wings, breast, backbone, thighs and drumsticks (ACMSF, 1996). Manual handling by workers during inspection for cosmetic defects in de-boned meat, such as chicken breast, can also increase the level of cross-contamination.

Data requirements

Data requirements for modelling transport and processing fall into two categories. First, data are needed to describe how the prevalence of contaminated birds, carcasses and products changes during each sequential step, and, second, data are needed to describe the corresponding changes in numbers of the pathogen per contaminated bird, carcass or product at each stage.

Change in prevalence and numbers during transport and processing will be variable in nature, due to varying conditions, handling practices and temperatures. In addition to variability, it is likely that there will be an extensive amount of uncertainty associated with each step. Therefore, ideally, data to quantify both variability and uncertainty would be useful to characterize these steps.

Many studies that have investigated the effect of processing on *Salmonella* contamination of broiler chicken only consider a single step or a few of the sequential steps. Consequently, if combining data to generate estimates of the magnitude of change, details of the sampling methods and tests used and the associated sensitivity and specificity is important. Several different methods have been employed by various researchers to determine the presence or numbers of *Salmonella*, and samples may range from carcass rinse fluids and carcass swabs, to neck skin, or intestinal contents for direct testing.

Summary of data available

Information collected for pathogen prevalence and concentration changes in and on birds during transport is limited. Studies in the late 1970s by Rigby et al. (1980b) indicated that *Salmonella* could be isolated from debris in live-haul trucks and crates before live poultry was loaded, after unloading, and after washing. In the United States of America, Jones et al. (1991a) reported that debris from 33.3% of live-haul trucks and crates were positive for *Salmonella*, and similar levels were reported by Carraminana et al. (1997) in Spain. However, these data do not provide sufficient quantitative information to use for risk modelling.

Tables 6.5 to 6.13 provide a summary of data collected for individual steps during processing, and give a snapshot of the *Salmonella* situation at the various processing steps. However, they do not monitor change directly. In Table 6.9, some data is included that shows changes that occur during one of the processing steps.

In general, most studies consider prevalence of positive birds or carcasses. Further, the extent of contamination in the surrounding environment is often investigated, such as the knife used for slaughter (Table 6.5), the scald tank water (Table 6.6), the de-featherer (Table 6.7) and the chill water (Table 6.9). Environmental data can be used to give an indication of the extent of cross-contamination and, in theory, could also be used to predict prevalence levels or numbers of organisms at a particular point. Such predictions would require appropriate mathematical techniques and might require a number of assumptions relating to, for example, the rate of transfer of organisms at different sites. However, the limited amount of available data would mean that any predictions would be very uncertain and thus should be undertaken with caution.

Differences in prevalence resulting from different practices are considered in several studies. In particular, differences between tanks (with and without additives) has been investigated for both scalding (Humphrey and Lanning, 1987) and chilling (Surkiewicz et al., 1969; Lillard, 1980; Campbell et al., 1983; Dougherty, 1974). The studies that look at the addition of chemicals show, in general, a reduction in prevalence (Table 6.10). In addition, variation during a day of processing is investigated for scalding (Abu-Ruwaida et al., 1994), plucking (Rigby et al., 1980a) and chilling (Rigby et al., 1980a). Also in relation to time, variation from day to day and from year to year is investigated for scalding (Abu-Ruwaida et al., 1994), evisceration (Baumgartner et al., 1992) and chilling (Rusul et al., 1996). Finally, plant-to-plant variation is considered for plucking (Chambers et al., 1998) and chilling (Lillard et al., 1990). Few of the studies on individual processing steps consider the number of organisms per bird. In fact, the only results relate to chilling (Surkiewicz et al., 1969; Dougherty, 1974; Waldroup et al., 1992). Although data for prevalence and numbers of organisms are available for individual processing steps, using these data to estimate levels of change requires additional assumptions because the data have been generated from different studies and thus there is no baseline value from which to commence estimation (Table 6.11).

Data relating to changes in prevalence and numbers of organisms are given in Table 6.9 and Table 6.11. Most of this data focuses on changes in prevalence; only one considers changes in numbers (Campbell et al., 1983). Of these studies, Abu-Ruwaida et al. (1994) and Lillard (1990) consider changes throughout the significant stages of processing. Abu-Ruwaida et al. (1994) also consider day-to-day variation, but their results give 100% prevalence at all points and thus would not be suitable for modelling change. The remaining studies in Table 6.11 commence later in processing and thus the problem of no baseline information from which to start, again arises. For example, the investigations by James et al. (1992a, b) commence after defeathering and so the prevalence level at the point of entry into the processing plant is unknown. These studies could, however, be used to look at change from one point to the next.

General conclusions on changes could be made from this data, but much of it is old and thus would require careful consideration within an exposure assessment. In particular, the effect of changes in practices and regulations would have to be investigated. Finally, Table 6.12 presents data on prevalence of *Salmonella* on finished products, at the end of processing. It is evident that it is difficult to combine these data for a risk assessment, as the different studies have used different sample types and analytical methods. Very few studies have quantified the numbers of *Salmonella*, and these are shown in Table 6.13 for whole carcass.

Table 6.5 Data collected at stun and kill processing stage.

Sample	No. tested	% *Salmonella*-Positive	Enumeration (average of positive samples	Reference (Country)
Throat-cutting knife	20	50		Carraminana et al., 1997 (Spain)
Feathers				Kotula and Pandya, 1995 (USA)
Breast	40	75	7.2 ±0.2 log CFU/g	
Thigh	40	53	6.5 ±0.2 log CFU/g	
Drum	40	55	6.5 ± 0.2 log CFU/g	
Skin				Kotula and Pandya, 1995 (USA)
Breast	40	45	6.3 ±0.2 log CFU/g	
Thigh	40	30	5.9 ±0.2 log CFU/g	
Drum	40	27	5.8 ±0.2 log CFU/g	
Foot	40	55	5.8 ±0.2 log CFU/g	

Table 6.6 Data collected at scalding processing stage.

Sample	Number tested	% *Salmonella-* positive	Enumeration (average of positive samples)	Reference (Country)
Tank Water	15	100	13.9 ±13.4 MPN/100 ml	Humphrey and Lanning, 1987 (UK)
Tank Water + NaOH	15	27	3.0 ±2.3 MPN/100 ml	
Tank Water - Entry	4	NS[1]	2.9 log CFU/ml	Abu-Ruwaida et al., 1994 (Kuwait)
Tank Water – Middle	4	NS	2.3 log CFU/ml	
	4	NS	2.1 log CFU/ml	
Tank Water – Exit	4	NS	2.3 log CFU/ml	
	4	NS	2.3 log CFU/ml	
Tank Water	20	75		Carraminana et al., 1997 (Spain)
Carcass, 52°C scald	NS		3.0 log MPN per carcass	Slavik, Jeong-Weon and Walker, 1995.
	NS		3.17 MPN per carcass	
	NS		3.09 MPN per carcass	
Carcass, 56°C	NS		3.16 MPN per carcass	
	NS		3.17 MPN per carcass	
	NS		3.34 MPN per carcass	
Carcass, 60°C	NS		3.50 MPN per carcass	
	NS		3.48 MPN per carcass	
	NS		3.36 MPN per carcass	

Note: (1) NS = not stated

Table 6.7 Data collected at de-feathering processing stage.

Sample	Number tested	% *Salmonella-* positive	Reference (Country)
De-featherer swabs			Rigby *et al.*, 1980a (Canada)
Before start-up	3	33.3	
Coffee break	3	100.0	
End of shift	3	66.7	
Crop swabs	273	2.2	Chambers *et al.*, 1998 (Canada)
(post-de-feathering)	362	5.8	
De-feathered carcass rinse	6	83.3	Fuzihara, Fernades and Franco, 2000 (Brazil)

Table 6.8 Data collected at evisceration processing stage.

Sample	Number tested	% *Salmonella*-positive	Reference (Country)
Carcass swabs			Morris and Wells, 1970
Pre-Evisceration	203	23.6	(USA)
Post- Evisceration	212	17.9	
Neck skin, 10-g sample[1]			Goren et al., 1988
Carcasses	3 099	11.7	(Netherlands)
Flocks (25 birds each)	124	62.9	
Neck skin, 50-g sample[1]			Baumgartner et al., 1992
Carcasses	485	19.2	(Switzerland)
Flocks (5 birds each)	97	47.4	

NOTES: (1) Sampled post-evisceration.

Table 6.9 Data collected at chilling.

Sample	Number tested	% *Salmonella*-positive[1]	Enumeration (average of +ve samples) if available	Reference (Country)
Carcass rinse				Lillard, 1990
Pre-chill A[2]	40	13		(USA)
Post-chill A	40	28		
Pre-chill B[3]	40	10		
Post-chill B	40	38		
Pre-chill	48	100		Izat et al., 1989
Post chill	103	58		(USA)
Carcass rinse				Campbell et al., 1983
Entry final wash	108	22	1–30 MPN – 17 samples	(USA)
			30–300 MPN – 4 samples	
			>300 MPN – 3 samples	
Entry chill tank	108	6	1–30 MPN – 5 samples	
			30–300 MPN – 0 samples	
			>300 MPN – 1 samples	
Exit chill tank	215	12	1–30 MPN – 24 samples	
			30–300 MPN – 1 samples	
			> 300 MPN – 0 samples	
Chill water 1st tank	71	20	< 1.1 MPN/ml – 14 samples)	Campbell et al., 1983
Final tank	71	3	>1 MPN/ml – 2 samples	(USA)

NOTES: (1) Percentages rounded. (2) Inside/outside bird washer used in facility. (3) Outside bird washer only

Table 6.10 Data collected at chilling processing stage: effects of chlorine addition (Lillard, 1980).

Concentration of ClO_2 (ppm)	Time of day	Number tested	% *Salmonella*-positive[1]	MPN/ml	Number tested	% Positive[1]	MPN/g
0	a.m.	30	43	<0.4–15.8	28	21	< 0.4–48
	p.m.	30	40		28	7	
3	a.m.	24	29	<0.4	24	4	< 0.4
	p.m.	24	21		24	0	
5	a.m.	24	0	0	48	0	< 0.4
	p.m.	24	0		48	2	
20	a.m.	26	15	<0.4	26	4	< 0.4
	p.m.	26	19		26	0	
34	a.m.	22	0	0	22	9	< 0.4
	p.m.	22	0		22	0	

NOTES: (1) Percentages have been rounded.

Table 6.11 Summary of data collected for changes during processing.

Sample and site	No. positive out of no. tested (%)[2]	Reference (Country)
Cloacal and pericloacal swabs, 5 pooled		Carraminana et al., 1997 (Spain)
Post-picking	11/20 (55%)	
Post-vent cutting	9/20 (45%)	
Post-evisceration	12/20 (60%)	
Post-spray washing	7/10 (70%)	
Post-air chilling	12/20 (60%)	
Overall change		
Neck skin		Abu-Ruwaida et al., 1994 (Kuwait)
Bleed (pre-scald)	11/11 (100%)[1]	
De-feathering	11/11 (100%)	
Carcass rinse		Dougherty, 1974 (USA)
Pre-evisceration	39/60 (65%)	
Final product	28/60 (47%)	
Carcass rinse		Fuzihara, Fernades and Franco, 2000 (Brazil)
Post-de-feathering	5/6 (83%)	
Post-evisceration	4/6 (66%)	
Post-immersion 1	5/6 (83%)	
Post-immersion 2	5/6 (83%)	
Carcass rinse		Lillard, 1990 (USA)
Pre-scald	16/84 (19%)	
Post-scald	10/84 (12%)	
Post-pick	10/84 (12%)	
Post-evisceration	12/84 (14%)	
Pre-chill (after wash)	12/84 (14%)	
Post-chill	31/84 (37%)	
Carcass rinse		James et al., 1992A (USA)
Pre-evisceration:	93/160 (58%)	
Pre-chill	77/160 (48%)	
Post-chill	114/158 (72%)	
Post-cut	119/154 (77%)	

Sample and site	No. positive out of no. tested (%)[2]	Reference (Country)
Carcass Rinse	33/99 (33%)	James et al., 1992b (USA)
Pre-evisceration	21/50 (43%)	
Pre-chill	23/50 (46%)	
Post-chill		
Carcass Rinse		James et al., 1992c (USA)
Pre-evisceration:	24/99 (24%)	
Pre-chill	28/99 (28%)	
Post-chill	24/49 (49%)	
Carcass Rinse		Jones et al., 1991b
After chilling	6/57 (11%)	
At packaging	3/14 (21%)	
Swab – post-scalding	Day 1: 0%	Patrick, Collins and Goodwin, 1973 (USA)
	Day 2: 0%	
	Day 3: 0%	
	Day 4: 4%	
	Day 5: 16%	
Swab – after de-feathering	Day 1: 12.5%	Patrick, Collins and Goodwin, 1973 (USA)
	Day 2: 0%	
	Day 3: 0%	
	Day 4: 4%	
	Day 5: 16%	
Swab – after chilling	Day 1: 19%	Patrick, Collins and Goodwin, 1973 (USA)
	Day 2: 4%	
	Day 3: 8%	
	Day 4: 4%	
	Day 5: 32%	
Carcass rinse/caeca cutting		McBride et al., 1980
Before scalding	129/330 (39%)	
After inspection	59/330 (18%)	
After chilling	73/330 (22%)	
Not stated		Rigby et al., 1980b
Unloading	311/331 (94%)	
After chilling	11/25 (44%)	

Table 6.12 Prevalence of *Salmonella* on finished carcasses and portions.

Country & year of sampling if known	Sample	Number sampled	Percentage positive	Data source
Argentina	Carcass surface swab	96	31.3	Terisotto et al., 1990
Argentina	Carcass rinse	86	2.3	Argentina – Call for
1994–98	Carcass rinse	39	15.4	data by FAO/WHO
Austria	NS[1]	1342	3.7	EC, 1998
Austria	NS	124	2.4	EC, 1998
Austria – 1998	Skin samples	1207	22.2	EC, 1998
Austria – 1997		80	62.5	
Austria – 1996		3485	20.9	
Belgium	NS	127	28.4	EC, 1998
Brazil	25 g of meat+skin	60[2]	42.0	Fuzihara, Fernandes and Franco, 2000

Country & year of sampling if known	Sample	Number sampled	Percentage positive	Data source
Canada – 1985–86	Carcass rinse	205 (46)[3]	80.5 (89.2)[4]	Lammerding et al.,
1984–85		180 (47)[3]	80.6 (76.6)[4]	1988
1983–84		140 (41)[3]	70.0 (68.3)[4]	
Denmark	Neck skin	4985	11.1	EC, 1998
Finland – 1998	NS	384	0.52	EC, 1998
1997		611	3.1	
Ireland – 1998	NS	2 695	16.6	EC, 1998
1997		2 218	22.6	
1996		1 632	22.2	
Malaysia	Carcass rinse – Plant A [5]	12	91.7	Rusul et al., 1996
		12	75	
		20	75	
Malaysia	Carcass rinse – Plant B [5]	20	30	Rusul et al., 1996
		20	0	
		20	55	
Netherlands – 1997	Neck skin	NS	53.4	EC, 1998
1998		NS	41-50	
Netherlands	10 g fillet[6]	10	0	EC, 1998
		10	1	
		10	90	
		10	80	
		10	10	
		10	80	
		10	60	
Norway – 1998	Neck skin	7 112	0.0	ARZN, 1998
1997		7 591	0.0	
Portugal	Swabs of surface and abdominal cavities	300	57	Machado and Bernardo, 1990
Sweden – 1998	NS	1 138	0.0	EC, 1998
1997		723	0.0	
1996		581	0.0	
Sweden	Neck skin	4 010	0.02	EC 1998
Thailand	Chicken meat [7]	353	181 (51%)	Jerngklinchan et al., 1994
USA		NS	3-4%	Lillard, 1989a
USA	Cloacal swabs, giblets, whole carcasses and parts	247	4.0%	Harris et al., 1986
USDA-FSIS	Carcass rinse	1 297	20% [8] 11.6% (MPN)	USDA-FSIS, 1996
USA	Carcass rinse[6]	14	21.4	
Venezuela		45	49	Rengel and Mendoza, 1984

NOTES: (1) NS = Not stated. (2) Sampled from 60 individual small poultry slaughterhouses (<200 birds per day). (3) Number of lots sampled, with 5 carcasses per lot. (4) Percentage of lots positive; one or more positive carcasses. (5) Samples not specified – some pre-chill, others post-chill;. (6) Sampled prior to packaging. (7) 25 g sample of raw chicken muscle. (8) Recovered using enrichment media.

Table 6.13 Numbers of *Salmonella* on finished carcasses.

Number of samples	%	MPN per carcass[1]	Source
136	79.5	< 1	Surkiewicz et al., 1969
28	16.4	1- 30	
1	0.6	30 -300	
6	3.5	> 300	
112	25.9	0.108 ±0.279	Waldroup *et al* , 1992
112	32.1	0.172 ±0.363	
112	77.3	0.736 ±0.672	
112	38.2	0.188 ±0.259	
112	30.4	0.085 ±0.226	
109	41.9	< 12	USDA-FSIS, 1996
118	45.4	12 - 120	
24	9.2	121 - 1200	
6	2.3	1201 - 12000	
3	1.2	>12000	
99	60.7	< 12	CFIA, 2000
60	36.8	12 - 120	
2	1.3	121 - 1200	
1	0.6	1201 - 12000	
1	0.6	>12000	

Notes: (1) MPN per carcass calculated from reported values (MPN per millilitre rinse fluid) × 400 ml total rinse fluid for USDA-FSIS and CFIA results.

Data gaps

The main data gaps for processing are:

- There is limited public information on the *processing practices* followed by different countries of the world (for example, scalding or chilling methods, including addition of chemicals).

- *Quantitative data* (i.e. numbers of organisms) are limited for several processing steps.

- Many studies are old, so *more recent information* on changes in prevalence and numbers would be beneficial.

6.2.5 Retail, distribution and storage

The aim of the retail, distribution and storage module is to estimate the change in numbers of *Salmonella* on broilers after processing and before preparation and consumption by the consumer.

Retail, distribution and storage steps

When considering distribution and storage of broilers, it is assumed that the broilers are already dressed, chilled or frozen, and ready for supply. Storage can mean storage at the processing plant prior to distribution, storage at the retail outlet or central distribution centre, and storage in the home.

The distribution and storage of processed broilers can influence the bacterial load on the meat. If broiler chickens are not packaged individually, cross-contamination can occur, increasing the prevalence of salmonellae within a batch. These bacteria can also multiply as a function of the temperature, the nutrient conditions, moisture content and pH of their environment. Hence there are several variables that can influence the contamination of an individual broiler by the time it is cooked in the home, including:

- The prevalence and numbers of salmonellae on finished broiler chickens.
- The conditions of storage, including:
  - storage temperature;
  - relative humidity and broiler moisture;
  - muscle pH;
  - whether pre-packed or unpacked; and
  - storage density.
- The conditions of distribution, especially
  - external temperature during:
    loading,
    transport, and
    delivery.

Data requirements and models available

There are several variables that may influence the prevalence and level of salmonellae on broiler chickens during retail, distribution and storage. For a general risk assessment framework, it is important to recognize the potential consequences of these variables in the production-to-consumer food chain. Factors such as likely temperature abuse conditions at any one stage can be utilized to model potential growth. For this, it is necessary to use predictive models that estimate the likely outcome of changes in the environmental conditions that the *Salmonella* experience. Data requirements for this purpose can be split into two main areas: choice of suitable predictive models, and the measurement of environmental changes during the retail, distribution and storage chain. In addition, studies that provide data on prevalence or numbers of organisms at retail are important in validating predictive modelling of the food chain.

Microbiological models can differ in mathematical complexity, but a complex model may not necessarily be the best choice to answer a particular risk management question (van Gerwen, 2000). The need for an accurate prediction needs to be offset by a consideration of whether the model is easy to use, whether it is robust and precise, and whether it has been validated against independent data. For example, if the objective of a risk assessment exercise is to demonstrate the most significant risk factors in a process, a simple model may have advantages over a complex model. However, if an accurate prediction of bacterial numbers is necessary, a more complex and accurate model may be preferable. In the choice of a suitable model, one must also consider the quality of the data that is going to be used to generate a prediction. If the temperature data on a process are poor, it may not be appropriate to use a complex model for the predictions. Often this can lead to a misinterpretation of the accuracy of the final prediction. The most appropriate model would be the simplest model possible for a given purpose and the given data quality, providing that it is validated and precise. A good model should also be subjected to an analysis method that quantifies the accuracy and bias of its predictions (Buchanan and Cygnarowicz, 1990). Ideally, a model should be both accurate and unbiased.

Models used in risk assessment must adequately reflect reality (Ross, Baranyi and McMeekin, 1999; Ross, Dalgaard and Tienungoon, 2000). Before predictive models are used in exposure assessment, their appropriateness to that exposure assessment and overall reliability should be assessed.

It is always possible to create a model that perfectly describes the data, simply by having a sufficiently complex model (Zwietering et al., 1991), but such models lack generality and would be *less* useful for predicting responses in other situations.

Two complementary measures of model performance can be used to assess the 'validity' of models (Ross, Baranyi and McMeekin, 1999; Ross, Dalgaard and Tienungoon, 2000). These measures have the advantage of being readily interpretable. The 'bias factor' (B_f) is a multiplicative factor by which the model, on average, over- or under-predicts the response time. Thus, a bias factor of 1.1 indicates that the prediction response exceeds the observed, on average, by 10%. Conversely, a bias factor less than unity indicates that a growth time model would, in general, over-predict risk, but a bias factor of 0.5 indicates a poor model that is overly conservative because it predicts generation times, on average, half of that actually observed. Perfect agreement between predictions and observations would lead to a bias factor of 1.

The 'accuracy factor' (A_f) is also a simple multiplicative factor indicating the *spread* of observations about the model's predictions. An accuracy factor of two, for example, indicates that the prediction, on average, differs by a factor of 2 from the observed value, i.e. either half as large or twice as large. The bias and accuracy factors can equally well be used for any time-based response, including lag time, time to an *n*-fold increase, death rate and D value. Modifications to the factors were proposed by Baranyi, Pin and Ross (1999). As discussed above, typically, the accuracy factor will increase by 0.10–0.15 for every variable in the model. Thus, an acceptable model that predicts the effect of temperature, pH and water activity on growth rate could be expected to have $A_f = 1.3$–1.5. Satisfactory B_f limits are more difficult to specify because limits of acceptability are related to the specific application of the model. Armas, Wynn and Sutherland (1996) considered that B_f values in the range 0.6–3.99 were acceptable for the growth rates of pathogens and spoilage organisms when compared with independently published data. te Giffel and Zwietering (1999) assessed the performance of many models for *Listeria monocytogenes* against seven datasets and found bias factors of 2–4, which they considered to be acceptable, allowing predictions of the order of magnitude of changes to be made. Other workers have adopted higher standards. Dalgaard (2000) suggested that B_f values for successful validations of seafood *spoilage* models should be in the range 0.8 to 1.3. Ross (1999) considered that, for pathogens, less tolerance should be allowed for $B_f >1$ because that corresponds to under-predictions of the extent of growth and could lead to unsafe predictions. That author recommended that for models describing *pathogen* growth rate, B_f in the range 0.9 to 1.05 could be considered good; be considered acceptable in the range 0.7 to 0.9 or 1.06 to 1.15; and be considered unacceptable if $<\sim 0.7$ or >1.15.

In another approach to assessing model performance, the group of researchers involved in the development of the predictive modelling program Food MicroModel™ proposed that validation could be split into two components: first, the model's mathematical performance (error$_1$), and second, its ability to reflect reality in foodstuffs (error$_2$) (Anon., 1998). They found that the error of a single microbiological concentration record was about 0.1–0.3 log$_{10}$ CFU/ml. Therefore, this could be considered the standard error obtained by fitting the model. If, during comparison of the predicted data with the measured data used to generate the primary model, the standard error was greater than 0.3–0.4 log$_{10}$ CFU/ml, then the authors suggested that the curve should only be used with caution for any secondary modelling stage. They went on to suggest that when a quadratic response surface was fitted to predicted kinetic parameters from the primary model to create the secondary model, the statistical tests should include a measure of goodness of fit. They suggested that the aim of a good model would be to achieve a standard error of no greater than 15-20%. Other suggested statistical tests were measures of parsimony (e.g. t-test), errors of prediction (e.g. least squares) and measures of robustness (e.g. bootstrap methods). The ability of a model to reflect reality in foodstuffs (error$_2$) is often assessed by conducting a review of the literature for measured data describing the kinetic parameter for prediction by the model. These data must not be the data used to generate the model. Ross (1999) suggested that validation data could be subdivided into sets that reflected the level of experimental control. Hence, data generated in a highly controlled broth system would be separated from data generated in a less controlled foodstuff. In this way, he argued that the performance of the model would not be undermined by evaluation against poor quality data or unrepresentative data. For examples of the limitations and difficulties of using validation data from the literature, see McClure et al., (1997); Sutherland and Bayliss, (1994); Sutherland, Bayliss and Roberts (1994); Sutherland, Bayliss and Braxton (1995); and Walls et al. (1996). The multiplicative factors of bias and accuracy discussed previously could be equally applied to quantification of both error$_1$ and error$_2$.

The selection of a model for a microbiological phenomenon must go further than the mathematics. It is all too easy to forget that a model is only as good as the data on which it is based. Bacteria are biological cells and as such the methodology used to enumerate their numbers greatly affects the count obtained. For this reason the predictive model should be based on replicate data using recognized enumeration methods. The use of resuscitation procedures for enumeration is particularly important when the organism has been growing near its physiological limits. Here, bacteria are often in a state of environmental stress and recovery is necessary to prevent the artificial depression of bacterial numbers. The method used to generate the data must be free from experimental artefacts that might artificially increase or decrease the bacterial count.

Growth

Bacteria multiply by a simple process of cell division, known as binary fission. A single bacterial cell reaches a stage in its growth when it undergoes a process that results in the single cell dividing into two daughter cells. The growth of bacterial populations therefore follows a predictable cycle that involves a period of assimilation – called the lag phase; a period of exponential growth – called the exponential phase; and a period of growth deceleration and stasis – called the stationary phase. Growth curves are often described kinetically by three variables: initial cell number (N_0), lag time (λ) and specific growth rate (μ), which can also be used to determine the generation time or doubling time of the population. Note that this simple description does not take the stationary phase into account. Prediction of the stationary phase is not always necessary for risk assessment, although a maximum population density parameter is often useful as an endpoint for the prediction of the exponential phase of growth. The values of these variables change with environmental conditions, including temperature, pH, water activity (a_w), nutrient state and the presence and concentration of preservatives. Studies of the growth of bacteria can generate different types of data. Kinetic data, involving the enumeration of bacteria during the growth cycle, describe the shape of the population growth curve in response to a specific set of growth conditions. Probabilistic data, involving measurement of simple growth or no-growth characteristics of the bacterial population, describe whether or not the bacteria will grow under certain growth conditions.

Growth Models

Microbiologists recognize that not all equations that are applied to bacterial processes can be considered models. A kinetic model should have a sound physiological basis (Baranyi and Roberts, 1995). This distinction has not always been made in the literature, and the word "model" has been invariably used to describe empirically-based curve fitting exercises.

Growth models increase in complexity from primary models that describe a population response, e.g. growth rate and lag time, to secondary models that describe the effect of environmental factors on the primary response, e.g. temperature and pH.

For the growth process of bacteria, an example of a simple primary model is shown in Equation 6.1.

$$N = N_0 \cdot \exp\left(\mu(t-\lambda)\right)$$
Equation 6.1

Where N = number of bacteria; N_0 = initial number of bacteria; μ = specific growth rate; and λ = lag time.

This type of model could be applied to growth data to determine the primary kinetic parameters for specific growth rate and lag time for the given set of environmental growth conditions under which the data was generated.

There are several primary models that have been used routinely to describe the growth of bacteria. Examples are the Gompertz equation (Gibson, Bratchell and Roberts, 1988; Garthright, 1991), which is an empirical sigmoidal function; the Baranyi model (Baranyi and Roberts, 1994), which is a differential equation; and the three-phase linear model (Buchanan, Whiting and Damart, 1997), which is a simplification of the growth curve into three linear components.

Secondary growth models based on primary models have been created by replacing the term for specific growth rate and the term for lag time with a function that described the change of these response variables with respect to environmental factors such as temperature, water activity and pH. Examples are the non-linear Arrhenius model – where the square root model relates the square root of the growth rate to growth temperature (Ratkowsky et al., 1982) – and the response surface model. In the case of the simple model example in

Equation 6.1, an example secondary model can be used to describe the growth of a bacterial population when temperature changes (Equation 6.2).

$$N = N_0 \cdot \exp(f_{TEMP} \mu(t - f_{TEMP} \lambda))$$
Equation 6.2

Where N = number of bacteria; N_0 := initial number of bacteria; μ = specific growth rate; λ = lag time; and f_{TEMP} = mathematical function for the effect of temperature, such as a quadratic equation.

This type of model could be applied to growth data at different temperatures and would allow the calculation of the number of bacteria after a given growth period when temperature changes during that growth period. Secondary models developed from primary models are more useful than primary models alone for the quantification of risk, providing that the environmental factors influencing growth can be measured dynamically.

Growth Models for Salmonella *in Chicken Meat*

An ideal growth model for *Salmonella* should take into account the general issues raised previously about model selection, but, in addition, it should be tailored for the product under study. The ideal growth model would aim to encompass the variable limits for temperature, pH and a$_w$ shown in Table 6.14, for which *Salmonella* are estimated to grow.

In the case of *Salmonella* in broilers, the model either should have been developed using data describing *Salmonella* growth in chicken meat, or at least be validated against real product data.

Table 6.14. Limits for growth of *Salmonella* (ICMSF, 1996)

Conditions	Minimum	Optimum	Maximum
Temperature (°C)	5.2	35–43	46.2
pH	3.8	7–7.5	9.5
Water activity (a$_w$)	0.94	0.99	>0.99

Table 6.15. Growth models for *Salmonella*

Salmonella serotype	Growth medium	Temp. range (°C)	pH range	Other conditions	Primary model	Secondary model	Reference
Typhimurium	Milk	10–30	4–7	a$_w$ 0.9–0.98. Glucose as humectant	Non-linear Arrhenius	Quadratic response	Broughall and Brown, 1984
Typhimurium	Laboratory media	19–37	5–7	Salt conc. 0–5%		Quadratic response	Thayer et al., 1987
Mixed Stanley, Infantis and Thompson)	Laboratory media	10–30	5.6–6.8	Salt conc. 0.5–4.5%	Gompertz	Quadratic response	Gibson, Bratchell and Roberts, 1988
Typhimurium	Laboratory media	15–40	5.2–7.4	Previous growth pH 5.7–8.6	2 phase linear	Quadratic response	Oscar, 1999a
Typhimurium	Cooked ground chicken breast	16–34		Previous growth temp. 16–34°C	2 phase linear	Quadratic response	Oscar, 1999b
Typhimurium	Cooked ground chicken breast	10–40		Previous growth salt 0.5–4.5%	2 phase linear	Quadratic response	Oscar, 1999c

Published growth models for *Salmonella* predict growth as a function of temperature, pH, water activity (a_w) and previous growth conditions. Table 6.15 summarizes the basis of several models.

The models of Broughall and Brown (1984) and Thayer et al. (1987) do not appear to have been validated by the authors. Validation is included for the other four models. Gibson, Bratchell and Roberts (1988) validated their model against growth data generated using pork slurry and data published in the literature. The model predictions were in good agreement with the observed data. The greatest variance was found at the extremes of the model, with low temperature or high salt concentration. This model has the advantage of being based on a considerable quantity of experimental observations and covers a wide selection of environmental growth conditions. However, the authors did not validate the work against observed data in chicken meat. The work reported by Oscar (1999a, b and c) concluded that previous growth temperature, pH and salt concentration had little effect on the estimates of specific growth rate and lag time for *Salmonella* Typhimurium. The author also demonstrated that it was possible to develop models in a food matrix including chicken meat, and hence these are useful for the purposes of this exposure assessment.

Survival

Under stress conditions, bacteria will either remain in a state of extended lag or may die slowly. Studies on the survival of *Salmonella* under stress conditions are limited. The number of *S.* Enteritidis was shown to remain constant during the storage of chicken breast at 3°C under a range of modified atmospheres over a 12-day study period (Nychas and Tassou, 1996). However, growth of enterobacteriaceae, including *Salmonella*, on naturally-contaminated chicken meat occurred at 2°C after 3 days in 30% CO_2, and after 5 days in 70% CO_2, with numbers increasing by 3 log cycles after 15 and 23 days, respectively (Sawaya et al, 1995). These investigators noted that *Salmonella* composed about 12% of the total enterobacteriaceae microflora, and the proportion remained constant throughout storage. It is possible that *Salmonella* growth is enhanced by the presence of competitive microflora. Hall and Slade (1981) carried out an extensive study of the effect of frozen storage on *Salmonella* in meat. In chicken substrate, the numbers of *S.* Typhimurium declined by 99.99% (4 log cycles) at –15°C over 168 days, and by 99.4% (2–3 log cycles) at -25°C over 336 days. Survival data for *Salmonella* have been summarized by ICMSF (1996).

Model selection for exposure assessment model

When considering broiler meat as a media for growth and survival of *Salmonella*, several factors can be simplified. At the surface of the meat, water activity might vary as a function of air moisture, chilling conditions and packaging method, but generally falls between a_w 0.98 and 0.99. The pH varies among muscle types, but is between pH 5.7 and 5.9 for breast meat and pH 6.4–6.7 for leg meat. The skin averages pH 6.6 for 25-week-old chickens (ICMSF, 1996). Poultry meat is also a rich source of nutrients such as protein, carbohydrate and fat, with essential minerals and vitamins. Consequently, it can be assumed that the growth of *Salmonella* will not be limited by the lack of available nutrients and hence the growth rate will be optimal for a given temperature within the pH and a_w limits of the poultry meat.

For the purposes of a simple exposure assessment model, the change in environmental conditions could be considered solely as a change in external temperature and chicken carcass temperature. It can be assumed that the pH of a broiler chicken will be pH 6.0 and that the water activity will be 0.99. Some appropriate models that could be used to predict changes in growth rate during retail, distribution and storage are:

- For temperatures between 10°C and 30°C, the growth model of Gibson, Bratchell and Roberts, 1988.
- For temperatures between 16°C and 34°C, the growth model of Oscar, 1999b.
- For temperatures between 4°C and 9°C, the survival model of Whiting, 1993.
- There are no appropriate models for temperatures below 4°C.

For the purposes of the current exposure assessment, the model developed by Oscar (1999b) was selected. The model was developed in chicken meat slurry and therefore took account of the interactions between the bacteria and the food matrix. In addition, the model was simple and easily applied. The author also assessed the accuracy and bias of the model by measuring the relative error of predictions against:

(i)   the data used to generate the model; and

(ii)   new data measured using the same strain and experimental conditions, but at intermediate temperatures not used in the data set used to develop the original model.

Median relative errors for lag time were given as 0.9% and -3% for comparisons (i) and (ii), respectively, and the median relative errors for growth rate were given as 0.3% and 6.8% for comparison (i) and (ii), respectively. The predictions for either parameter were unbiased. The accuracy of the model was deemed to be within accepted guidelines, as discussed above.

Temperature data characterizing retail, distribution and storage

Providing that suitable secondary kinetic models are available, it is necessary to examine the change in the environmental conditions with time during the retail, distribution and storage chain. The most common studies involve the use of temperature probes to measure the changes in product temperature during a process. For broiler chickens, the measurement of external surface and deep muscle temperatures may be used to characterize the growth or survival of *Salmonella* at these locations. Sampling can be used to measure pH and water activity changes with time, but these types of study are rarely conducted. Alternatively, thermodynamic models can be used to predict the temperature of a product given the external temperature and time. To ensure the predictions are consistent with measured data, caution must be exercised when using this approach.

Temperatures in the retail, distribution and storage chain tend to become less controlled from processor to consumer. Temperature and time studies of storage at the processing plant, distribution to the retailer and storage at the retailer often remain the unpublished property of the broiler industry or retailers. Few studies, if published, carry detailed data. Temperature and time studies of transport and storage by the consumer tend to be carried out by food safety organizations and are also largely unpublished. This presents problems for risk assessment unless access to these data can be arranged. Even with access to data in commercial organizations, it is often unlikely that data will be released that characterizes poor practice.

Data requirements and the data available

Growth modelling

Calibrated equipment should always be used for measuring time and temperature profiles of processes. Studies can be of a single step, such as storage at the retail stage, or be of multiple steps. In both cases, it is important to measure the environmental temperature, the external product temperature and the internal product temperature. Profiles should be measured in more than one product and, in the case of multi-step measurements, careful notes on the start and end times of the individual steps must be kept. It is important, where possible, to follow the same product throughout a multi-step process so that measurements from one step to the other can be related. Wherever possible, data should be analysed statistically to determine the within-step and step-to-step variability. If continuous measurement is not possible using a temperature data logger, then as many real-time measurements as possible should be made using a temperature probe.

Few thermal profile data for retail, storage and distribution were provided by FAO/WHO member countries as a result of the call for data. No actual data were found in the literature, although profiles were shown in graphic form in some studies. As an example, time and temperature data were kindly provided on whole broilers by Christina Farnan (Carton Group, Cavan, Republic of Ireland). These data are summarized in Tables 6.16 and 6.17.

When carrying out a quantitative exposure assessment, it is important to access national data. Data should be requested from national broiler processors and retailers.

Table 6.16. Summary of chilled chain data from Carton Group.

Location of product (probed chicken in box of 5 carcasses)	Trial 1: 1000-g broilers			Trial 2: 2300-g broiler		
	Time (minutes)	Average temperature (°C)		Time (minutes)	Average temperature (°C)	
		surface	muscle		surface	muscle
Primary chill	0	–	36	0	–	41
Packing hall	43	–	7.0	80	–	10.2
Boxed	55	–	7.0	85	–	10.2
Blast chill	57	–	7.0	100	–	10.2
Storage chill	75	1.1	2.0	155	5.0	6.2
Dispatch lorry	717	1.1	1.1	230	4.0	4.0
Depart plant	755	1.1	1.1	315	3.0	2.4
Arrival at retailer	945	1.7	1.1	500	3.0	0.7
Storage at retailer (back chill)	968	2.3	1.1	505	3.0	0.7
Storage at retailer	>48 hours	Max. 3.7	Max. 3.3	N/A	N/A	N/A

SOURCE: Data supplied by Christina Farnan, Carton Group, Cavan, Republic of Ireland.

Table 6.17. Summary of frozen chain data from Carton Group.

Location of product (probed chicken in box of 5 carcasses)	Time (minutes)	Trial 2: 2300-g broiler	
		Average temperature (°C)	
		surface	muscle
Boxed	0	19.5	2.8
Into blast freezer	1	19.5	2.8
Out of blast freezer	3925	-34.7	-32.8
Into cold store	3930	-33.9	-32.8
Depart plant	4140	-32.1	-32.3
Arrive central distribution	4180	-32.0	-31.6

SOURCE: Data supplied by Christina Farnan, Carton Group, Cavan, Republic of Ireland.

Transport and storage temperatures during consumer handling of products can vary greatly. In the United States of America, a study was carried out in 1999 to quantify this process (Audits International, 1999). This work is a good template for carrying out similar research in other nations. Data were generated on retail backroom storage temperature, display case temperature, transit temperature, ambient temperature in the home, home temperature and home temperature after 24 hours. Tables 6.18 and 6.19 summarize the data. These example data were not generated in chicken but may be used as a guide.

These data can be useful to estimate growth or survival, or both, in a deterministic assessment, or as a basis for probability distributions for time and temperature in stochastic modelling.

Table 6.18. Summary of consumer transport and storage study on chilled products including meat

Location	Average time (minutes)	Average temperature (°C)	Maximum time (minutes)	Maximum temperature (°C)
Retail backroom cold store air	N/A	3.3	N/A	15.5
Product in retail backroom cold store	N/A	3.3	N/A	16.6
Product in retail display refrigerator	N/A	4.0	N/A	14.4
Product from retail to home	65	10.3	>120	(max. 36.6 at home)
Product in home refrigerator (after 24 h)	N/A	4.0	N/A	21.1
Home ambient temp	N/A	~27.0	N/A	>40.5

NOTES: N/A = Not available. SOURCE: Audits International, 1999.

Table 6.19: Summary of consumer transport and storage study on frozen dairy products

Location	Average time (minutes)	Average temperature (°C)	Maximum time (minutes)	Maximum temperature (°C)
Product in retail display freezer	N/A	-12.9	N/A	6.6
Product from retail to home	51	-8.4	>120	20
Product in home refrigerator (after 24 h)	N/A	-15.9	N/A	8.9
Home ambient temp	N/A	~27.0	N/A	>40.5

NOTES: N/A =Not available. SOURCE: Audits International, 1999.

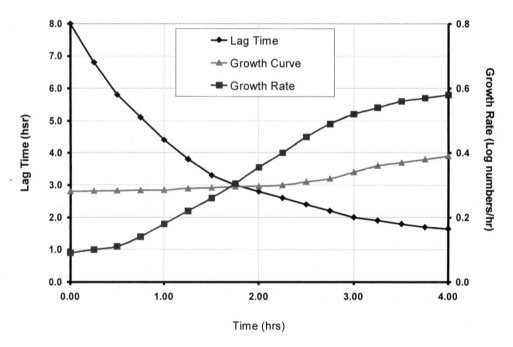

Figure 6.4. Relationship of lag time and growth rate with increasing temperature as a function of time.

To illustrate a deterministic approach, the data in Table 6.19 can be used to demonstrate the predicted effect on the growth of *Salmonella* in a product during transport from the retail store to the consumer's home. For this example, let the number of salmonellae on the product be 1000 CFU at the start and assume that the temperature increases linearly over the transport period. It is also assumed that the growth of the organism starts at the beginning of the transport period rather than in the store. The Oscar growth model (1999b) can be used to calculate the predicted growth pattern. The model calculates the lag time and specific growth rate for salmonellae as a function of time and temperature. The organism cannot grow until the elapsed time exceeds the lag period. As temperature increases, the lag period decreases and the specific growth rate increases. This is shown in Figure 6.4. Until the elapsed time is equal to the lag period the numbers of bacteria are fixed at the starting number (in this case 1000 CFU). Figure 6.4 shows that after 2.5 hours the lag period has been exceeded and the organism is allowed to grow at a rate set by the specific growth rate.

To calculate the relationship shown in Figure 6.4, the steps followed were:

*   The thermal profile was divided into equal time and temperature blocks of 15 minutes.
*   For each block, the model was used to calculate the lag time and specific growth rate.
*   The growth curve was fixed at the starting cell number until the elapsed time was greater than the lag period (2.5 hours).
*   After completion of the lag period, the growth at each time and temperature block was calculated by dividing the specific growth rate by the growth period.
*   The increases in bacterial numbers predicted at each time and temperature block were summed to give the final increase in numbers after completion of the thermal profile.

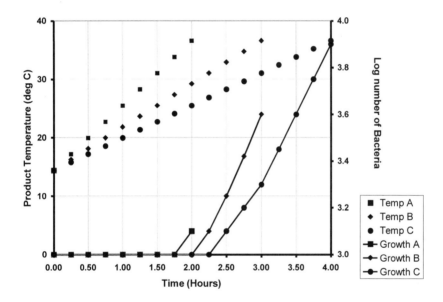

Figure 6.5. The predicted effect on the growth of *Salmonella* of temperature increase during consumer transport of product to home.

Data in Table 6.18 suggest that in a worst case scenario, a product at 14.4°C in the store could reach 36.6°C during transport over a period greater than 2 hours. Using the same approach, the effect of journey time on the growth of salmonellae can be demonstrated. Figure 6.5 shows the predicted consequences of a journey that results in a product at 14.4°C reaching 36.6°C over a 2-, 3- or 4-hour journey time.

The Oscar model (1999b) has a temperature range of 16°C to 34°C and calculations were only performed within this temperature range. It must be emphasized that predictive models should not be extrapolated beyond their boundaries.

Retail level prevalence and concentration data

Data on concentration and prevalence at the retail level could be useful as a starting point for an exposure assessment. Tables 6.20a, 6.20b and 6.20c summarize the data reported and collected to date. It is important to note, however, that study design details are lacking and the future collation of such details should be recommended.

Table 6.20a. Reported prevalence of *Salmonella* in poultry at retail.

Type of Product	Number sampled	Percentage positive	Reference (Country), and year of sampling, if reported
Fresh or frozen poultry (NS)[1], domestic and imported	322	7.8	Kutsar, 2000 (Estonia), FAO/WHO call for data. No year.
Imported frozen	151	7.3	Al Busaidy, 2000 (Sultanate of Oman), FAO/WHO call for data. No year.
Broiler chicken and hens	1186	17.3	BgVV, 2000 (Germany) – 1999
Supermarket, frozen	52	2.0	Wilson, Wilson and Weatherup,
Supermarket, chilled	58	5.0	1996 (Northern Ireland, UK). No
Butcher, frozen	6	0.0	year.
Butcher, chilled	24	25.0	
Giblets, skin and carcass samples			ACMSF, 1996 (UK)
Chilled	281	33.0	– 1994
Frozen	281	41.0	– 1994
Chilled	143	41.0	– 1990
Frozen	143	54.0	– 1990
Chilled	103	54.0	– 1987
Frozen	101	64.0	– 1987
Frozen	100	79.0	– 1979/80
Poultry products (NS)			EC, 1998
	1931	17.5	Austria – 1998
	286	10.6	Denmark – 1998
	404	5.7	– 1997
	462	9.5	– 1996
	114	0.88	Finland – 1998
	100	3.0	– 1996
	1207	22.2	Germany – 1998
	3062	22.2	– 1997
	3979	27.2	– 1996
	198	5.6	Greece – 1998
	69	0	– 1997
	51	47.1	Ireland – 1998
	104	14.4	Italy – 1997
	1010	20.2	Netherlands – 1998
	1314	29.2	– 1997
	1196	32.8	– 1996
	31	0	Northern Ireland (UK) – 1998
	314	12.1	– 1996

Type of Product	Number sampled	Percentage positive	Reference (Country), and year of sampling, if reported
	73	34.3	Portugal – 1998
	34	23.5	– 1997
	562	36.8	UK – 1996
Poultry breast meat			Boonmar et al., 1998 (Bangkok,
5 traditional open markets	50	80	Thailand). No year.
5 supermarkets	50	64	
Carcasses, at distribution centre for	123	24.4	Uyttendaele et al., 1998 (Belgium) 1996
large food chain	131	17.6	– 1995
[Positive if >1CFU/100 cm^2 or /25g]	114	27.2	– 1994
	81	19.7	– 1993
Chicken portions	153	49.0	– 1996
[Positive if >1CFU/100 cm^2 or /25g]	117	39.3	– 1995
	112	41.1	– 1994
	101	35.0	– 1993
Carcasses, retail markets. [Positive if	133	33.8	Uyttendaele, de Troy and Debevere,
>1 CFU/100 cm^2 or /25 g]			1999 (Belgium, France, Italy, the
Chicken products	41	82.9	Netherlands, UK). No year.
Chicken portions	225	51.1	
Carcasses, cuts, processed			
with skin	183	47.0	
without skin	182	34.6	
Carcasses, cuts, processed	279	54.0	Belgium. No year.
	434	33.6	France. No year.
	13	30.8	Italy. No year.
	2	0.0	Netherlands. No year.
	44	47.7	UK. No year.
Wet market – carcasses	445	35.5	Rusul et al., 1996 (Malaysia). No
– intestinal content	54	11.0	year.
Open Market – chicken meat	164	87.0	Jerngklinchan et al., 1994
gizzard	14	86.0	(Thailand). No year.
liver	94	91.0	
heart	8	88.0	
Supermarket – chicken meat	188	77.0	
gizzard	31	77.0	
liver	36	28.0	
heart	38	87.0	
Chicken meat, supermarkets	41	7.3	Swaminathan, Link and Ayers, 1978 (USA). No year.
Chicken meat	283	10.6	ARZN, 1998 (Denmark). No year.
Products (drumsticks, wings, livers, fillets, etc.)	81	54	de Boer and Hahn, 1990 (the Netherlands). No year.

Type of Product	Number sampled	Percentage positive	Reference (Country), and year of sampling, if reported
Products (drumsticks, wings, livers, fillets, etc.)	822	33.3	Mulder and Schlundt, in press (the Netherlands) – 1995
	907	32.5	– 1994
	840	32.1	– 1993

NOTES: NS = not stated.

Table 6.20b. Prevalence and concentration.

Sample	Country	Year of Sampling	No. positive/ No. sampled	Numbers on positive carcasses	Reference
Frozen thawed carcasses	USA		2/12 (16.7%)	0.23 MPN/ml	Izat, Kopek and McGinnis, 1991; Izat et al., 1991
			3/12 (25%)	0.06 MPN/ml	
			3/12 (25%)	0.09 MPN/ml	
			3/12 (25%)	0.07 MPN/ml	
			6/12 (50%)	0.34 MPN/ml	
			4/12 (33.3%)	0.05 MPN/ml	
Carcasses, after chill[1]	Canada	1997-98	163/774 (21.1%) C.I. 18 –24	<0.03MPN/ml: 99 0.03 – 0.30: 60 0.301 – 3.0: 2 3.0 1 – 30.0: 1 >30.0: 1	CFIA, 2000
Carcass rinse, after chill[2]	USA	1994-95	260/1297	Per cm^2	USDA-FSIS, 1998
Carcass rinse, after chill	USA	[1992]	29/112 (25.9%)		Waldrop et al., 1992

Notes: (1) Immersion, no chlorine. (2) Immersion, unspecified level of chlorine present in chill water.

Table 6.20c. Numbers of *Salmonella* on whole carcasses at retail.

Type of product	Number of samples	%	MPN[1]	Direct count/10 cm^2
Fresh	40	89	0 – 10	<100
	4	9	11 – 100	
	0	0	101 – 1100	
	1	2	> 1100	
Frozen	30	68	0 – 10	
	10	23	11 – 100	
	2	4	101 – 1100	
	1	2	> 1100	
	1	2	No MPN	

Notes: (1) MPN = Most probable number per carcass. Source: Dufrenne et al., 2001.

6.2.6 Preparation

The aim of the preparation module is to estimate the numbers of salmonellae in broiler chicken meat prior to consumption.

Preparation steps

The preparation process begins at the point the chilled or frozen broiler chicken, whole or portions, is removed from the refrigerator or freezer, respectively. Frozen whole broilers and portions must be thawed, but then preparation steps for both frozen and chilled whole broiler are essentially the same. Figure 6.6 summarizes common preparation steps. In the following module description, the case of whole broilers is considered. However, a similar approach can be applied to chicken portioned, provided that time and temperature data are available to characterize the storage, thawing, preparation and cooking pathways.

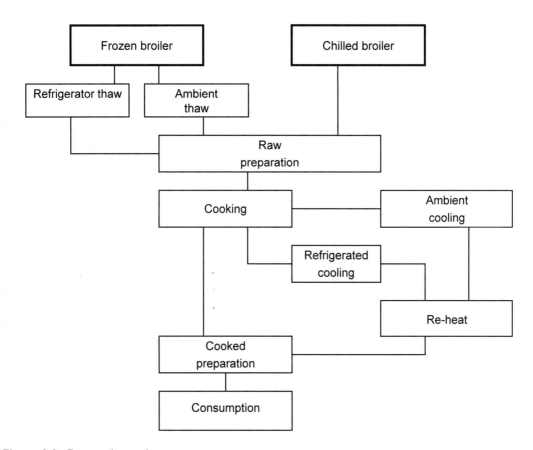

Figure 6.6. Preparation pathways

Thawing

Retailers of frozen poultry recommend that a frozen broiler chicken should be thawed overnight in a refrigerator. This is to maintain the surface of the broiler at a low enough temperature to prevent the growth of bacteria. However, in reality, broilers are often thawed outside a refrigerator or in an oven or microwave. If thawed at ambient temperature, the surface of the broiler can approach moderate ambient temperatures and because thawing often requires several hours there is potential for bacteria to grow on the surface. Thawing a frozen broiler in a heated oven takes a shorter period but surface temperatures are higher and bacteria grow more quickly. Thawing a broiler in a microwave creates an uneven heating pattern that again raises temperature and growth rate. This is offset by the reduced time-scale, although uneven thawing can leave frozen areas of the meat that can prevent adequate cooking. The thawing process also causes drip loss and this contaminated fluid can be an additional hazard during raw preparation.

Raw Preparation

Raw preparation involves a considerable amount of handling and hence cross-contamination is a serious hazard. Bacteria present on the meat can be transferred to hands, cloths, utensils and surfaces during the process. These items then become a source of contamination for ready-to-eat food in the kitchen. The preparation of the broiler chicken will also influence the cooking step. For example, a stuffed bird may take longer to cook than one where the cavity is open.

Cooking

Cooking is a critical step in the process. Thorough cooking should kill all the bacteria on and in a broiler. However, low cooking temperatures or short cooking times can result in undercooked meat with potentially live bacteria. The probability that bacteria will survive in these circumstances depends on the degree of undercooking and the bacterial load on the raw broiler. If cooked correctly, the meat should be free from bacteria at the point of consumption.

Cooling and re-heating

It is not uncommon for cooked meat portions, or even the whole cooked broiler chicken, to be cooled, stored, then eaten later, either re-heated or not. If carried out correctly, this process should not be hazardous. However, if the cooked meat is not cooled in a refrigerator but left to cool at ambient temperature, then any bacteria that survive cooking or are transferred by cross-contamination can grow, often quickly. If the meat is not re-heated then there is no opportunity to reduce the bacterial load. If the meat is re-heated thoroughly, then these bacteria will again be killed and the product should be safe. Obviously any less than thorough re-heating, as with cooking, may fail to kill all the bacteria. If the product is cooled quickly to refrigerator temperatures and stored chilled, then the product should be safer than cooling at ambient temperatures. There are numerous documented cases of food poisoning attributable to poorly controlled cook and chill processes.

Meal preparation

Preparation of the cooked broiler can involve processes like carving and dressing. The main hazard here is the potential to contaminate the meat with bacteria. Cross-contamination caused by poor hygienic practices may introduce bacteria onto a product that should be free from them following a correct cooking process.

...ta requirements, models available and data

...neral hygienic practice studies

...ble 6.21 indicates research into general food safety practices in the home. These studies ...e an indication of how many consumers may handle food in an unsafe manner. The ...dies selected for Table 6.21 are a cross-section of the types of studies that have been ...nducted. Data from this type of work can be used in an exposure assessment to evaluate ...e probability of unsafe practices occurring in the home.

...ble 6.21. General quantitative surveys of hygiene in the home.

Study	Subject numbers	Data type	Comments
...orsfold and Griffith, ...95	NA[1]	Riskscores	Model for assessing food safety behaviour
...ekruse et al., 1995	1620	% respondents to food safety questions	Telephone survey
...ott, 1996	NA	Various	Review article
...orsfold and Griffith, ...97a	108	% subjects displaying unhygienic practices	Direct observation *in situ*
...y, Komar and ...ovenlock, 1999	40	% subjects displaying unhygienic practices	Direct observation via video
...chutze et al, 1999	NA	*Salmonella* serotype, culture sources %	Investigation follow-up after salmonellosis diagnosis.

...otes: (1) NA = not applicable.

...awing

...r an exposure assessment model, any changes in the number of salmonellae during the ...awing process can be predicted using the survival and growth models discussed in ...ction 6.2.5, provided that adequate data are available to describe the temperature changes.

Studies on the thawing of broilers are often carried out by broiler chicken processors and ...ailers for the development of safe thawing instructions. These data can often be obtained ...risk assessors on application to the company. Unlike freezing and chilling where the ...armest part would be the deep muscle, the reverse is true of the thawing process. It is ...portant therefore to measure the thermal profile at the surface of the broiler as well as in ...e deep muscle (Table 6.22). Unfortunately, these measurements are rarely taken. Such is ...e emphasis for developing thermal profiles for cooking where the coldest spot is measured ...ie geometric centre), that workers often use the same approach to measure thawing. In ...ese studies the emphasis is on whether thawing is complete, which is essential for the ...bsequent cooking process. However, few data in the literature are available to describe the ...rface temperature where *Salmonella*, if present, can begin to multiply. It is possible to use ...ermodynamic models for thermal diffusivity to calculate a surface temperature given air ...nperature (Brown et al., 1998).

Table 6.22. Example of data on thawing of a 2300-g raw, frozen broiler chicken carcass.

Process step	Deep muscle temp. (°C)	Surface temp. (°C)
Start thaw in packaging at ambient temperature	-17.3	-16.0
After 24 hours in packaging	-1.9	1.4
After 29 hours (with 5 hours in ambient conditions, removed from packaging)	0	11.3

SOURCE: Provided by Christina Farnan, Carton Group, Cavan, Republic of Ireland.

Bryan and McKinley (1974) studied the preparation process for whole frozen turkey a produced detailed time and temperature profiles for all processes, including thawi However, they reported only deep muscle temperatures and the air temperature. For a 20 [9 kg] turkey, they found that after 40 hours thawing in a refrigerator at ~4°C the de muscle temperature was only –2.8°C. At ambient temperature (~24°C), the deep mus temperature was 0°C after 9 hours and 10°C after 18 hours. The surface temperature in t latter case was 10°C after 5 hours and 16.6°C after 22 hours. This demonstrates that surface temperatures can be relatively high by the time the turkey is thawed. For broil where weights are lower, the thawing time would be reduced, but the surface temperatu after similar periods are likely to be the same or slightly higher, due to the reduced mass the bird.

It is important to validate any predictions of growth during thawing and at least c suitable study is reported in the literature. Data on the growth of *Salmonella* followi thawing was generated in minced chicken substrate (White and Hall, 1984). Such data co be used to develop a model for frozen storage, periods of freeze-thaw and thawing, but t type of model development is outside of the scope of the current exposure assessment. T White and Hall data show that the numbers of *S.* Typhimurium decreased during fro storage by approximately 99% after 168 days of storage, but by only 90% for *S.* Hadar i similar period. They also showed that the numbers of *S.* Typhimurium increased by 1.8 cycles after 24 hours thawing at 20°C and by 2.93 log cycles after the same period at 27 *S.* Hadar grew by 2.87 log cycles after 24 hours thawing at 20°C and by 5.4 log cycles af the same period at 27°C. These data on thawing can be used to validate the growth mod selected, given the thawing profiles reported.

Preparation handling of raw chicken

Handling which is typically carried out at ambient temperatures can transfer bacteria cross-contamination of the hands and food preparation environment and especially prolonged, this is another factor that may lead to growth of salmonellae.

There are few data available in the literature on time and temperature studies duri preparation. Data on the time taken to prepare poultry and the temperature changes w reported by Garey and Simko (1987).

Several studies of cross-contamination have been conducted, but these consider gene contamination of the food environment rather than the contamination attributable to specific process such as preparation of chicken (Scott and Bloomfield, 1990; Josephs bino and Pepper, 1997). Others have quantified the incidence of cross-contamination due specific processes (Humphrey, Martin and Whitehead, 1994; Cogan, Bloomfield and mphrey, 1999). However, few have quantified the numbers of bacteria that are transferred ing cross-contamination. Cross-contamination resulting from the preparation of broilers been studied (de Wit, Broekhuizen and Kampelmacher, 1979). In an elegant study, they d naladixic-acid-resistant *E. coli* K12 as a marker organism to artificially contaminate ilers. The spread of this organism during preparation was studied. The cross-tamination rates show that the more direct the contact between broiler and item, the ater the percentage of positive samples from that item. Washing reduces the incidence of ss-contamination, but not completely. In the preparation process, other surfaces, such as ter taps and spice jars, also become contaminated, but to a lesser extent, indicating lirect contamination from hands.

For a quantitative exposure assessment model, these data could be used to calculate the bability of cross-contamination by direct and indirect means, which would be more ctical than separate calculations for surfaces, utensils and hands.

A measure of the probability of cross-contamination is not sufficient for an exposure del without an idea of the quantity of bacteria involved. Zhao et al. (1998) developed a del system to enumerate bacteria transferred during common food preparation practices. ey found that chicken meat and skin inoculated with 10^6 CFU bacteria transferred ^{5} CFU to a chopping board and hands, and then 10^3–10^4 CFU to vegetables subsequently opped on the unclean board. Disinfection of the chopping board and hand washing luced the numbers of bacteria by 1–2.8 logs and reduced the incidence of cross-ntamination of the cut vegetables (52%: no bacteria; 33%: 10–50 bacteria; 5%: 100–200 cteria).

These data can be used as the basis to estimate the numbers of bacteria transferred to a od by cross-contamination. From the work of Zhao et al. (1998), it appears that bacteria nsfer at a rate of approximately 10% between items, e.g. between raw chicken and the opping board. Direct cross-contamination involves two steps, e.g. raw meat to chopping ard, and then to another food item. Hence, the direct cross-contamination bacterial load r the second food item should be a maximum point estimate of 1% of the numbers of lmonellae on the broiler chicken. Indirect cross-contamination involves a minimum of ree steps, e.g. broiler chicken to hands, to plate, and then to another food item. Hence, for direct cross-contamination, the bacterial load transferred to another food item would be a aximum of 0.1% of the salmonellae on the broiler.

Cross-contamination can also occur from inadequate hand washing. Studies on hand ashing have shown that numbers of bacteria on the hands influences the number of samples at are contaminated through finger contact (Pether and Gilbert, 1971). Reviews of hand ashing practices are available in the literature (Snyder, 1999; Restaino and Wind, 1990; eybrouck, 1986).

ooking and thermal death models

acteria die when subjected to the elevated temperatures found during cooking. It is widely cepted by microbiologists that bacteria die in a predictable, logarithmic way. This is ferred to as first-order inactivation kinetics. The physiological assumption is that there is only one heat target per cell that is responsible for the death of the whole cell. Classica‖ the death of bacteria has been described by the Arrhenius equation that was developed first-order chemical reaction kinetics (Equation 6.3):

$$K = A \cdot e^{\left(\frac{E_a}{RT}\right)} \qquad \text{and} \qquad \log_{10} N = \log_{10} N_o - \frac{(K \cdot time)}{2.303} \qquad \text{Equation 6.3}$$

where: E_a = activation energy (J mol^{-1}); A = pre-exponential factor, R = gas const‖ (8.31 JK^{-1} mol^{-1}), N_0 = initial cell number, N = cell number after time at T, and T = absol‖ temperature (Kelvin).

However, deviations from the first-order death kinetic model have been observe‖ Shoulders and tails to the survivor curves are reported. This area has been review‖ extensively (Clark, 1933; Withell, 1942; Rhan, 1945; Cerf, 1977; Casolari, 1994). Seve‖ models that have characterized non-linear thermal death curves have used a log-logis‖ function to describe the data (Cole et al., 1993; Little et al., 1994; Ellison et al., 1994; Duf‖ et al., 1995; Anderson et al., 1996; Blackburn et al., 1997).

The equation for the log-logistic curve, with a shoulder and a tail, is shown in equati‖ 6.4.

$$\log_{10} N = \alpha + \frac{(\varpi - \alpha)}{1 + e^{\frac{4\sigma(\tau - \log_{10} time)}{\varpi - \alpha}}} \qquad \text{Equation 6.4}$$

where N = cell number after time at the temperature studied, α = upper asymptote of t‖ curve, ϖ = lower asymptote of the curve, σ = maximum inactivation rate, and τ = time to t‖ point of maximum inactivation rate.

For the present exposure assessment, the traditional log-linear-death kinetic model will ‖ considered, for simplicity. Many investigators do not show the inactivation data for the‖ studies and merely quote D-values (i.e. time for a 90% reduction in the numbers of bacter‖ at a given temperature). Generally, these workers will use regression analysis of da‖ showing log$_{10}$ bacteria numbers vs heating time. The equation of the regression line can ‖ used to calculate a D-value over 1 log cycle reduction in the numbers of bacteria. Wh‖ D-values are calculated for a number of different temperatures, a relationship between t‖ D-value and the temperature can be calculated. Data expressed as the reciprocal of t‖ D-value vs temperature of the D-value can be analysed by regression to give a straight-li‖ equation. This equation can be used to calculate a z-value, which is the temperature chang‖ required to bring about a 90% change in D-value. Hence, if the z-value = 10°C and a D-val‖ at 70°C = 1min, then by applying the z-value to the D-value we can see that the D-value ‖ 80°C = 0.1 minute and the D-value at 60°C = 10 minutes. Therefore, with a D-value at ‖ given temperature and a z-value for a bacterium in a given heating medium it is possible ‖ calculate the reduction in the numbers of that bacterium at any other temperature.

Secondary models can be constructed that relate the change in D-value to parameters suc‖ as pH and water activity. A model describing the death of *S.* Enteritidis was developed b‖ Blackburn et al. (1997). This model is comprehensive, covering the effects of temperatur‖ pH and salt on survival. In addition, the model validated well against D-value data derive‖

in whole foods. Unfortunately, this model is incorporated into the Food MicroModel™ software, which is proprietary.

An alternative approach that has been used in other exposure assessments is to take published D-values for *Salmonella* in foodstuffs, analyse the data, and determine an average D-value and *z*-value using the method described earlier (Buchanan and Whiting, 1997). Table 6.23 shows data used in this exposure assessment model for the calculation of an average D-value and *z*-value (Figure 6.7).

Table 6.23. Data on the inactivation of *Salmonella*

Serotype	Meat	D-value (minutes)	Temperature (°C)	Reference
Salmonella	Chicken	0.176	70	Murphy et al., 1999
Salmonella	Chicken	0.286	67.5	Murphy et al., 1999
S. Typhimurium	Ground Beef	0.36	63	Goodfellow and Brown, 1978
Salmonella	Ground Beef	0.7	62.76	Goodfellow and Brown, 1978
S. Thompson	Minced Beef	0.46	60	Mackey and Derrick, 1987
Salmonella	Ground Beef	4.2	57.2	Goodfellow and Brown, 1978
S. Typhimurium	Ground Beef	2.13	57	Goodfellow and Brown, 1978
S. Typhimurium	Ground Beef	2.67	57	Goodfellow and Brown, 1978
S. Typhimurium	Skin macerate[1]	61.72	52	Humphrey, 1981
Salmonella	Ground Beef	62	51.6	Goodfellow and Brown, 1978

NOTES: (1) from chicken neck.

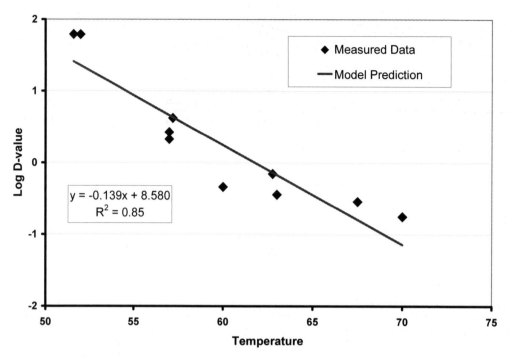

Figure 6.7. Plot of D-values from Table 6.23 with linear regression model used to subsequently calculate D- and z-values.

The D-value can be calculated using Equation 6.5:

$$D_{value} = 10^{(-0.139 \cdot Temp)+8.580}$$ Equation 6.5

and the z-value is the reciprocal of the slope of the line, Equation 6.6:

$$z_{value} = \frac{1}{0.139} = 7.19$$ Equation 6.6

To utilize the linear model for the thermal death of salmonellae in an exposure assessment model, it is necessary to measure the time and temperature profile for the cooking step. For conventional conduction-limited cooking (i.e. oven roasting, boiling, steaming), measurements are normally taken at the coldest spot, which is the deep muscle tissue of a broiler chicken carcass. However, this does not give information about the temperature at the surface of the carcass, where salmonellae may also be located.

For microwave cooking, where the thermal profile may be uneven, measurements must be taken in a number of places. An exposure assessment should account for differences in cooking methodology and the heterogeneity of temperature that this may cause. Models for microwave cooking are very complex and often require the use of thermodynamic modelling techniques to generate the time and temperature distributions.

Many studies reported in the literature do not contain the thermal profiles. Some report end-product temperatures and cooking time (Baker, Poon and Vadehra, 1983; Schnepf and Barbeau, 1989). Table 6.24 shows some publications where appropriate data are given.

Example data have also been supplied by a manufacturer of cooked chicken products (personal communication). Data for cooking of chicken drumsticks are summarized in Table 6.25.

Table 6.24. Studies on thermal profiles for cooking of poultry.

Study	Item cooked	Cooking method	Parameters measured
Bryan, 1971	Whole turkey	Boiling and steaming	Deep muscle, surface and external temperature
Bryan and McKinley, 1974	Whole turkey	Oven roast	Deep muscle, surface and oven temperature
Lyon et al., 1975	Chicken thighs	Boiling	Internal temperature
Ibarra et al., 1999	Chicken breast	Oven roast	Infrared surface and internal temperature
Chen and Marks, 1997	Chicken patties	Oven roast	Surface, interior and oven temperature
Chang, Carpenter and Toledo, 1998	Whole turkey	Oven roast	Various points

Table 6.25. Example thermal profile data on roasting chicken drumsticks.

Time (minutes)	Time of the temperature block (minutes)	External temp. (°C) (mean of 2 measurements)	Internal temp. (°C) (mean of 6 measurements)
0	5	12.6	14.9
5	5	13.0	14.2
10	5	136.6	13.7
15	5	161.2	27.8
20	5	150.0	43.2
25	5	150.7	56.2
30	5	164.0	68.6
35	5	166.1	78.0
40	5	168.6	85.8
45	5	166.3	83.7
50	5	176.3	93.3
55	5	161.4	94.9
60	5	49.2	82.1

To calculate the lethal effect of the process shown in Table 6.25, the following approach can be applied:

1. Break the profile up into time and temperature blocks as shown in Table 6.25.

2. Using Equation 6.5, calculate a D-value at a suitable reference temperature within the range of the profile.

3. Use Equation 6.7 to calculate the equivalent process time at the reference temperature for each time and temperature block:

$$Etime_{T_{ref}} = \frac{(10^{\frac{(T \cdot T_{re})}{z}})}{time}$$ Equation 6.7

where: *Etime* = equivalent time at the reference temperature; T_{ref} = reference temperature; T = temperature (°C) of the time and temperature block; z = temperature change resulting in a 90% change in D-value, calculated from Equation 6.6; and time = time period of the time and temperature block (in minutes).

4.   Use Equation 6.8 to calculate the equivalent reduction in log numbers of bacteria for each time and temperature block.

$$\log red = \frac{Etime_{T_{ref}}}{D_{T_{ref}}}$$ Equation 6.8

where: Log *red* = reduction in log numbers of bacteria; *Etime* = equivalent time at the reference temperature; T_{ref}= reference temperature; and D = D-value.

5.   Subtract each reduction from the starting log number of bacteria to determine the number of bacteria surviving the process.

Figure 6.8 shows the application of this approach to the data given in Table 6.25, using the model generated in Equations 6.7 and 6.8 and an assumed starting number of salmonellae of 10 million.

Cooling and re-heating

Providing suitable time and temperature profiles are available, the growth and thermal death models can be used to predict the numbers of salmonellae that may be present after a process. Published time and temperature profiles can be found (Bryan, 1971; Bryan and McKinley, 1974) but, as with all such profiles, data are scarce.

Meal preparation

Meal preparation can involve re-contamination of the cooked chicken from salmonellae present on hands, utensils and surfaces. This can be accounted for in the exposure assessment in a similar manner to the modelling of the raw preparation step. An assumption based on data can be made for the probability of cross-contamination and the numbers of salmonellae transferred (see above: *Preparation handling of raw chicken*).

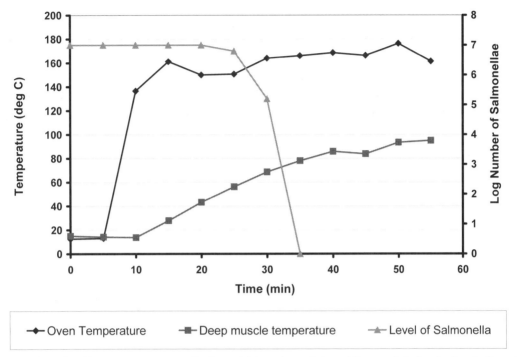

Figure 6.8. Cooking temperature profile for a chicken drumstick and the predicted reduction in the number of salmonellae in the deep muscle tissue.

6.2.7 Consumption

The aim of the consumption module is to quantify the frequency with which broiler meat is consumed in the form specified in the preparation module, and to quantify the portion size.

Consumption studies

One aim of an exposure assessment model is to provide quantitative data to input into the dose-response model. To do this, it is necessary for the exposure assessment to predict the likelihood of human exposure to a bacterial pathogen, and the numbers of the pathogen to which a person may be exposed. So far in this report, tools have been described that enable a quantitative prediction of the number of salmonellae on ready-to-eat broiler chicken meat, and the likelihood that the chicken meat will be contaminated with salmonellae. However, to become exposed to the bacteria, a person must consume broiler meat. Therefore, the number of bacteria that enter the person's body also depends on the amount of the meat they eat and possibly the frequency of consumption. The final stage in the exposure assessment model is a determination of consumption patterns for broiler chicken meat.

Food consumption patterns vary from country to country, by demographic group, and by age group. Therefore, ideally, countries should determine their own national consumption patterns. Additionally, consumption studies are often undertaken for purposes other than exposure assessment, e.g. nutrition studies. The design of these studies is not necessarily appropriate for determination of exposure to microorganisms from consumption of a product.

Data requirements and available data

Data required for a consumption module would relate to the products specified for the risk assessment, and in the exposure assessment. In this work, consumption data are required for a single serving from whole or portioned broiler chicken, prepared in the home according to the methods used in the previous module.

Commodity consumption data have been compiled and published by WHO (1998). For countries without national studies, this work is a good reference. Unfortunately, because of its general nature, it gives consumption data for chicken meat from all sources on a regional basis. The consumption of chicken meat per day per capita was reported as follows:

- Middle Eastern diet   30.5 g
- Far Eastern diet   11.5 g
- African diet   5.5 g
- Latin American diet   25.3 g
- European diet   44.0 g

These data include meat from whole cooked broilers, but in addition also include servings of cooked minced chicken preparations, pre-prepared commercial meals, and other sources outside the scope of the present exposure assessment model.

A more detailed breakdown of food consumption can often be gained from national nutrition surveys. For example, in Australia, a national survey conducted in 1995 (McLennon and Podger, 1995) classified consumption of whole muscle poultry meat *inter alia* by age group, sex and socioeconomic group. Table 6.26 summarizes the relevant data.

Table 6.26. Mean daily intake of poultry muscle meat per person in Australia.

Male age group (years)									
2–3	4–7	8–11	12–15	16–18	19–24	25–44	45–64	65+	19+
11.3 g	19.2 g	26.8 g	48.4 g	51.4 g	73.1 g	66.7 g	62.6 g	45.4 g	63.3 g
Female age group (years)									
2–3	4–7	8–11	12–15	16–18	19–24	25–44	45–64	65+	19+
8.8 g	12.5 g	23.6 g	29.4 g	32.3 g	33.7 g	31.5 g	34.2 g	29.7 g	32.2 g
SEIFA quintile of relative socioeconomic disadvantage[1], 19 years +									
1st		2nd		3rd		4th		5th	
47.3		48.1		45.6		47.1		48.9	

NOTES: (1) Based on the characteristics of an area where the person lives. People in the first quintile live in the most disadvantaged areas, whereas people in the fifth quintile live in the least disadvantaged areas. SOURCE: McLennon and Podger, 1995.

These data allow exposure predictions to be targeted to vulnerable groups, such as the very young and the elderly. The study also showed that, in Australia, the consumption of poultry muscle meat was not influenced to any great degree by socioeconomic group.

The single drawback to these data is that the only value reported is the mean daily intake. Reporting the standard deviations of the mean values would allow estimation of the distribution range of size of meals consumed.

In Ireland, the Irish Universities Nutrition Alliance (IUNA) have recently completed a food consumption survey. The primary aim of the survey was to establish a database of the habitual food and drink consumption of Irish adults between the ages of 18 and 64. The Republic of Ireland section of the database contains entries for 958 subjects, but as the data were collected *per eating occasion*, there are 159 091 entries in the database. The total food consumption for each subject must be taken to represent his or her habitual weekly intake of a given food. The IUNA database was searched for meals where chicken was casseroled, grilled, stir-fried, deep-fried or roasted. Prepared chicken dishes – chicken Kiev, chicken vindaloo, etc. – were excluded.

Of the 159 091 eating occasions entered in the database, 1289 referred to chicken muscle. In real terms, 633 subjects out of a possible 958 consumed chicken muscle at least once per week (66%). Of those consuming at least once a week, it was found that the chicken muscle was consumed on average 2.04 times per week (maximum 7 times; minimum once).

It is worth noting that consumption data is very country specific as consumption patterns may be very different in different parts of the world. Thus, any national exposure assessment should use data specific for that country rather than data from any other country.

6.2.8 Review of models available

Overview

To date, no full exposure assessments of *Salmonella* in broiler chicken products have been presented, i.e. an exposure assessment that includes all the steps outlined in Figure 6.1 for the production-to-consumption pathway. However, exposure models have been developed for subsections of this pathway. Oscar (1997) considers levels of exposure throughout processing (Module 2 in Figure 6.1) while Oscar (1998) and Oscar (in press) developed models to describe exposure from the point of packaging to the point of consumption (Modules 3 and 4 in Figure 6.1). Brown et al. (1998) consider changes in the numbers of organisms on contaminated raw chicken products following cooking (Module 4 in Figure 6.1).

In contrast, a full exposure assessment has been described for *Campylobacter jejuni* in fresh poultry (Fazil et al., unpublished; A.M. Fazil, personal communication). Although there are key differences between *Salmonella* and *Campylobacter jejuni*, this model can be used as a basis for review.

These models are summarized here with respect to the objectives of the work, and the various methodologies used. It is noted that several of the models consider, to some extent, dose-response and hazard characterization as well as exposure assessment. In such cases, only the exposure assessment part is reviewed. Following each summary, the methodologies are discussed with respect to a full exposure assessment of *Salmonella* in broiler chicken products.

The models of Oscar (1997), Oscar (1998) and Oscar (in press)

The model of Oscar (1997) is essentially a demonstration tool to illustrate the use of simulation modelling in food safety decision-making. Consequently, real data are not used within the model and hence results do not represent actual estimates of exposure.

The demonstration model considers the prevalence of *Salmonella*-positive broiler chicken carcasses and the number of organisms per contaminated carcass following each sequential step of processing, as outlined in Figure 6.3. Each step is characterized by two quantitative parameters, the prevalence and the extent of a specific pathogen event. Pathogen events correspond to either an increase or reduction in numbers of organisms per carcass, depending on the step-specific factors. In the model, increases reflect only cross-contamination, hence bacterial growth is not included.

A similar pathogen event approach is used in subsequent models (Oscar, 1998; Oscar, in press). These models commence at packaging of raw chicken and describe changes in the prevalence of *Salmonella*-positive products and the numbers of organisms per positive product until the point of consumption. Pathogen events again refer to either an increase or decrease in pathogen load, but these subsequent models also consider bacterial growth.

These three models provide simple assessments of exposure. The underlying methodology involves simulation of a random chicken product through various exposure steps.

In general terms, the framework presented in the Oscar models could be used as a basis for development of a full exposure assessment of *Salmonella* in broiler products. Indeed, in the first instance, it may be possible to combine the processing model (Oscar, 1997) with either of the packaging-to-consumption models (Oscar, 1998; Oscar, in press). However, there are important points that would need further consideration before such use.

First, the model framework describes the inherent variability of the sequential exposure steps. In particular, the probabilities (p_i) represent the randomness associated with whether or not the particular pathogen events will occur while the distributions for changes in numbers of organisms describe all possible magnitudes of change. However, it is possible that for *Salmonella* we may not know the exact values for p_i and all possible magnitudes of change. Therefore it is likely that there will be uncertainty associated with these parameters. As is, the model framework does not account for such uncertainty and thus may produce inaccurate estimates of exposure.

The second feature of this framework that should be addressed focuses on the notion of cross-contamination. The model framework for packaging through to consumption does not include cross-contamination to other products or the environment. However, during preparation, for example, such cross-contamination might be very important. Consequently, this approach could underestimate exposure. For the processing model (Oscar, 1997), the methodology used to account for cross-contamination is not explicitly stated, hence it cannot be determined whether or not this would be appropriate in a full exposure assessment.

A final point worth considering is the representation of growth and survival within the packaging-to-consumption models. Growth and reduction due to temperature abuse and cooking, respectively, are not given as time-dependent processes within the models. Rather, the overall change following a period of abuse or cooking is modelled. Although this gives a mechanism for estimating changes in exposure, the effect of different temperature profiles and product specific parameters cannot be investigated. Consequently, investigation of control strategies would be difficult.

In summary, these provide a basis for the development of a full exposure assessment, but issues concerning uncertainty, cross-contamination, growth and decline would have to be addressed before further use.

The model of Brown et al. (1998)

This model considers the prevalence of *Salmonella* on raw chicken portions and the numbers of organism per contaminated portion. Prevalence is estimated by a point value while a probability distribution is used to describe the variability in the numbers of organisms per contaminated portion. Given the initial level of pathogen on the raw product, the final level of exposure is then determined by modelling the effects of cooking.

The approach used within this model is deterministic in nature. In particular, point values are used for model parameters such as prevalence of contaminated chicken portions, and the heat transfer coefficient. Estimates of exposure are determined by integration over all parameters that are inherently variable, more specifically time, microbial distribution, and measurement of depth into the product. Although this approach accounts for inherent variability, it does not incorporate uncertainty in parameter values. As a result, it does not facilitate the derivation of confidence intervals for estimates of exposure. The authors present several suggestions for including uncertainty that could be incorporated in a full exposure assessment.

There are two main exposure steps in this model: first, the level of raw chicken contamination and, second, the effect of cooking. Cross-contamination within the kitchen, prior to cooking, is not considered. As discussed previously, cross-contamination within the kitchen could be a very important pathway for exposure to *Salmonella* from raw poultry and thus should be included within a full exposure assessment.

Overall, the framework presented in this model will be very useful for the development of Module 4 (Preparation) (see Figure 6.1) of any full exposure assessment. The framework could also be further enhanced by including uncertainty in model parameters and attempting to model cross-contamination in the kitchen.

The model of Fazil et al. (unpublished)

This assessment is still in-progress (A.M. Fazil, personal communication.). The review presented here considers the information that was available at the time of this review. It is expected that this model will be refined in the future, thus the comments made here may require appropriate modification.

The preliminary model provides a full exposure assessment for *Campylobacter jejuni* in fresh chicken. All stages from on-farm production to consumption are considered. At each stage, the 'fate' of *C. jejuni* on chickens is estimated with particular reference to surface contamination and the numbers of organisms per contaminated unit (carcass). In this way, changes in prevalence and numbers are described and a final estimate of exposure is derived. As this model considers fresh chicken products, the framework presented provides a basis for the development of an exposure model for *Salmonella* in the same commodity.

In a similar manner to the model pathway outlined in Figure 6.1, the exposure assessment commences with estimation of farm-level parameters. More specifically, the number of organisms on the skin and feathers of birds is calculated. Estimation is undertaken by determining the number of organisms excreted in the faeces and then assuming that a proportion of these contaminate the external parts of the bird. Consequently, it is assumed that feather, skin, etc., (i.e. surface) contamination arises directly from the birds. Given that within-flock prevalence of *Campylobacter* is generally very high (Hartnett et al., 2001), this would appear to be a valid assumption. However, for *Salmonella,* within-flock prevalence is much more variable and it may be more appropriate to consider other sources of contamination.

From the initial concentration of organisms on the exterior of birds at the farm level, changes in numbers during transport and subsequent processing are modelled. The modelling approach considers each step in turn and determines the magnitude of change in terms of either a log increase or decrease, depending on the particular step. The magnitude of change is estimated from several data sets that provide this specific type of information, hence particular reasons for change, such as cross-contamination or wash-off, are accounted for. If equivalent data were available for *Salmonella* spp, a similar modelling approach could be used. It is of course important to point out one key difference between *Salmonella* and *Campylobacter*, that is that conditions during processing that may be favourable for the growth of *Salmonella* would probably not result in multiplication of *Campylobacter*.

As changes in concentration are modelled, changes in prevalence of contaminated birds, carcasses or products from farm to the end of processing are also described. The initial prevalence estimate relates to prevalence of contaminated birds on entry into the processing plant, and this estimate essentially describes the probability that any random bird is contaminated. During processing, changes in prevalence have been modelled by initially ranking the different stages according to the extent to which cross-contamination is likely to occur. Based on this ranking, a cross-contamination factor is then assigned to each step. For each step, the resulting prevalence is a function of the prevalence at the start of the step and the cross-contamination factor. Given the generality of the cross-contamination factor approach, it is likely that a similar methodology could be used to model changes in *Salmonella* prevalence during processing.

Following processing, the time between processing and preparation of the chicken in the home is considered. This period covers both storage and transit. It is assumed that the chicken product remains at refrigerated temperatures and reduction in the number of organisms per day is calculated. This approach, which essentially models survival, is appropriate for *Campylobacter*. However, for organisms such as *Salmonella*, growth during storage and transit may be important, depending on whether or not temperature abuse occurs. Consequently, growth as well as survival would have to be considered.

The final step of the exposure assessment models consumer handling and preparation. It is assumed that exposure to *C. jejuni* occurs via two independent routes: consumption of undercooked chicken and through the raw chicken fluids that may be subsequently ingested through cross-contamination. The models presented for these steps could be adapted for a *Salmonella* exposure assessment by incorporating species-specific data.

The *Campylobacter* exposure assessment is a stochastic model. The stochastic component of the model framework describes the variability in changes in prevalence and numbers of organisms throughout the sequential exposure steps. However, as yet, the uncertainty associated with these distributions of change is not accounted for. Given the limited quantitative information relating to changes in prevalence and numbers, inclusion of uncertainty will be important for an exposure assessment of *Salmonella*.

Overall, the general framework on which the preliminary model of Fazil and collaborators is based could provide a basis for the development of a *Salmonella* exposure assessment. However, other factors would also have to be included, particularly growth during storage and transport and uncertainty associated with probability distributions to describe magnitudes of change.

6.2.9 Recommendations

To date, no full exposure assessments have been undertaken for *Salmonella* in broilers. This present report has considered:

- What is required for undertaking such assessments.
- What information is available.
- How the available information meets the requirements.

The following recommendations for directing future work can be made.

(i) Reporting of prevalence at different steps of the full exposure pathway should be encouraged in all regions of the world.

(ii) Reported data should give full details of study methodology, including sampling site, sampling time, how the sample relates to the overall population, and microbiological methods.

(iii) Determination of quantitative data should be encouraged, and, if it becomes available, then full exposure assessments could be developed to investigate mitigation strategies (e.g. use of chlorine in chill water) or to compare alternative practices (e.g. air chilling versus immersion chilling).

(iv) Cross-contamination during processing and handling operations should be studied quantitatively and methodologies for modelling this process should be developed. Cross-contamination during these stages is a critical factor, which is often associated with outbreaks.

(v) At the national level, the collection of consumption data should be promoted. The design of these studies should accommodate the data requirements for exposure assessments. These requirements include population variability, portion size and frequency of consumption.

(vi) In predictive microbiology, the area of survival has been less well studied than growth or death. There are few predictive models that describe survival at chill and frozen temperatures. Further development of these models is essential.

6.3 EXPOSURE ASSESSMENT MODEL, MODEL PARAMETERS AND ASSUMPTIONS

6.3.1 Introduction

Previous sections examined the data and models available to generate a production-to-consumption risk model. Although there is a substantial amount of literature relating to *Salmonella* in poultry-rearing operations and during processing, the existing data have severe limitations for usefulness in quantitative (or semi-quantitative) risk assessment. Very few investigations have enumerated *Salmonella* either on-farm or at processing, or measured how the populations change, for example in a specific stage during processing. Evidence suggests that numbers of *Salmonella* on poultry carcasses during processing are generally low, at the limits of detection using current enumerative methods, and even then, the commonly used MPN method is very labour and cost intensive. Hence, for practicality, only detect/non-detect (prevalence) investigations are commonly carried out. This results, therefore, in a critical data gap because without enumeration data, risk cannot be estimated. In addition, for both prevalence and the few enumerative investigations, there is a wide diversity in conditions of sampling (sample type, site, size, unit, etc.) and of laboratory testing methods, as well as other confounding factors introduced by the original purposes of the studies and their experimental design. Accommodating these variations and assessing the validity, sensitivity and specificity in each individual report would probably be an exercise in futility. Furthermore, when temporal (if considering data from the early 1980s together with more recent information) and geographical factors are considered, a comprehensive risk model would not be very informative. However, the foregoing sections provide guidelines for the type of information and approach that might be used to develop a production-to-consumption risk model that could be applied to data that represent an individual processing operation, country or region.

Given the lack of use of enumeration data for stages prior to processing, the Exposure Assessment model for purposes of this risk assessment therefore begins at the end of commercial processing, with survey data for contamination levels on chilled broiler carcasses. The subsequent changes in contamination due to storage, handling and preparation were modelled based on information that was presented in detail in the previous section. The construct of the exposure assessment model is summarized in the following model description, and the parameters are shown in Section 6.4

6.3.2 Model overview

The exposure assessment considered fresh, whole broilers that are purchased at retail, then prepared and consumed in the home. The exposure model was analysed using Monte-Carlo simulation facilitated by @RISK software (©Palisade). Each iteration of the model tracks a randomly selected broiler carcass from the time of exit from processing, through storage, preparation and cooking, to consumption. Thus, each run represents a random serving of cooking chicken and the exposures (including cross-contamination) that arise as a result of preparing this serving.

At the start of each iteration, the carcass is assigned to either the *Contaminated* or *Not Contaminated* state according to the prevalence of contaminated carcasses. If the carcass is contaminated, the number of *Salmonella* assumed to be present is selected from the range of values specified by a custom distribution of reported data. If the carcass it is not contaminated, the concentration of organisms is set to zero and this value is held constant for the remainder of the model. For contaminated carcasses, following the start-up step of the model the changes in the level of contamination through storage, preparation (including cross-contamination) and cooking are modelled.

Changes in the level of contamination during the various stages from chilling to consumption occur as a result of a number of variable processes, including storage times and temperatures, practices during preparation, and cooking times and temperatures. This variability is described by probability distributions derived from published and unpublished data and, where necessary, expert opinion.

The model is defined in terms of a number of parameters that describe the processes of broiler carcass distribution and storage, preparation, cooking and consumption. Many of these parameters can be considered general in that they can be used to describe the situation in many countries, such as cooking temperatures and duration of storage. In contrast, some parameters are country specific, such as prevalence of contaminated carcasses exiting processing, and thus to obtain results for individual countries, country-specific data must be input. In addition to the scope for generalization, the model parameters can be modified to determine their influence on the final estimates of exposure.

6.3.3 Processing

Prevalence of Salmonella-*contaminated carcasses*

Prevalence immediately after primary processing was set as a "user" input to the model. In the reports available, prevalence can vary widely from lot to lot, among different processing operations, and among regions and countries, particularly if national standards have been established or *Salmonella* control programmes operate. Thus, if using this model to describe the situation in a specific country, the likely ranges of local prevalence should be used to generate the risk estimate. Reductions to this value can then demonstrate the effect of prevalence reduction strategies, no matter how they are implemented, on the risk of illness. This would be an important preliminary investigation, prior to determining the best options for reduction, because an idea of the magnitude of the benefit can be realized.

For the purposes of this assessment, a baseline model was first developed, using as the initial input a fixed prevalence level of 20% *Salmonella*-contaminated carcasses after chilling. The predicted relative change in risk associated with higher or reduced levels of prevalence were modelled for comparison, using fixed values from 1 to 90% contaminated carcasses, and the results compared with the baseline risk estimate.

Numbers of Salmonella *on contaminated carcasses*

Few studies report on concentrations of *Salmonella* on broilers. Five studies reporting pathogen numbers at the end of processing (chilling) were summarized in Tables 6.9, 6.10, 6.12 and 6.13 (Surkiewicz et al., 1969; Dougherty, 1974; Lillard, 1980; USDA-FSIS, 1996; Campbell et al., 1983). Since then, data from Canada (CFIA, 2000), shown in Table 6.13, has been made available. All of these studies report MPN values rather than $\log_{10}$ values and all consider immersion chilling.

Some of these studies have characteristics that mean that they are of limited use for inclusion within this exposure assessment. The studies of Surkiewicz et al. (1969) and Campbell et al. (1983) report combined distributions of MPNs from carcasses randomly selected from a number of processing plants. These processing plants differ in their practices relating to the use of chlorine. As chlorine has been reported as having an influence on counts of pathogenic organisms on carcasses (Waldroup et al. 1992), the combined distributions would only be representative if, at a national level, chlorine were used in the same proportion of plants in which it was used in these studies. In addition, these studies are old (published in 1969 and 1983, respectively) and practices affecting concentrations are likely to have changed. Thus, the distributions may not be representative of the current situation. For these reasons, it was decided not to include these studies in the example.

The results reported by Dougherty (1974) and Lillard (1980) give only the mean MPN values, with no information about the distributions of the data. These are therefore of limited use for describing the inherent variability of this parameter. Further, the studies are again old, and may not be representative of current practices. As a consequence, it was decided to exclude them from the example. It is noted that, in the future, more details about unreported original data might be obtainable by contacting the investigators.

National chicken broiler baseline surveys have been conducted in the United States of America in 1994-95 (USDA-FSIS, 1996) and in Canada in 1997-98 (CFIA, 2000) (Table 6.13). These surveys employed statistically based sampling plans, and the same sample collection and laboratory procedures. In the USDA study, carcasses were collected from federally inspected processing plants responsible for approximately 99% of all chickens slaughtered in the United States of America. Similarly, the processing plants from which carcasses were sampled in Canada were federally registered and produced 99.9% of broilers. Both studies report MPN distributions for levels of *Salmonella* on chilled carcasses.

Although the USDA and CFIA studies are similar in nature, and both reported similar prevalence of *Salmonella* on chilled carcasses (20% and 21.1%, respectively, by qualitative enrichment of carcass rinse samples), the resulting MPN distributions cannot be combined. This is because practices relating to the use of chlorine differed between the two countries at the time the baseline surveys were conducted. In the United States of America, the addition of chlorine at levels sufficient to maintain 1–5 ppm free chlorine in the overflow was the norm, while in Canada this was not general practice. However, in isolation, the two studies provide good data sets for characterizing the concentration on carcasses after chilling; they are recent, representative and all sampling methods are clearly described. Of course, neither study reports *Salmonella* concentrations prior to chilling, therefore careful consideration would have to be given if incorporating either data set into a specific processing model.

For the baseline risk model in this assessment, the levels of contamination on chilled broiler carcasses in Canada were used as inputs. This can probably be considered a general data input rather than a country-specific one.

Estimating numbers of Salmonella *on contaminated carcasses*

Carcass rinses (400 ml) were obtained for 774 broiler carcasses (CFIA, 2000). From each rinse fluid, a sample was tested for the presence or absence of salmonellae using a qualitative enrichment method. Of these, 163 tested positive. Positive rinse fluids were tested by the

MPN method, and the MPN per millilitre calculated. The frequency of positive carcasses in five ranges was recorded. These data are shown in Table 6.27. The MPN per carcass was calculated by making two assumptions: first, all organisms on the carcass would be recovered during the shaking procedure, and, second, these organisms would be uniformly distributed within the rinse fluid. Based on these assumptions, the estimated MPN/carcass is equal to 400×MPN/ml (Table 6.28).

Table 6.27. Canadian national baseline data for *Salmonella* on chicken broiler carcasses

Range (MPN/ml)	Range (MPN/carcass)	Frequency
<0.03[1]	<12	99
0.03–0.3	12–120	60
0.301–3.0	121–1200	2
3.01–30	1201–12 000	1
>30.0[2]	>12 000	1

NOTES: (1) Positive by qualitative method, negative by quantitative MPN method.
(2) Maximum reported value was 110 MPN/ml.   SOURCE: CFIA, 2000.

The distribution for the 163 positive carcasses in Table 6.27 gives a description of the variability in the MPN/carcass. However, as the data is from a sample of carcasses, there will be uncertainty concerning the true variability. The cumulative distribution (Table 6.28) set the minimum value as 1 MPN/carcass and the maximum equal to 110% of the maximum observed MPN, i.e. 110% of 110 MPN/ml = 121 × 400 ml.

The resulting distribution for log MPN/carcass is shown in Figure 6.9. These distributions were used to characterize the variability in the numbers of *Salmonella* on contaminated carcasses at the end of processing.

The assumptions concerning the calculation of MPN/carcass from the data reported in the Canadian study require thoughtful consideration. In particular, there is uncertainty and variability relating to the MPN method, which has not been accounted for here. Further, it is likely that the carcass rinse method will not recover all organisms from the carcass. Indeed, it has been reported that on successive carcass rinses of the same bird, aerobic bacteria and enterobacteriaceae can still be recovered after 40 rinses (Lillard 1988, 1989b). These issues should be addressed in future refinements of the exposure assessment.

Table 6.28. Cumulative distribution for carcass concentration, with assumed minimum and maximum concentrations.

MPN/carcass	Log$_{10}$ MPN/carcass	Cumulative probability
1	0.00	0.00
12	1.08	0.60
120	2.08	0.97
1200	3.08	0.98
12 000	4.08	0.99
44 000	4.64	0.99
48 400	4.68	1.00

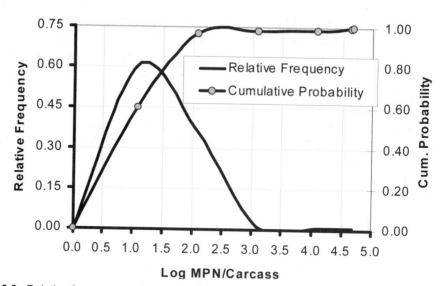

Figure 6.9. Relative frequency and cumulative distribution of Log$_{10}$MPN/carcass

6.3.4 Distribution and storage

After processing and packaging, poultry carcasses are distributed to retail stores. It was assumed that between processing plant and retail there would be no change in the prevalence of contaminated carcasses or in numbers of *Salmonella* on those carcasses. The latter was based on assuming controlled refrigerated transportation conditions (see below for growth at <10°C). Although in the current exercise it was assumed that transportation was well controlled, this needs to be determined on a case-by-case basis. For that reason, a module was created, although not simulated, to illustrate how this step might be potentially modelled and is summarized in Section 6.2.5, together with the other modules.

Three opportunities for *Salmonella* to multiply on the raw chicken were considered, from the time it enters the retail chain to the point at which the consumer prepares the chicken for cooking. These were (i) during retail storage and display, (ii) in transport from retail to the home, and (iii) during storage in the home. Survival and growth models currently available for estimating population changes during these stages were reviewed in Section 6.2.5. There are no suitable models to estimate survival and die-off for salmonellae in or on broilers, and therefore for the purposes of this risk assessment it was assumed that the salmonellae either grow given suitable conditions, or the population remains static on poultry, but does not decrease.

Several growth models for salmonellae were evaluated for their relevance and 'ease of use' for this assessment. The growth model selected was developed by Oscar (1999b) for *S.* Typhimurium (Equation 6.9).

$$LGR = \exp \left(\begin{array}{l} -6.225 - [0.0114 \times NaCl] + [0.3234 \times Temp] + \\ [0.002 \times \{NaCl \times Temp\}] - [0.0085 \times NaCl^2] - [0.0045 \times Temp^2] \end{array} \right) \qquad \text{Equation 6.9}$$

The equation parameters were developed using ground chicken breast meat as the growth medium (rather than laboratory media), and the model has a wider growth temperature range than others (10°C–40°C). The author also validated the model satisfactorily. The growth model takes account of the influences of temperature and salt concentration (including any previous exposure to NaCl, as in the case of pre-culturing inocula in the laboratory) on the growth of *S.* Typhimurium. The author's opinion was sought on the likely water activity of broiler meat used in the development of the model and a value of 0.99 or 1.9% salt was advised. Therefore, the salt concentration parameters were both fixed at 1.9%, and the external temperature remained a variable that determined the growth rate. A final assumption was that there was no lag phase in the growth phases modelled. This is reasonable given that salmonellae on or in broiler meat would have had ample time to adapt to their environmental conditions prior to retail delivery, and it would be unlikely that the cells would experience significant lag time before commencing growth once storage temperatures rose. It has been emphasized elsewhere that predictive models can only be used for interpolation within their boundaries. The growth model has a lower temperature bound of 10°C and hence it was assumed that there was no growth below this. The upper temperature bound (40°C) was assumed not to be exceeded under normal storage conditions. The lower temperature bound assumption may underestimate some growth at <10°C.

Section 6.2.5 of the Exposure Assessment discussed the modelling of non-isothermal temperature profiles. It was noted that time-temperature storage (retail display, home, etc.) profiles are generally not available for raw poultry. Therefore, in this assessment, the observations of Ross (1999) were used, namely that microbial growth during isothermal temperature conditions could be reasonably predicted using the average temperature of the isothermal profile. Hence, any growth of salmonellae in broilers during storage was based on distributions around the reported average storage temperatures.

Note that while the growth model can be considered general, the time and temperature profiles used within it must be country specific.

Retail storage

A study in the United States of America (Audits International, 1999) reported survey data on the variability of average retail storage temperatures. These may or may not reflect similar conditions in other countries, but in the model these values can be readily replaced with other, more representative, temperatures if appropriate. Temperatures were recorded for 975 fresh meat products. The overall average temperature recorded was 4°C with a standard deviation of 2.8°C. The maximum temperature reported was 10°C and the minimum was – 7.2°C. For this exposure assessment model, the variability in retail storage temperatures was represented by a truncated Normal distribution using these data. Hence, during the simulation, values could not be selected which were above the maximum or below the minimum recorded temperatures. Therefore, as 10°C represents the lower temperature bound of the growth model, growth was achieved only when an average retail storage temperature of 10°C or above was selected at random during the Monte Carlo simulation.

The growth model calculated a specific growth **rate** for *S.* Typhimurium at the average storage temperature. The **extent** of growth was determined by the length of storage time. Advice from retailers in Ireland was sought to estimate the minimum and maximum length of time that fresh chicken broilers were kept at retail. It would be preferable to obtain the information in a much more structured manner, or through a commissioned study, but, as first step, this may be appropriate. The minimum value was estimated as 2 days and the maximum as 7 days. It was assumed that all values were equally likely and therefore the retail storage duration was represented by a uniform distribution. However, a correlation factor of –0.75% was used to ensure that, in the simulation, combinations of high storage temperatures and long storage times were unlikely (resulting in detectable spoilage and the product discarded before consumption, as would be the case in reality). The specific growth rate calculated by the growth model was multiplied by the storage time in days to give a value for the $\log_{10}$ increase in numbers of salmonellae.

Transport from retail to home

Data describing the variability of temperatures for foods during transport from the retail store to the home have been collected in the United States of America (Audits International, 1999). Variability in transport times from store to home was also measured during this study.

Given the parameters of the growth model used in this assessment, a product temperature of 10°C must be exceeded before a specific growth rate is calculated for *Salmonella*. Therefore it was assumed that if the external temperature that was experienced during transport to the home were below 10°C, no growth would occur. Hence, an estimate of the external temperature during transport was important to determine microbial growth. For the purposes of this assessment, typical northern European temperatures were applied, with the temperature variability represented using a PERT distribution centred on the most likely temperature value. These were a minimum of 0°C, a maximum of 24°C, and with a most likely value of 13°C.

The United States of America study reported changes in product temperature during transport from the retail store to the home for 975 fresh meat products. The overall average was 3.72°C with a standard deviation of 2.82°C. The maximum temperature change was defined as the difference between the external (ambient) air temperature and the minimum growth temperature of the model (10°C). The minimum temperature change was taken as 0°C (no change). This variability was represented by a truncated Normal distribution in a similar way to that described previously for retail storage.

The maximum product temperature during transport was calculated as the retail temperature of the product plus the temperature change if any change occurred. The average product temperature was calculated as the mean of the maximum product temperature plus the retail product temperature. This average temperature was used to calculate the specific growth rate for salmonellae in the growth model.

The data for transport time were reported as the frequency of measurements in 15-minute time intervals. A cumulative distribution was fitted to these values and used to represent the variability in these data (Table 6.29 and Figure 6.10). The increase in $\log_{10}$ numbers of salmonellae in or on a simulated broiler was calculated by multiplying the specific growth rate by the transport time.

Table 6.29. Transportation time from retail to home.

Time (minutes)	Frequency	Cumulative
–	0.000	0.000
15	0.005	0.005
30	0.050	0.055
45	0.180	0.235
60	0.250	0.485
75	0.220	0.705
90	0.160	0.865
105	0.070	0.935
120	0.030	0.965
240	0.035	1.000

SOURCE: Data from Audits International, 1999.

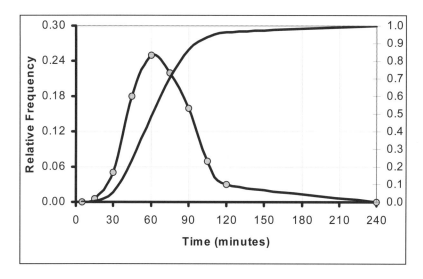

Figure 6.10. Probability distribution for transportation time from retail to home.

Home storage

Data on the variability of product temperatures during domestic refrigerated storage are available for the United States of America (Audits International, 1999). Temperatures during domestic refrigerated storage were recorded for unspecified food products, being chilled at an average of 4°C with a standard deviation of 2.65°C, with a maximum temperature of 21.1°C and a minimum of –6.1°C. The variability of reported home storage temperatures was represented by a truncated Normal distribution using these temperatures. Values could not be selected during the simulation above the maximum or below the minimum temperatures recorded. Again, because 10°C represented the lower temperature bound of the growth model, pathogen growth occurred only when a home storage temperature of 10°C or above was randomly selected during the Monte Carlo simulation.

The specific growth rates for *S.* Typhimurium were calculated for the average home storage temperatures and the extent of growth was determined by the length of storage time in the home.   Based on adherence to the "use-by" date, expert opinion estimated the minimum value would be no storage time (immediate use), the maximum would be 5 days with a most likely value of 2 days in the refrigerator.   The variability in storage time was represented with a PERT distribution centred on the most likely value of 2 days. A correlation factor of –0.75% was used to ensure that combinations of high temperature and long storage time were unlikely (detectable spoilage and discard of product). The specific growth rate calculated by the growth model was multiplied by the storage time to give a value for the $\log_{10}$ increase in numbers of salmonellae.

6.3.5 Preparation and consumption

Cross-contamination

Cross-contamination of foods during the handling and preparation of raw meats is a recognized hazard in the home. If this parameter were ignored in a risk assessment model, it is likely that the risk would be underestimated. To estimate the risk of illness attributable to the cross-contamination of other foods during preparation of raw poultry, it is necessary to have information about the likelihood that cross-contamination will occur, and what is the likely number of bacteria transferred from raw to a ready-to-eat food. Section 6.2.6 described investigations that have studied these aspects. Estimates of the probability of cross-contamination were available from observational studies of food preparation behaviours (Worsfold and Griffith, 1997b). Estimates for the proportion of bacteria transferred from a raw food to hands or cutting board, and subsequently to other foods, were obtained from studies by Zhao et al. (1998). For the present risk assessment, two pathways of potential cross-contamination were modelled: from raw poultry via hands, and from raw poultry to cutting boards to other foods.

The probability of a person not washing their hands after handling raw poultry was estimated to be 0.6 (Worsfold and Griffith, 1997b).   The proportion of salmonellae transferred from the raw broiler to the hands was estimated to be 10% (Zhao et al., 1998). It was assumed that if salmonellae were present, the number of salmonellae on the broiler and the proportion transferred would determine the numbers transferred. If hands were then washed, no further cross-contamination occurred.   Hand washing was described by a Binomial distribution with a probability based on the values returned from the uncertainty model, as described previously. Based on this, if the model returned that hands were not washed, then salmonellae would be transferred to other foods. The numbers of salmonellae contaminating the other food was then calculated to be a function of the number of organisms on the hands and the proportion transferred.

Cross-contamination via cutting boards was simulated in the same way as cross-contamination from hands. However, here the probability estimate for the board being used for other food was 0.6 (Worsfold and Griffith, 1997b).

Cooking module

Preparing the food for consumption was modelled following an approach described by Fazil et al. (unpublished) and A.M. Fazil (personal communication) in a risk assessment for *Campylobacter* in poultry. Briefly, adequate cooking will destroy salmonellae and therefore it is only the broilers that are inadequately cooked that may still contain salmonellae at time of consumption (for the purposes of this module alone, post-cooking contamination is not considered). However, even with undercooking, it was assumed that salmonellae present on the external surfaces of the carcass will be inactivated, and that only some proportion of the total number – those more protected from heat penetration –would survive. The survival of the 'protected' bacteria will then depend on their heat resistance, and the length of time at some final temperature. The work of Fazil et al. (unpublished) and A.M. Fazil (personal communication) modelled this scenario based on published data for thermal profiles during cooking and on expert opinion, which were included in the example model for *Salmonella* in broilers.

The input variables in this module are sources of uncertainty in the example model. Table 6.30 shows the variables and their associated probabilities (Fazil et al., unpublished; A.M. Fazil, personal communication).

Table 6.30. Variables used to describe cooking of broilers.

Variable	Probabilities		
	Minimum	Most Likely	Maximum
Proportion of broilers not adequately cooked	0.05	0.1	0.15
Proportion of salmonellae in protected areas	0.1	0.16	0.2
Temperature exposure of protected bacteria (°C)	60	64	65
Time exposure of protected bacteria (minutes)	0.5	1	1.5

SOURCE: Fazil et al., unpublished.

The probability that a randomly selected broiler would be undercooked was determined by a Binomial distribution. If the simulation determined that the broiler was adequately cooked, the broiler was considered *Salmonella*-negative. If the simulation determined that the broiler was inadequately cooked, then the number of salmonellae surviving was calculated as described previously in this section. Having determined the number of salmonellae in protected areas and the time and temperature they may experience, a D-value was used to calculate the numbers of salmonellae surviving. The D-value applied was dependent on temperature and was developed and described in Equation 6.5 (see Section 6.2.6).

Consumption

Section 6.2.7 discussed the consumption data requirements for microbial risk assessment. For the purposes of this assessment, consumption data collected in Ireland by IUNA were used to estimate the range of amounts of chicken in a single serving that might be consumed by individuals 18 to 64 years of age. Note, however, that use of this model in a national setting will require country-specific data to be used. These data are shown as frequency and cumulative distributions in Figure 6.11. The amounts consumed were for meals consisting of whole portions of chicken meat; recipes in which the chicken was present as an ingredient were not considered. For chicken on the bone, the intake was calculated by correcting the weight to reflect the edible portion. The consumption database showed that over the 7 days of recording for 1379 subjects, 65.5% of subjects (903) consumed chicken on 1695 eating occasions. For the purposes of this assessment, the risk estimations were based on one serving of chicken every two weeks (specifically as whole portions, prepared from fresh whole carcasses in the home).

The cumulative frequency distribution was used during simulation to randomly generate serving weights for broiler meat. A United Kingdom retailer supplied data on the likely weights of broilers. Minimum weight was estimated to be 1.1 kg, maximum weight was estimated to be 2.5 kg, with a most likely value of 1.5 kg. These data were fitted with a PERT distribution, which was used during the simulation to randomly generate a broiler weight. Expert opinion from a United Kingdom producer estimated that 30% of the weight of a chicken was edible meat. Therefore the broiler weight was reduced by 70% to generate an edible meat weight. Finally, the edible meat weight was divided by the serving size to calculate the number of servings per broiler.

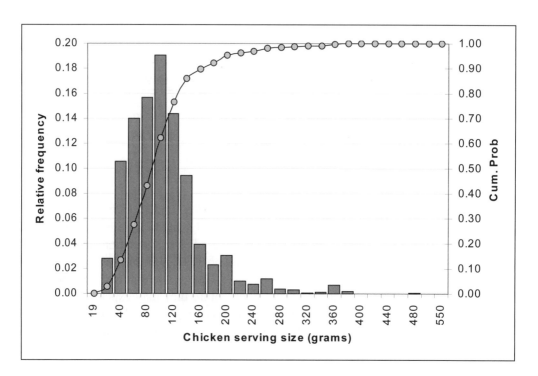

Figure 6.11. Frequency distribution for the consumption of chicken meat per eating occasion.

6.3.6 Calculation of the number of salmonellae consumed

The outcomes of exposure assessment are probability of ingestion and number of salmonellae ingested in a serving. The number of organisms ingested from undercooked poultry was calculated by dividing the number of organisms on a broiler by the number of servings from the broiler. The ingestion of *Salmonella* from raw poultry via a cross-contamination event was treated as a separate pathway with a separate risk estimate. By running the model through Monte Carlo simulation, distributions were generated of the number of salmonellae that a consumer might be exposed to per serving of cooked broiler meat, and per cross-contamination exposure event.

6.4 MODEL DESCRIPTION AND PARAMETERS

Table 6.31. End of processing.

Description	Variable	Unit	Distribution or Equation		
Prevalence	Prev			Min	Max
			Fixed value	0	1
Concentration	Conc	MPN/bird	Cumulative		

Table 6.32. Transport from processing plant to retail [not simulated in current model]

Description	Variable	Unit	Distribution or Equation		
Transport temperature	T_pr	degree C		Min	Max
			Uniform		
Transport time	t_pr	hours		Min	Max	CF
			Correlated uniform			-0.75
Minimum growth temperature	Tmin_pr	degree C	Constant	10	
Salt concentration	Slt_pr	%	Constant	1.9	
Log growth per hour	LGR_pr	log/hr	$=EXP(-6.2251-(0.0114*Slt\_pr) +(0.3234*T\_pr) +(0.002*(Slt\_pr*T\_pr)) -(0.0085*(Slt\_pr*Slt\_pr)) -(0.0045*T\_pr*T\_pr)))$		
Total log growth at retail	LG_pr	log	$=IF(T\_pr<Tmin\_pr,0,t\_pr*LGR\_pr)$		

Table 6.33. Storage at retail

Description	Variable	Unit	Distribution or Equation			
Retail temperature	Rtl_Temp	degree C		Mean	SD	Min	Max
			Truncated Normal	4	2.8	-7.2	10
Retail time	Rtl_Time	days		Mean	Max	CF
			Correlated Uniform	2	7	-0.75
Minimum growth temperature	MGT	degree C	Constant	10		
Salt concentration	NaCl	%	Constant	1.9		
Log growth per hour	LogSGR_Rtl	log/hr	$=EXP(-6.2251 -(0.0114*NaCl) +(0.3234*Rtl\_Temp) +(0.002*(NaCl*Rtl\_Temp)) -(0.0085*(NaCl*NaCl)) -(0.0045*(Rtl\_Temp*Rtl\_Temp)))$			
Total Log growth at retail	Rtl_growth	log	$=IF(Rtl\_Temp<MGT.0.Rtl\_Time*24*LogSGR\_Rtl)$			

Table 6.34. Transport from retail to home

Description	Variable	Unit	Distribution or Equation			
Ambient temperature during transport	Trans_Temp	degree C	Pert	Min 0	ML 13	Max 24
Maximum change in temperature during transport	TransMax	degree C	= Trans_Temp -Rtl_Temp			
Potential change in temperature during transport	Trans_DTemp1	degree C	Truncated Normal	Mean 3.72	SD 2.82	Min 0	Max TransMax
Change in temperature during transport	Trans_Dtemp2	degree C	=IF(Trans_Temp -Rtl_Temp<=0,0,Trans_DTemp1			
Chicken temperature after transport	Post_Trans_Temp	degree C	=Rtl_Temp +Trans_DTemp2			
Average transport temperature	Avg_Trans_Temp	degree C	=Average(Rtl_Temp, Post_Trans_Temp)			
Transport time	Trans_Time	Minutes	Correlated Cumulative	Min 5	Max 240	CF -0.75
Log growth per hour	LogSGR_Trans	log/hr	=EXP(-6.2251 -(0.0114*NaCl) +(0.3234*Avg_Trans_Temp) +(0.002*(NaCL*Avg_Trans_Temp (0.0085*(NaCL*NaCL)) -(0.0045*(Avg_Trans_Temp* Avg_Trans_Temp)))			
Total log growth during transport	Trans_growth	log	=IF(Avg_Trans_Temp<MGT,0,Trans_Time/60*LogS GR_Trans)			

Table 6.35. Storage at home

Description	Variable	Unit	Distribution or Equation			
Home storage temperature	Home_Temp	degree C	Truncated Normal	Mean 4	SD 2.65	Min -6.1	Max 21.1
Home storage time	Home_Time	days	Correlated PERT	Min 0	ML 2	Max 5	CF -0.75
Log growth per hour	LogSGR_Home	log/hr	=EXP(-6.2251 -(0.0114*NaCL) +(0.3234*Home_Temp) +(0.002*(NaCl*Home_Temp)) -(0.0085*(NaCl*NaCL)) -(0.0045*(Home_Temp*Home_Temp)))			
Total log growth in home	Home_growth	log	+IF(Home_Temp<MGT,0,Home_Time*24*LogSGR_Ho me)			
Total log growth in storage, transport and home	Growth	log	Rtl_growth + Trans-growth + Home_growth			

Table 6.36. Cross-contamination during preparation

Description	Variable	Unit	Distribution or equation			
Number of organisms on bird	Num	cells	=IF(Conc=0,0,10^Conc)			
Chickens ⇒ Hands						
Transfer from chicken to hands?	XCH	—	=IF(Num=0,0,1)			
Proportion trans-ferred from chicken	Pop_CH	pro-portion	Pert	Min 0	ML 0.1	Max 0.15
Number on hands	Num_H	cell	=IF(XCH=0,0,Num*Prop_CH)			
Number left on chicken	Num_C1	cell	=Num -Num_H			
Hands ⇒ Other food						
Probability that hands are not washed	HW_Prob	—	Beta	alpha 64	beta 46	
Hands not washed?	HW	—	=binomial(1,HW_Prob)			
Proportion trans-ferred from hands	Prop_HF	—	Pert	Min 0.00	ML 0.10	Max 0.15
Number on other foods via hands	Num_OF1	—	=IF(HW=0,0,Num_H*Prop_HF)			
Chickens ⇒ Board						
Transfer from chicken to board	XCB	—	=IF(Num=0,0,1)			
Proportion trans-ferred from chicken to board	Prop_CB	pro-portion		Min	ML	Max
			Pert	0	0.1	0.15
Number on board	Num_B	cell	=IF(XCB=0,0,Num*Prop_CB)			
Number left on chicken		cell	=NUm_C1 -Num_B			
Board ⇒ Other food						
Probability that board is used for other foods	Brd_use_Prob	—	Beta	alpha 66	beta 44	
Boards used for other food?	Brd_use	—	=binomial(1,Brd_use_Prob)			
Proportion trans-ferred from board	Prop_BF	—	Pert	Min 0.00	ML 0.10	Max 0.15
Number on other foods from chicken via board	Num_OF2	—	=IF(Brd_use=0,0,Num_B*Prop_BF)			
Number ingested via-cross-contamination	Num_XC	cell	=Num_OF1 +Num_OF2			
Ingestion via cross-contamination?	—	—	+IF(Num_XC=0,0,1)			

Table 6.37. Cooking

Description	Variable	Unit	Distribution or equation			
Probability of in-adequate cooking	Prob_AC	—		Min	ML	Max
			Pert	0.05	0.10	0.15
Adequately cooked?	AC	—	=binomial(1,1-Prod_AC)			
Proportion o0f cells in areas that permit a chance of survival	Prop_Prot			Min	ML	Max
			Pert	0.10	0.16	0.20
Log number of cells with chance of survival	Num_Prot	log cells	=IF(Conc=0,0,LOG10(10^Conc*Prop_Prot))			
Exposure time at exposure temp-erature for cells in "protected area"	Time_Prot	minutes		Min	ML	Max
			Pert	0.50	1.00	1.50
Exposure temp-erature during cooking in "protected areas"	Temp_Prot	degree C		Min	ML	Max
			Pert	60	64	65
D-value (at this temperature)	D_Prot	minutes	=10^(-0.139*Temp_Prot +8.58)			
log reduction in "protected area"	Prto_LR	log	=IF(AC=1,"death",Time_Prot / D_Prot)			

Table 6.38. Consumption.

Description	Variable	Unit	Distribution or equation			
Weight of a broiler carcass	Broiler_WT	gram		Min	ML	Max
			Pert	1100	1500	2500
Proportion of edible meat	Prop_edible	—	Fixed. 0.3			
Weight of edible meat	Edible_WT	gram	=Broiler_WT*Prop_Edible			
Serving size	Serve_size	gram		Min	Max	
			Cumulative	19	550	
Number of servings per broiler	Num_Serve	—	=IF(Edible_WT<Serve_Size,1, ROUND(Edible_WT/Serve_Size,0))			

6.5 REFERENCES CITED IN CHAPTER 6

Abu-Ruwaida, A., Sawaya, W., Dashti, B., Murad, M., & Al-Othman, H. 1994. Microbiological quality of broilers during processing in a modern commercial slaughterhouse in Kuwait. *Journal of Food Protection*, **57**: 887–892.

ACMSF [Advisory Committee on the Microbiological Safety of Food]. 1996. Report on Poultry Meat. London: HMSO.

Al Busaidy, S. [2000]. Information supplied by the Ministry of Health, Sultanate of Oman in response to the FAO/WHO Call for Data.

Altekruse, S., Street, D., Fein, S., & Levy, A. 1995. Consumer knowledge of foodborne microbial hazards and food-handling practices. *Journal of Food Protection*, **59**: 287–294.

Anderson, R., & May, R. 1991. *Infectious diseases of humans: dynamics and control*. Oxford, UK: Oxford University Press.

Anderson, W., McClure, P., Baird-Parker, A., & Cole, M. 1996. The application of a log-logistic model to describe the thermal inactivation of *Clostridium botulinum* 213B at temperatures below 121.1°C. *Journal of Applied Bacteriology*, **80**

Angen, O., Skov, M., Chriel, M., Agger, J., & Bisgaard, M. 1996. A retrospective study on *Salmonella* infection in Danish broiler flocks. *Preventative Veterinary Medicine*, **26**: 223–237.

Anon. 1998. Validation of predictive models of microbial growth. MAFF predictive microbiology group. MSG 97/4/4

Armas, A.D., Wynne, A., & Sutherland, J.P. 1996. Validation of predictive models using independently published data. Poster/Abstract. *2nd International Conference of Predictive Microbiology, Hobart, Tasmania*.

ARZN. 1998. Trends and sources of zoonotic agents in animals, feedstuffs, food and man in the European Union and Norway. Data provided pusuant to Article 5, Paragraph 1, of Council Directive 92/117/EEC). Collated and published by the Federal Institute for Health Protection of Consumers and Veterinary Medicine (BgVV), Berlin, Germany.

Audits International. [1999]. Information on United States cold temperature evaluation. Submitted to FAO/WHO in response to 2000 call for data.

Baker, R., Poon, W., & Vadehra, D. 1983. Destruction of *Salmonella typhimurium* and *Staphylococcus aureus* in poultry products cooked in a conventional and microwave oven. *Poultry Science*, **62**: 805–810.

Baranyi, J., & Roberts, T. 1994. A dynamic approach to predicting bacterial growth in food. *International Journal of Food Microbiology*, **23**: 277–294.

Baranyi, J., & Roberts, T. 1995. Mathematics of predictive food microbiology. *International Journal of Food Microbiology*, **26**: 199–218.

Baranyi, J., Pin, C., & Ross, T. 1999. Validating and comparing predictive models. *International Journal of Food Microbiology,* **48**: 159–166.

Baumgartner, A., Heimann, P., Schmid, H., Liniger, M., Simmen, A. 1992. *Salmonella* contamination of poultry carcasses and human salmonellosis. *Archiv fur Lebensmittelhygiene*, **43**: 121–148.

BgVV [Bundesinstitut für gesundheitlichen Verbraucherschutz und Veterinärmedizin]. 2000. Gericht über die epidemiologiche Situation der Zoonosen in Deutschland für 1999. Herausgeben von M. Hartung. BgVV-Hefte 08/2000. Berlin, Germany.

Blackburn, W., Curtis, L., Humpheson, L., Billon, C., & McClure, P. 1997. Development of thermal inactivation models for *Salmonella* Enteritidis and *Escherichia coli* 0157:H7 with temperature, pH and NaCl as controlling factors. *Journal of Food Microbiology*, **38**: 31–44.

Boonmar, S., Bangtrakulnonth, A., Pornrunangwong, S., Marnrim, N., Kaneko, K., & Ogawa, M. 1998a. Predominant serovars of *Salmonella* in humans and foods from Thailand. *Journal of Veterinary Medical Science*, **60**: 877–880.

Broughall, J., & Brown, C. 1984. Hazard analysis applied to microbial growth in foods: development and application of three-dimensional models to predict bacterial growth. *Journal of Food Microbiology*, **1**: 13–22.

Brown, M., Davies, K., Billon, C., Adair, C., & McClure, P. 1998. Quantitative microbiological risk assessment: principles applied to determining the comparative risk of salmonellosis from chicken products. *Journal of Food Protection*, **61**: 1446–1453.

Bryan, F. 1971. Use of time-temperature evaluations in detecting the responsible vehicle and contributing factors of foodborne disease outbreaks. *Journal of Milk Food Technology*, **34**(12): 576–582.

Bryan, F., & McKinley, W. 1974. Prevention of foodborne illness by time-temperature control of thawing, cooking, chilling, and preheating turkeys in school lunch kitchens. *Journal of Milk Food Technology*, **37**: 420–429.

Buchanan, R.L., & Cygnarowicz, M. 1990. A mathematical approach toward defining and calculating the duration of the lag-phase. *Food Microbiology*, **7**: .

Buchanan, R.L., & Whiting, R.C., 1997. Rish assessment – a means for linking HACCP plans and public health. *Journal of Food Protection*, **61**: 1531–1535.

Buchanan, R.L., Whiting, R.C., & Damart, W. 1997. When is simple good enough: a comparison of the Gompertz, Baranyi and three-phase linear models for fitting bacterial growth curves. *Journal of Food Microbiology*, **14**: 313–326.

Campbell, D., Johnston, R., Campbell, G., McClain, D., & Macaluso, J. 1983. The microbiology of raw, eviscerated chickens: a ten-year comparison. *Poultry Science*, **62**: 437–444.

Carraminana, J., Yanguela, J., Blanco, D., Rota, C., Agustin, A., Arino, A., & Herrera, A. 1997. *Salmonella* incidence and distribution of serotypes throughout processing in a Spanish poultry slaughterhouse. *Journal of Food Protection*, **60**: 1312–1317.

Casolari, A. 1994. About basic parameters of food sterilization technology. Letter to the editor. *Journal of Food Microbiology*, **11**: 75–84.

Cassin, M., Lammerding, A., Todd, E., Ross, W., & McColl, R. 1998. Quantitative risk assessment for *Escherichia coli* 0157:H7 in ground beef hamburgers. *International Journal of Food Microbiology*, **41**: 21–44.

Cerf, O. 1977. Tailing of survival curves of bacterial spores. *Journal of Applied Bacteriology*, **42**: 1–19.

CFIA [Canadian Food Inspection Agency]. 2000. Canadian Microbiological Baseline Survey of Chicken Broiler and Young Turkey Carcasses, June 1997 – May 1998. (On Internet at http://inspection.gc.ca/english/anima/meavia/mmopmmhv/chap19/baseline-e.pdf)

Chambers, J., Bisaillon, J., Labbe, Y., Poppe, C., & Langford, C. 1998. *Salmonella* prevalence in crops of Ontario and Quebec broiler chickens at slaughter. *Poultry Science*, **77**: 1497–1501.

Chang, H., Carpenter, J., & Toledo, R. 1998. Temperature histories at critical points and recommended cooking time for whole turkeys baked in a conventional oven. *Journal of Food Science*, **62**: 262–266.

Chen, H., & Marks, B. 1997. Evaluating previous thermal treatment of chicken patties by visible/near-infrared spectroscopy. *Journal of Food Science*, **62**: 753–756.

Clark, A. 1933. *The Significance of Time Action Curves. The Mode of Action of Drugs on Cells.* pp. 163–188. London: Edward Arnold and Co.

Cogan, T., Bloomfield, S., & Humphrey, T. 1999. The effectiveness of hygiene procedures for prevention of cross-contamination from chicken carcasses in the domestic kitchen. *Letters in Applied Microbiology*, **29**: 354–358.

Cole, M., Davies, K., Munro, G., Holyoak, C., & Kilsby, D. 1993. A vitalistic model to describe the thermal inactivation of *Listeria monocytogenes. Journal of Industrial Microbiology*, **12**: 232–239.

Corrier, D.E., Nisbet, D.J., Scanlan, C.M., Hollister, A.G., Caldwell, D.J., Thomas, L.A., Hargis, B.M., Tompkins, T., & Deloach, J.R. 1995. Treatment of commercial broiler chickens with a characterized culture of cecal bacteria to reduce salmonellae colonization. *Poultry Science*, **74**(7): 1093-1101.

CVM [Centre for Veterinary Medicine, US Food and Drug Administration]. 2001. Risk assessment on the human health impact of fluoroquinolone resistant *Campylobacter* associated with consumption of chicken. Published 18 October 2000, and revised as of January 2001. Available at www.fda.gov/cvm/antimicrobial/Risk_asses.htm.

Dalgaard, P. 2000. Fresh and lightly preserved seafood. pp. 110-139, *in:* C.M.D. Man and A.A. Jones (eds). *Shelf Life Evaluation of Foods.* 2nd Edition. Maryland, USA: Aspec Publishing Incorporated.

de Boer, E., & Hahne, M. 1990. Cross-contamination with *Campylobacter jejuni* and *Salmonella* spp. from raw chicken products during food preparation. *Journal of Food Protection*, **53**: 1067–1068.

de Wit, J., Broekhuizen, G., & Kampelmacher, E. 1979. Cross-contamination during the preparation of frozen chickens in the kitchen. *Journal of Hygiene*, **83**: 27–32.

Dougherty, T. 1974. *Salmonella* contamination in a commercial poultry (broiler) processing operation. *Poultry Science*, **53**: 814–821.

Duffy, G., Ellison, A., Cole, W., & Stewart, G. 1994. Use of bioluminescence to model the thermal inactivation of *Salmonella typhimurium* in the presence of a competitive microflora. *Journal of Food Microbiology*, **23**: 467–477.

EC [European Commission]. 1998. Trends and sources of zoonotic agents in animals, feedstuffs, food and man in the European Union in 1998. Berlin.

Eley, A.R. 1996. *Microbial Food Poisoning.* London: Chapman and Hall.

Ellison. A., Anderson, W., Cole, M., & Stewart, G. 1994. Modelling the thermal inactivation *Salmonella typhimurium* using bioluminescence data. *International Journal of Food Microbiology*, **23**: 467–477.

FAO-OIE-WHO. 1995. Animal Health Yearbook. FAO, Rome.

Fazil, A.M., Lowman, R., Stern, N., & Lammerding, A.M. Unpublished. A quantitative risk assessment model for *C. jejuni* in chicken. Abstract CF10. p.65, *in: Proceedings of the 10th International Workshop on CHRO.* Baltimore, Maryland.

Fuzihara, T.O., Fernandes, S.A., & Franco, B.D.G.M. 2000. Prevalence and dissemination of Salmonella serotypes along the slaughtering process in Brazilian small poultry slaughterhouses. *Journal of Food Protection*, **63**: 1749–1753.

Garey, J., & Simko, M. 1987. Adherence to time and temperature standards and food acceptability. *Journal of the American Dietetic Association*, **87**: 1513–1518.

Garthright, W. 1991. Refinements in the prediction of microbial growth curves. *Food Microbiology*, **8**: 239–248.

Geornaras, G., & von Holy, A. 1994. Bacterial contamination in poultry processing. *Food Industries of South Africa*, 1994: 31–34.

Gibson, A., Bratchell, N., & Roberts, T. 1988. Predicting microbial growth: growth responses of salmonellae in a laboratory medium as affected by pH, sodium chloride and storage temperature. *International Journal of Food Microbiology*, **6**(2): 155–178.

Goodfellow, S., & Brown, W. 1978. Fate of *Salmonella* inoculated into beef for cooking. *Journal of Food Protection*, **41**: 598–605.

Goren, E., de Jong, W.A., Doornenbal, P., Bolder, N.M., Mulder, R.W.A.W., & Jansen, A. 1988. Reduction of *Salmonella* infection of broilers by spray application of intestinal microflora: a longitudinal study. *Veterinary Quarterly*, **10**(4): 249–255.

Haas, C., Rose, J., & Gerba, C. 1999. *Quantitative microbial risk assessment.* Chichester, UK: John Wiley & Sons,

Hadad, J., & Mohammed Ali, K. 1986. Isolation of *Salmonella* from poultry and poultry meat. *Journal of Veterinary Medicine*, **34**: 189–199.

Hall, L.B., & Slade, A. 1981. *Food poisoning organisms in food - effect of freezing and cold storage III (final report).* Campden Food Preservation Research Association, Chipping Campden, United Kingdom. *Technical Memorandum,* No. 276.

Harris, N., Thompson, D., Martin, D., & Nolan, C. 1986. A survey of *Campylobacter* and other bacterial contaminants of pre-market chicken and retail poultry and meats, Kings County, Washington. *American Journal of Public Health*, **76**: 401–406.

Hartnett, E., Kelly, L., Newell, D., Wooldridge, M., & Gettinby, G. 2001. A quantitative risk assessment for the occurrence of *Campylobacter* in chickens at the point of slaughter. *Epidemiology and Infection,* **127**(2): 195-206.

Hartung, M. [1999]. Salmonellosis – sources of infection in Germany. Information provided as a personal communication, summarizing in English the material published in German as *Bericht uber die epidemiologische Situation der Zoonosen in Deutschland fur 1998.* Bundesinstitut für gesundheitlichen Verbraucherschutz und Veterinärmedizin, Berlin. BgVV-Hefte, 09/1999.

Huis in 't Veld, J., Mulder, R., & Snijders, J. 1994. Impact of animal husbandry and slaughter technologies on microbial contamination of meat: monitoring and control. *Meat Science,* **36**: 123–154.

Humbert, F. 1992. *Salmonella* and poultry production: epidemiology and incidence on human health. *Le Point Veterinaire,* **24**.

Humphrey, T. 1981. The effects of pH and levels of organic matter on the death rates of *Salmonellas* in chicken scald-tank water. *Journal of Applied Bacteriology,* **51**: 27–39.

Humphrey, T., & Lanning, D. 1987. *Salmonella* and *Campylobacter* of broiler chicken carcasses and scald tank water: the influence of water pH. *Journal of Applied Bacteriology,* **63**: 21–25.

Humphrey, T.J., Martin, K.W., & Whitehead, A. 1994. Contamination of hands and work surfaces with *Salmonella enteritidis* PT4 during the preparation of egg dishes. *Epidemiology and Infection,* **113**(3): 403–409.

Ibarra, J., Tao, Y., Walker, J., & Griffis, C. 1999. Internal temperature of cooked chicken meat through infrared imaging and time series analysis. *Transactions of the American Society of Agricultural Engineers,* **42**: 1383–1390.

ICMSF [International Commission for the Microbiological Specifications of Foods]. 1996. *Micro-organisms in food.* Vol. 5: *Characteristics of Microbial Pathogens.* London: Blackie Academic and Professional. 513 pp.

IUNA [Irish Universities Nutrition Alliance]. 2000. North/South Ireland Food Consumption Survey. Report. Dublin: Food Safety Promotion Board.

Izat, A., Druggers, C., Colberg, M., Reiber, M., & Adams, M. 1989. Comparison of the DNA probe to culture methods for the detection of *Salmonella* on poultry carcasses and processing waters. *Journal of Food Protection,* **52**: 564–570.

Izat, A., Kopek, J., & McGinnis, J. 1991. Research note: incidence, number and serotypes of *Salmonella* on frozen broiler chickens at retail. *Poultry Science,* **70**: 1438–1440.

Izat, A.L., Yamaguchi, W., Kaniawati, S., McGinnis, J.P., Raymond, S.G., Hierholzer, R.E., & Kopek, J.M. 1991. Research note: use of consecutive carcass rinses and a most probable number procedure to estimate salmonellae contaminations. *Poultry Science,* **70**(6): 1448-1451.

Jacobs-Reitsma, W.F., Bolder, N.M., & Mulder, R.W.A.W. 1994. Cecal carriage of *Campylobacter* and *Salmonella* in Dutch broiler flocks at slaughter: a one-year study. *Poultry Science,* **73**(8): 1260–1266.

Jacobs-Reitsma, W.F., Bolder, N.M., & Mulder, R.W.A.W. 1991. *Salmonella* and *Campylobacter-* free poultry products: an utopia? Prevention and control of potentially pathogenic microorganisms in poultry and poultry meat processing. 4. Hygienic Aspects of Processed Poultry Meat.

James, W., Williams, W., Prucha, J., Johnston, R., & Christensen, W. 1992a. Profile of selected bacterial counts and *Salmonella* prevalence on raw poultry in a poultry slaughter establishment. *Journal of the American Veterinary Medical Association,* **200**: 57–59.

James, W., Brewer, R., Prucha, J., Williams, W., Parham, D. 1992b. Effects of chlorination of chill water on the bacteriologic profile of raw chicken carcasses and giblets. *Journal of the American Veterinary Medical Association,* **200**: 60–63.

James, W., Prucha, J., Brewer, R., Williams, W., Christensen, W., Thaler, A., & Hogue, A. 1992c. Effects of counter-current scalding and post-scald spray on the bacteriologic profile of raw chicken carcasses. *Journal of the American Veterinary Medical Association*, **201**: 705–708.

Jay, L., Comar, D., & Govenlock, L. 1999. A video study of Australian domestic food-handling practices. *Journal of Food Protection*, **62**: 1285–1296.

Jernglinchan, J., Koowatananukul, C., Daengprom, K., & Saitanu, K. 1994. Occurrence of *Salmonella* in raw broilers and their products in Thailand . *Journal of Food Protection*, **57**: 808–810.

Jones, F., Axtell, R., Rives, D., Scheideler, S., Tarver, F., Walker, R., & Wineland, M. 1991a. A survey of *Salmonella* contamination in modern broiler production. *Journal of Food Protection*, **54**: 502–507.

Jones, F., Axtell, R., Tarver, F., Rives, D., Scheideler, S., & Wineland, M. 1991b. *Environmental factors contributing to Salmonella colonization of chickens. Colonization Control of Human Bacterial Entopathogens in Poultry.* pp. 3–20. San Diego, CA: Academic Press.

Josephson, K., Rubino, J., Pepper, I. 1997. Characterization and quantification of bacterial pathogens and indicator organisms in household kitchens with and without the use of disinfectant cleaner. *Journal of Applied Microbiology*, **83**: 737–750.

Kahneman, D., Slovic, P., Tversky, A. 1982. *Judgement under uncertainty: Heuristics and biases.* New York, NY: Cambridge University Press.

Kotula, K., & Pandya, Y. 1995. Bacterial contamination of broiler chickens before scalding. *Journal of Food Protection*, **58**: 1326–1329.

Kutsar, K. [2000]. Information from the Health Protection Inspectorate, Estonia, submitted in response to the FAO/WHO Call for Data.

Lahellec, C., Colin, P., & Bennejean, G. 1986. Influence of resident *Salmonella* on contamination of broiler flocks. *Poultry Science*, **65**: 2034–2039.

Lammerding, A., & Fazil, A. 2000. Hazard identification and exposure assessment for microbial food safety risk assessment. *International Journal of Food Microbiology*, **58**: 1–11.

Lammerding, A., Garcia, M., Mann, E., Robinson, Y., Dorward, W., Truscott, R., & Tittiger, F. 1988. Prevalence of *Salmonella* and thermophilic *Campylobacter* in fresh pork, beef, veal and poultry in Canada. *Journal of Food Protection*, **51**: 47–52.

Lillard, H. 1980. Effect on broiler carcasses and water of treating chilled water with chlorine and chlorine dioxide . *Poultry Science*, **59**: 1761–1766.

Lillard, H. 1988. Comparison of sampling methods and implications for bacterial decontamination of poultry carcasses by rinsing . *Journal of Food Protection*, **51**: 405–408.

Lillard, H. 1989a. Factors affecting the persistence of *Salmonella* during the processing of poultry. *Journal of Food Protection*, **52**: 829–832.

Lillard, H. 1989b. Incidence and recovery of salmonellae and other bacteria from commercially processed poultry carcasses at selected pre- and post-evisceration steps. *Journal of Food Protection*, **52**: 88–91.

Lillard, H. 1990. The impact of commercial processing procedures on the bacterial contamination and cross-contamination of broiler carcasses. *Journal of Food Protection*, **53**: 202–204.

Little, C., Adams, M., Anderson, W., & Cole, M. 1994. Application of a logistic model to describe the survival of *Yersinia enterocoiliica* at sub-optimal pH and temperature. *International Journal of Food Microbiology*, **22**: 63–71.

Lyon, C, Lyon, B., Klose, A., & Hudspeth, J. 1975. Effects of temperature-time combinations on doneness and yields of water-cooked broiler thighs. *Journal of Food Science*, **40**: 129–132.

Machado, J., & Bernardo, F. 1990. Prevalence of *Salmonella* in chicken carcasses in Portugal. *Journal of Applied Bacteriology*, **69**: 447–480.

Mackey, B., Derrick, C. 1987. The effects of prior heat shock on the thermal resistance of *Salmonella thompson* in foods. *Letters in Applied Microbiology*, **5**: 115–118.

McBride, G., Skura, B., Yoda, R., & Bowmer, E. 1980. Relationship between incidence of *Salmonella* contamination among pre-scalded, eviscerated and post-chilled chickens in a poultry processing plant. *Journal of Food Protection*, **43**: 538–542.

McClure, P.J., Beaumont, A.L., Sutherland, J.P., & Roberts, T.A. 1997. Predictive modelling of growth of *Listeria monocytogenes*: the effects on growth of NaCl, pH, storage temperature and NaNO$_2$. *International Journal of Food Microbiology*, **34**: 221–232.

McLennon, W., & Podger, A. 1995. *National Nutritional Survey of Foods Eaten in Australia.* Australian Bureau of Statistics.

McMeekin, T.A., Olley, J., Ross, T., & Ratkowsky, D.A. 1993. *Predictive microbiology. Theory and application.* Taunton, UK: Research Studies Press.

Mead, G. 1992. Food poisoning *Salmonellas* in the poultry-meat industry. *The Meat Hygienist*, 1992: 6–12.

Morris, G., & Wells, J. 1970. *Salmonella* contamination in a poultry-processing plant. *Applied Microbiology*, **19**: 795–799.

MSF [Microbiological Safety of Food]. 1990. Report of the Committee on the Microbiological Safety of Food, Part 1. London: HMSO.

Mulder, R., & Schlundt, J. (in press) Control of foodborne pathogens in poultry meat.

Murakami, K., Horikawa, K., Ito, T., & Otsuki, K. 2001. Environmental survey of salmonella and comparison of genotypic characters with human isolates in Western Japan. *Epidemiology and Infection,* **126**: 159–171.

Murphy, R., Marks, B., Johnson, E., & Johnson, M. 1999. Inactivation of *Salmonella* and *Listeria* in ground chicken breast meat during thermal processing. *Journal of Food Protection*, **62**: 980–985.

Notermans, S., & Kampelmacher, E. 1974. Attachment of some bacterial strains to the skin of broiler chickens. *British Poultry Science*, **15**: 573–585.

Notermans, S., Kampelmacher, E. 1975. Further studies on the attachment of bacteria to skin. *British Poultry Science*, **16**: 487–496.

Nychas, G., & Tassou, C. 1996. Growth/survival of *Salmonella enteritidis* on fresh poultry and fish stored under vacuum or modified atmosphere. *Letters in Applied Microbiology*, **23**: 115–119.

Oscar, T.P. (in press) A risk assessment for *Salmonella* spp., *Campylobacter jejuni*, and chicken.

Oscar, T.P. 1997. Use of computer simulation modelling to predict the microbiological safety of chicken. *Proceedings of the 32nd National Meeting of Poultry and Health and Processing*.

Oscar, T.P. 1998. The development of a risk assessment model for use in the poultry industry. *Journal of Food Safety*, **18**: 371–381.

Oscar, T.P. 1999a. Response surface models for effects of temperature, pH, and previous growth pH on growth kinetics of *Salmonella typhimurium* in brain heart infusion broth. *Journal of Food Protection*, **62**: 106–111.

Oscar, T.P. 1999b. Response surface models for effects of temperature, pH, and previous temperature on lag-time and specific growth rate of *Salmonella typhimurium* on cooked ground chicken breast. *Journal of Food Protection*, **62**: 1111–1114.

Oscar, T.P. 1999c. Response surface models for effects of temperature and previous growth sodium chloride on growth kinetics of *Salmonella typhimurium* on cooked chicken breast. *Journal of Food Protection*, **12**: 1470–1474.

Patrick, T., Collins, J., & Goodwin, T. 1973. Isolation of *Salmonella* from carcasses of steam- and water-scalded poultry. *Journal of Milk Food Technology*, **36**: 34–36.

Pether, J., & Gilbert, R. 1971. The survival of *Salmonellas* in finger-tips and transfer of the organisms to foods. *Journal of Hygiene*, **69**: 673–681.

Poppe, C., Irwin, R., Messier, S., Finley, G., & Oggel, J. 1991. The prevalence of *Salmonella enteritidis* and other *Salmonella* spp. among Canadian registered commercial chicken broiler flocks. *Epidemiology and Infection*, **107**: 201–211.

Ratkowsky, D., Olley, J., McMeekin, T., & Ball, A. 1982. Relationship between temperature and growth rate of bacterial cultures. *Journal of Bacteriology*, **149**: 1–5.

Rengel, A., & Mendoza, S. 1984. Isolation of *Salmonella* from raw chicken in Venezuela. *Journal of Food Protection*, **47**: 213–216.

Restaino, L., & Wind, C. 1990. Antimicrobial effectiveness of handwashing for food establishments. *Dairy, Food and Environmental Sanitation*, **10**: 136–141.

Reybrouck, G. 1986. Handwashing and hand disinfection. *Journal of Hospital Infection*, **8**: 5–23.

Rhan, O. 1945. *Injury and death of bacteria by chemical agents*. Biodynamica Monograph, No. 3. Biodynamica Normandy

Rigby, C., Pettit, J., Bentley, A., Salomons, M., & Lior, H. 1980a. Sources of salmonellae in an uninfected commercially-processed broiler flock. *Canadian Journal of Comparative Medicine*, **44**: 267–274.

Rigby, C., Pettit, J., Baker, M., Bently, A., Salomons, M., & Lior, H. 1980b. Flock infection and transport as sources of salmonellae in broiler chickens and carcasses. *Canadian Journal of Comparative Medicine*, **44**: 324–337.

Rose, N., Beaudeau, F., Drouin, P., Toux, J., Rose, V., & Colin, P. 1999. Risk factors for *Salmonella enterica* subsp. *enterica* contamination in French broiler-chicken flocks at the end of the rearing period. *Preventative Veterinary Medicine*, **39**: 265–277.

Ross, T. 1999. *Predictive food microbiology models in the meat industry*. Sydney, Australia: Meat and Livestock Australia. 196 pp.

Ross, T., & McMeekin, T.A. 1994. Predictive microbiology – a review. *International Journal of Food Microbiology*, **23**: 241–264.

Ross, T., Baranyi, J. & McMeekin, T.A. 1999. Predictive Microbiology and Food Safety. *In:* R. Robinson, C.A. Batt and P. Patel (eds). *Encyclopaedia of Food Microbiology.* London: Academic Press.

Ross, T., Dalgaard, P., & Tienungoon, S. 2000. Predictive modelling of the growth and survival of Listeria in fishery products. *International Journal of Food Microbiology*, **62**: 231–246.

Rusul, G., Khair, J., Radu, S., Cheah, C., & Yassin, R. 1996. Prevalence of *Salmonella* in broilers at retail outlets, processing plants and farms in Malaysia. *International Journal of Food Microbiology*, **33**: 183–194.

Sawaya, W., Elnawawy, A., Abu-Ruwaida, A., Khalafawi, S., & Dashti, B. 1995. Influence of modified atmosphere packaging on shelf-life of chicken carcasses under refrigerated storage conditions. *Journal of Food Safety*, **15**: 35–51.

Schnepf, M., & Barbeau, W. 1989. Survival of *Salmonella typhimurium* in roasting chickens cooked in o microwave, conventional microwave, and a conventional electric oven. *Journal of Food Safety*, **9**: 245–252.

Schutze, G., Sikes, J., Stefanova, R., & Cave, M. 1999. The home environment and salmonellosis in children. *Paediatrics*, **103**.

Scott, E. 1996. Foodborne disease and other hygiene issues in the home. *Journal of Applied Bacteriology*, **80**: 5–9.

Scott, E., & Bloomfield, S. 1990. The survival and transfer of microbial contamination via cloths, hands and utensils. *Journal of Applied Bacteriology*, **68**: 271–278.

Slavik, M., Jeong-Weon, K., & Walker, J. 1995. Reduction of *Salmonella* and *Campylobacter* on chicken carcasses by changing scalding temperature. *Journal of Food Protection*, **58**: 689–691.

Snyder, O.P., Jr. 1999. A 'safe hands' hand wash program for retail food operations: a technical review. Available at : www.hi-tm.com/documents/handwash-FL99.html.

Soerjadi-Liem, A., &Cumming, R. 1984. Studies on the incidence of *Salmonella* carriers in broiler flocks entering a poultry processing plant in Australia. *Poultry Science*, **63**: 892–895.

Sutherland, J.P., & Bayliss, A.J. 1994. Predictive modelling of growth of *Yersinia enterocolitica*: the effects of temperature, pH and sodium chloride. *International Journal of Food Microbiology*, **21**: 197–215.

Sutherland, J.P., Bayliss, A.J., & Roberts, T.A. 1994. Predictive modelling of growth of *Staphylococcus aureus*: the effects of temperature, pH and sodium chloride. *International Journal of Food Microbiology*, **21**: 217–236.

Sutherland, J.P., Bayliss, A.J., & Braxton, D.S. 1995. Predictive modelling of growth of *Escherichia coli* O157:H7: the effects of temperature, pH and sodium chloride. *International Journal of Food Microbiology*, **25**: 29–49.

Surkiewiez, B., Johnston, R., Moran, A., & Krumm, G. 1969. A bacteriological survey of chicken eviscerating plants. *Food Technology*, **23**: 80–83.

Swaminathan, B., Link, M., & Ayers, J. 1978. Incidence of *Salmonella* in raw meat and poultry samples in retail stores. *Journal of Food Protection*, **41**: 518–520.

te Giffel, M.C. & Zwietering, M.H. 1999. Validation of predictive models describing the growth of *Listeria monocytogenes*. *International Journal of Food Microbiology*, **46**: 135–149.

Terisotto, S., Tiburzi, M.C., Jiménez, S.M., Tessi, M.A., & Moguilevsky, M.A. 1990. Prevalencia de *Salmonella* encanales de pollos evisceradas. *La Industria Cárnica Latinoamericana*, **18**: 40–46.

Thayer, D., Muller, W., Buchanan, R.L., & Phillips, J. 1987. Effect of NaCl, pH, temperature, and atmosphere in growth of *Salmonella typhimurium* in glucose-mineral salts medium. *Applied and Environmental Microbiology*, **53**: 1311–1315.

USDA-FSIS [United States Department of Agriculture, Food Safety and Inspection Service]. 1996. Nationwide broiler chicken microbiological baseline data collection program. pp. 1-34. Washington DC: Government Printing Office.

USDA-FSIS. 1998. *Salmonella* Enteritidis Risk Assessment. Shell Eggs and Egg Products. Final Report. Prepared for FSIS by the *Salmonella* Enteritidis Risk Assessment Team. 268 pp. Available on Internet as PDF document at: www.fsis.usda.gov/ophs/risk/contents.htm.

Uyttendaele, M., de Troy, P., & Debevere, J. 1999. Incidence of *Salmonella, Campylobacter jejuni, Campylobacter coli,* and *Listeria monocytogenes* in poultry carcasses and different types of poultry products for sale on the Belgian retail market. *Journal of Food Protection*, **62**: 735–740.

Uyttendaele, M., Debevere, J., Lips, R., & Neyts, K. 1998. Prevalence of *Salmonella* in poultry carcasses and their products in Belgium. *International Journal of Food Microbiology*, **40**: 1–8.

Van Gerwen, S.J.C., te Giffel, M.C., van 't Riet, K., Beumer, R.R., & Zwietering, M.H. 2000. Stepwise quantitative risk assessment as a tool for characterization of microbiological food safety. *Journal of Applied Microbiology*, **88**: 938–951.

Vose, D. 2000. *Risk analysis: a quantitative guide.* Chichester, UK: John Wiley & Sons,

Waldroup, A., Rathgeber, B., Forsythe, R., & Smoot, L. 1992. Effects of six modifications on the incidence and levels of spoilage and pathogenic organisms on commercially processed post-chill broilers. *Journal of Applied Poultry Research*, **1**: 226–234.

Walls, I., Scott, V.N. & Bernard, D. 1996. Validation of predictive mathematical models describing growth of *Staphylococcus aureus*. *Journal of Food Protection*, **59**: 11–15.

White, C., & Hall, L. 1984. The effect of temperature abuse on *Staphylococcus aureus* and salmonellae in raw beef and chicken substrates during frozen storage. *Food Microbiology*, **1**: 29–38.

White, P., Baker, A., & James, W. 1997. Strategies to control *Salmonella* and *Campylobacter* in raw poultry products. *Revue scientifique et technique de l'Office international des epizooties*, **16**: 525–541.

Whiting, R.C. 1993. Modelling bacterial survival in unfavourable environments. *International Journal of Food Microbiology*, **12**: 240-246.

Wilson, I., Wilson, T., & Weatherup, S. 1996. *Salmonella* in retail poultry in Northern Ireland.

Withell, E. 1942. The significance of the variation on shape of time-survivor curves. *Journal of Hygiene*, **42**: 124–183.

Worsfold, D., & Griffith, C. 1995. A generic model for evaluating consumer food safety behaviour. *Food Control*, **6**: 357–363.

Worsfold, D., & Griffith, C. 1996. Cross-contamination in domestic food preparation. *Hygiene and Nutrition in Foodservice and Catering*, **1**: 151–162.

Worsfold, D., & Griffith, C. 1997a. Food safety behaviour in the home. *British Food Journal*, **99**: 97–104.

Worsfold, D., & Griffith, C. 1997b. Assessment of the standard of consumer food safety behaviour . *Journal of Food Protection*, **60**: 399–406.

Zhao, P., Zhao, T., Doyle, M., Rubino, J., & Meng, J. 1998. Development of a model for evaluation of microbial cross-contamination in the kitchen. *Journal of Food Protection*, **61**: 960–963.

Zwietering, M.H., de Koos, J.T., Hasenack, B.E., de Wit, J.C. & van't Riet, K. 1991. Modelling of bacterial growth as a function of temperature. *Applied and Environmental Microbiology*, **57**: 1094–1101.

7. RISK CHARACTERIZATION OF *SALMONELLA* IN BROILERS

7.1 SUMMARY

In this section, the results from the exposure assessment are used within the hazard characterization to estimate two quantities: the risk per serving and the risk from cross-contamination as a result of preparing that serving. As before, for the exposure assessment, the risk characterization is not specific to any country and thus comparison with surveillance data is not appropriate. Following calculation of the baseline model, the effect of a number of mitigation strategies is investigated.

7.2 RISK ESTIMATION

7.2.1 Results

The risk estimate for probability of illness was first simulated using a set prevalence for the presence of *Salmonella* in chilled, raw broiler chickens. At a prevalence of 20% contaminated carcasses, and based on the other model parameters, including the probability that the product will be undercooked, approximately 2% of the broilers prepared for consumption in the home could potentially contain viable cells of *Salmonella*. Figure 7.1 shows the distribution of average doses (colony-forming units, CFUs) per serving for contaminated chicken that is subsequently undercooked.

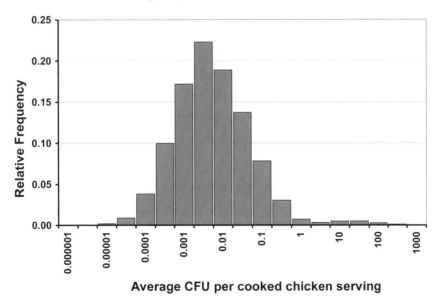

Figure 7.1. Average dose (CFU *Salmonella*) per serving in meals prepared from contaminated broilers.

Note that in Figure 7.1 the interpretation of values of less than 1 CFU per serving is 1 CFU per multiple servings, e.g. an average dose of 0.01 cells per serving translates to one in 100 servings contains a single cell.

Assuming a 20% prevalence of contaminated broilers, the estimated frequency and cumulative distribution of average risk per serving are shown in Figure 7.2. The expected risk per serving is 1.13E-5, or 1.13 illnesses per 100 000 servings. This value represents the average risk for all individuals in the population that consume servings of chicken that are stored, transported and prepared in the manner described in the model, and also accounts for the probabilities that the serving was from a chicken contaminated with *Salmonella*, and that the meal was undercooked. It should be recognized that some individuals consuming a serving on certain occasions would experience a much higher risk than others who may be consuming servings with no salmonellosis risk at all, since the serving is free of the pathogen.

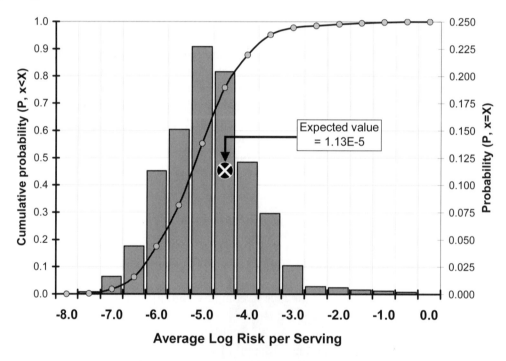

Figure 7.2. Distribution of average risk per serving.

The expected risk per serving can be extended to the expected risk over multiple servings, such as meals consumed in a year. If it is assumed that the risk posed by one exposure (serving) is statistically independent from any other exposure (serving), then the overall risk following a series of exposures can be estimated using Equation 7.1:

$$P_A = 1 - \prod_{j=1}^{i} \left(1 - P_{Dj}\right)$$
                                                                                          Equation 7.1

where P_A is the risk of infection following a series of exposures (annual risk) and P_{Di} is the risk of infection per exposure (daily or serving risk). In order to estimate the annual risk of infection, two pieces of information are required: the risk of infection per serving, and the number of servings consumed in a year. The calculation of annual risk based on the estimated average per serving risk and the assumptions for this baseline scenario are illustrated in Table 7.1.

Table 7.1. Calculation of expected annual risk.

Prevalence of contaminated carcasses	20%
Expected risk per serving	1.13E-05
Number of servings in year	26
Annual expected risk	2.94E-04
Rate of illness per 100 000	29.38

Illustrative calculation for annual expected number of illnesses for a country or region with this annual expected risk:

Population	20 000 000
Proportion of population that eats chicken	0.75
Potentially exposed population	15 000 000
Expected number of cases in the year	4406

The assumption inherent in the calculation above for the expected annual risk is that each of the servings consumed during the year has the same expected risk per serving and that the risk from each exposure is independent of every other exposure. The number of servings used to estimate the annual risk is assumed to be 26 meals, or once every 2 weeks. For illustration, a population risk for 20 million individuals was assumed to be under consideration, with 75% of that population eating chicken. In this example, the total expected number of salmonellosis cases arising from the model assumptions is estimated to be 4400, equivalent to a rate of 29 cases per 100 000 population. Obviously, these statistics need to be tailored for a specific country or region.

In addition to estimating the risk per serving based on consumption of undercooked poultry, the assessment also modelled the risk from cross-contamination. The sequence and nature of events that need to occur in order for the bacteria on raw chicken to be disseminated and ingested via other pathways is complex and difficult to model completely. There is a lack of information to adequately describe cross-contamination, but it is acknowledged that this is an important route for food-borne illness. The following estimates offer an approximation for the magnitude of the problem, although not all potential pathways were modelled that could result in exposure and illness.

In the baseline scenario, the expected risk from cross-contamination (transfer from raw chicken to hands to non-cooked foods, or from raw chicken to cutting board to non-cooked foods) was estimated to be 6.8E-4, or 6.8 illnesses per 10 000 exposures to contaminated material. This is more than an order of magnitude higher than the expected risk from a serving. This estimate is a function of two factors (conditional probabilities) in the current model: first, the expected risk when the event occurs, and, second, the expected probability that the event occurs.

The expected probability that the event occurs is driven by the prevalence of contamination **and** the probability of undercooking in the case of consumption, versus the prevalence of contamination **and** the probability of not washing hands or not washing cutting boards in the case of cross-contamination. Given the assumptions made in the model, the expected risk from this cross-contamination pathway is equivalent to approximately 60 chicken consumption exposures. Although the frequency with which people do or do not wash their hands can be debated, the ultimate risk from cross-contamination could in fact be even higher than that estimated here since there are multiple cross-contamination opportunities that exist in the home preparation environment.

7.2.2 Validation of model results

Validation of results derived from microbiological risk assessments (MRAs) is often difficult, primarily due to the large uncertainties that are commonly associated with predictions. Surveillance data can be used for this purpose, but such use should account for sensitivity of detection and reporting methods. Downstream validation can also be used. In this case, intermediate results can be compared with data not used for model development. For example, predictions for the prevalence of contaminated products at the point of retail can be validated using data from retail surveys. The recognized problems associated with validation strengthen the fact that other outputs from risk assessment, for example the identification of data gaps and the ranking of control strategies, are often more useful than the predicted values.

The model developed here does not estimate the risk for a specific country and therefore it was not possible to attempt to validate the predicted results.

7.2.3 Impact of uncertain parameters on risk estimates

In generating the model, some of the input parameters were modelled as variable while others inputs were considered uncertain, so uncertainty and variability were not explicitly separated. Variability is a property of the phenomenon and the variations that are described are a reflection of what could be expected in nature. Uncertainty is driven by the lack of knowledge about the nature and behaviour of a phenomenon. Inputs that are derived from large representative data sets generated by scientifically sound methods are less uncertain than inputs that are based on sparse data, small sample sizes, or poor scientific methods, or a combination. Good data sets can be regarded as representing the actual variability of phenomenon. In contrast, uncertainty arises when assumptions must be made to generate a distribution around a single data point that is reported in the scientific literature (e.g. when only a mean value is available), or when little or no data exist. Although it is recommended that uncertainty and variability should be explicitly separated within the MRA, this would lead to a complex model. For this reason, the effect of uncertainty was investigated by considering the uncertain parameters in a separate analysis.

Several of the parameters in the cooking module were considered uncertain and are listed in Table 7.2. The impact of uncertainty in these parameters was investigated in order to evaluate their influence on the risk estimate. To do this, the model was re-simulated using a fixed single value for each of the uncertain parameters while allowing the other parameters of the model to vary within their defined distributions. Three simulations were performed: in the first, the parameters listed in Table 7.2 were set at their mean value. The fixed values used for the second simulation were those that would generate a "worst case" scenario, i.e. the maximum value for probability that the chicken was undercooked, the maximum value for proportion of cells in a protected region, the minimum heat exposure time, and the minimum value for the temperature reached in a protected region (0.15, 0.2, 0.5 minutes and 60°C, respectively). It is recognized that such a scenario may not occur in reality, but it gives an upper bound to the range of possible values. The third simulation used the values that would give a "best case" scenario, i.e. minimum value for probability undercooked, etc. This approach allowed the extremes in the risk distribution, driven by the uncertain parameters, to be highlighted. The results of performing the analysis on the uncertain parameters influencing consumption risk are shown in Figures 7.3 and 7.4.

Table 7.2. Uncertain parameters in the cooking module.

Consumption relationship	Mean	Min.	Max.
Probability not adequately cooked	0.1000	0.0500	0.1500
Proportion in protected area	0.1567	0.1000	0.2000
Exposure time to cooking temperature of cells in protected areas	1.00	1.50	0.50
Cooking temperature reached in protected areas.	63.50	65.00	60.00

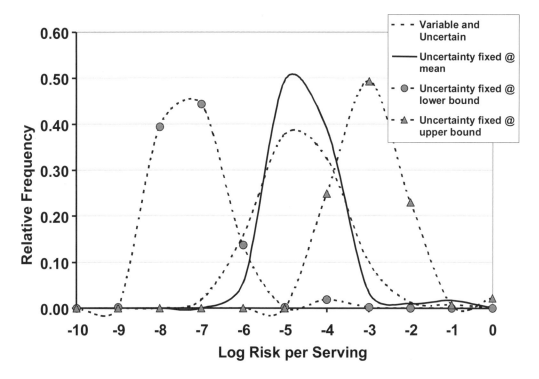

Figure 7.3. Effects of uncertain parameters on per serving risk distribution.

When the uncertain parameters were fixed at their mean values (Uncertainty fixed @ mean) and compared with the risk distribution generated by the model when all parameters were allowed to vary (Variable and Uncertain), it appears that within the range of uncertainty that was assumed to define the parameters, the impact of variation is not very large. The resulting risk distributions are similar and the tails of the currently defined uncertainty distributions do not have a dramatic impact on the overall risk uncertainty distribution. In other words, the range and shape of the distributions defining uncertainty do not influence the risk uncertainty significantly. Alternatively, if the assumptions made were incorrect and the uncertain parameters actually spanned a different range, e.g. if the true values are centred nearer to the min. or max. values rather than at the value assumed to be the mean, the distribution of risk would approach the extreme distributions shown. In these situations, the expected risk would be dramatically different. It should be noted that the extreme risk distributions shown in Figures 7.3 and 7.4 are truly bounds on the uncertainty range since the worst case or best case scenario has been compounded through the model. For example, the worst case scenario was defined by assuming that all of the uncertain parameters would simultaneously take on the values that give the worst outcome.

A complete quantitative uncertainty analysis of the model and all input parameters was beyond the scope of this work. This type of analysis is time consuming and not necessarily more informative for the purposes of this document. Many of the inputs are generic approximations in order to provide a representative risk scenario. Nevertheless, it is important to recognize these two characteristics – uncertainty and variability – in the probability distributions used in quantitative risk assessments. It is also readily recognizable that several input parameters in this model are *both* variable and uncertain, and, if the individual parameters are important in determining the magnitude of the risk estimate, it may be necessary to separate the uncertainty and variability in the quantitative analysis in order to understand their impacts and arrive at proper risk estimations (Nauta, 2000).

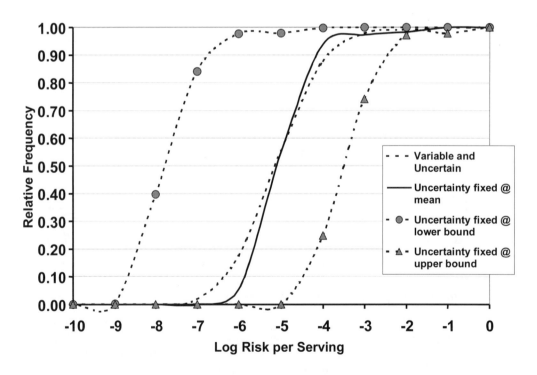

Figure 7.4. Effects of uncertain parameters on per serving cumulative risk distribution.

7.3 RISK MANAGEMENT OPTIONS USING ALTERNATIVE ASSUMPTIONS

7.3.1 Reducing prevalence

A change in the prevalence of contaminated raw product affects the risk to the consumer by altering the frequency of exposure to risk events, i.e. exposure to the pathogen. The change in risk as a result of a change in the prevalence of *Salmonella*-contaminated broilers was estimated by simulating the model using a range of initial prevalence levels. Seven different prevalence levels were investigated: 0.05%, 1%, 5%, 10%, 20%, 50% and 90%. If the prevalence of contaminated chickens leaving processing is altered, through some management practice either at the farm level or at the processing level, the expected risk per serving is altered. The magnitude of the changes in risk per serving and risk per cross-contamination event as a result of changes in prevalence are summarized in Table 7.3.

Table 7.3. Change in prevalence impact on risk.

	Prevalence						
	0.05%	1.0%	5.0%	10.0%	20.0%	50.0%	90.0%
Consumption							
Expected risk per serving	2.81E-08	5.63E-07	2.81E-06	5.63E-06	1.13E-05	2.81E-05	5.07E-05
Number of servings	26	26	26	26	26	26	26
Annual expected risk	7.32E-07	1.46E-05	7.32E-05	1.46E-04	2.93E-04	7.31E-04	1.32E-03
Rate of illness per 100 000	0.07	1.46	7.32	14.63	29.26	73.14	131.61
Calculation of expected number of cases in the year based on assumed population size and exposed population							
Population			20 000 000				
Proportion of population that eats chicken			0.75				
Potentially exposed population			15 000 000				
Expected number of cases in the year	11	219	1 097	2 195	4 389	10 970	19 741
Cross-contamination							
Expected risk per event	1.70E-06	3.41E-05	1.70E-04	3.41E-04	6.81E-04	1.70E-03	3.07E-03

A reduction of 50% in the number of cases of salmonellosis was estimated if a 20% contamination rate at the retail level was reduced to 10% contamination. The relationship between prevalence and expected risk is largely a linear one, specifically a percentage change in prevalence, assuming everything else remains constant, can be expected to reduce the expected risk by the same percentage.

The effectiveness of specific mitigations, either on-farm or treatments during processing, were not evaluated in the present risk model because of lack of representative data to analyse changes in either or both prevalence and level of contamination that might be attributable to a specific intervention. See Section 7.3.4 for a summary of poultry processing treatments. However, the influence of reducing prevalence can be interpreted, although with a high degree of uncertainty given our current state of knowledge, in the context of chlorine addition to the chill tanks during processing. There is little evidence that the addition of chlorine at levels of 50 ppm or less actually decreases the numbers of the pathogen attached to the skin of poultry carcasses. However, available data suggest that chlorine prevents an increase in the prevalence of contaminated carcasses, i.e. a reduction in cross-contamination (Table 7.4), although one study observed a substantial reduction in prevalence. The factor listed in the last column of the table is a ratio of the prevalence after chilling to the prevalence before chilling. A ratio greater than 1 indicates an increase in prevalence of contaminated carcasses.

Table 7.4. Experimental data for effects of chlorine on *Salmonella* prevalence after immersion chill tank.

Ref.	Amount	Prevalence before chilling			Prevalence after chilling			Ratio[1]
		Total	Positive	Prevalence	Total	Positive	Prevalence	
With Chlorine
[1]	20–50 ppm (tank)	48	48	100%	103	60	58%	0.58
[2]	4–9 ppm (overflow)	50	21	42%	50	23	46%	1.10
[3]	1–5 ppm (overflow)?	90	18	20%	90	17	19%	0.94
[4]	15–50 ppm (tank)	48	4	8%	96	7	7%	0.88
								0.87
Without Chlorine
[5]	–	160	77	48%	158	114	72%	1.50
[6]	–	99	28	28%	49	24	49%	1.73
[7]	–	40	5	13%	40	11	28%	2.20
[7]	–	40	4	10%	40	15	38%	3.75
[7]	–	84	12	14%	84	31	37%	2.58
[8]	–	60	2	3%	120	18	15%	4.50
								2.71

NOTES: (1) Ratio of prevalence after chilling to prevalence before chilling. A ratio >1 indicates an increase in prevalence of contaminated carcasses.
DATA SOURCES: [1] Izat et al., 1989. [2] James et al., 1992a. [3] Cason et al., 1997. [4] Campbell 1983. [5] James et al., 1992a. [6] James et al., 1992a. [7] Lillard, 1980. [8] Campbell, 1983.

7.3.2 Reduction in numbers of *Salmonella* on contaminated carcasses

The effect was assessed of reducing the numbers of *Salmonella* on poultry carcasses without changing the prevalence of contaminated carcasses. The values of the cumulative concentration distribution used in the baseline scenario were reduced by 50% (approximately 0.3 logCFU per carcass; Figure 7.5). The model was run using the reduced level of conta-mination while maintaining the prevalence at 20% and with no changes in any of the other parameters. Figure 7.6 shows a comparison of the per serving risk estimates for the modified simulation against the original data.

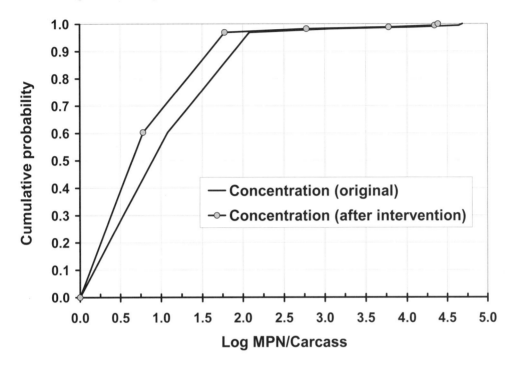

Figure 7.5. Original and post-intervention concentration distributions.

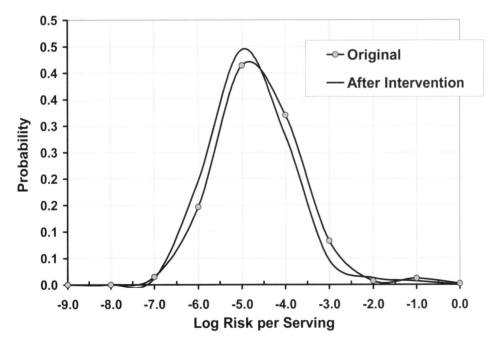

Figure 7.6. Risk per serving distribution before and after concentration-changing intervention.

Unlike a change in prevalence, a change in concentration of the pathogen does not necessarily have a linear relationship with the risk outcome. The distribution of risk shown in Figure 7.6, similar to the distribution of risk per serving shown earlier, is the risk per serving when contaminated. The servings were estimated to be contaminated and potentially undercooked approximately 2% of the time. That statistic remains unchanged if the level of contamination is reduced.

The expected risk per serving, which incorporates the prevalence of contaminated servings and the probability of undercooking, was estimated to be 1.13E-5 (1.13 illnesses per 100 000 servings) in the original case, and 4.28E-6 (4.28 per 1 000 000 servings) in the situation when the level of contamination is reduced. The expected risk per serving is therefore reduced by approximately 62%. A summary of the results is shown in Table 7.5.

Table 7.5. Risk summary before and after intervention to change concentration.

	Original	After Intervention
Prevalence	20%	20%
Expected risk per serving	1.13E-05	4.28E-06
Number of servings in year	26	26
Annual expected risk	2.94E-04	1.11E-04
Rate of illness per 100 000	29	11

Illustrative calculation for annual expected number of illnesses for a country/region with this annual expected risk

	Original	After Intervention
Population	20 000 000	20 000 000
Proportion of population that eats chicken	0.75	0.75
Potentially exposed population	15 000 000	15 000 000
Expected number of cases in the year	4406	1670

The risk from cross-contamination events is also affected when the level of contamination is reduced.

7.3.3 Change in consumer behaviour and the impact on risk

Finally, a change in consumer practices can have an impact on risk. The consumer represents the final intervention in mitigating the risk. However, the effectiveness of strategies aimed at changing consumer behaviour is difficult to anticipate, and difficult to measure. For purposes of this assessment, the potential impact on risk by modifying food preparation practices was investigated by running the simulation assuming that a strategy is implemented which changes consumer behaviour. The assumed changes were as follows:

– probability that product is not adequately cooked:

(OLD):	Min = 5%,	Most likely = 10%,	Max = 15%
(NEW):	Min = 0%,	Most likely = 5%,	Max = 10%

– exposure time (minutes):

(OLD):	Min = 0.5,	Most likely = 1.0,	Max = 1.5
(NEW):	Min = 1.0,	Most likely = 1.5,	Max = 2.0

The changes are thus assumed to reduce the probability of the consumer not adequately cooking their food, and, for those that do tend to undercook, the degree of undercooking is less.

If the simulation model is re-run with these assumptions, the expected risk is reduced to 2.22E-6 (2.22 illnesses per 1 000 000 servings) from 1.13E-5 (1.13 illnesses per 100 000 servings). As a result, the changes in consumer practices reduce the expected risk per serving by almost 80%. The changes in consumer practices have an impact on the frequency with which a potentially contaminated product remains contaminated prior to consumption (probability of undercooking) and reduces the risk when the potentially contaminated product reaches the consumer as well (longer cooking time). The distribution of risk per serving before and after the intervention is shown in Figure 7.7.

It is important to note that the mitigation strategy to alter cooking practices does not address the cross-contamination risk. In the baseline scenario, the expected risk per cross-contamination event was shown to be much larger than the risk from consumption. As a result, the strategy to change the consumers cooking practices needs to be tempered by the fact that cross-contamination may in fact be the predominant source of risk and the nature of cross-contamination in the home is still a highly uncertain phenomenon.

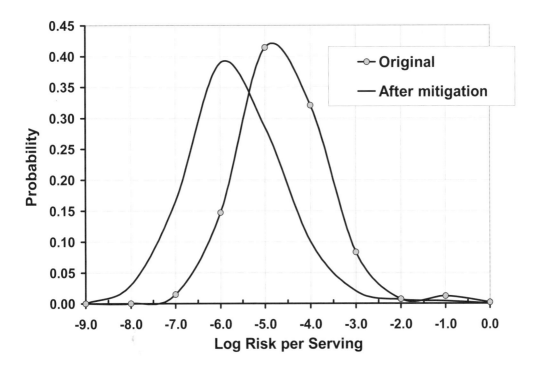

Figure 7.7. Risk distribution per serving before and after consumer behaviour altering intervention.

7.3.4 Intervention methods for controlling *Salmonella* on poultry

SUBSTRATE	CONTROL LEVEL	REDUCTION REPORTED	CONDITIONS	REFERENCE
CHEMICAL TREATMENT – Acetic acid				
Broiler carcasses	0.6%	Significant reduction: 96% of controls positive while treated carcasses were only 8% positive	Used with air injection in a 10-minute pre-chill at 10°C	Dickens and Whittemore, 1994
Chicken carcasses	0.6%	Reduction of 0.34 $\log_{10}$, darkened the carcasses and caused the feather follicles to protrude	1 hour static ice slush in chill tank	Dickens and Whittemore, 1995
Chicken carcasses	0.6%	Reduction of 0.62 $\log_{10}$, darkened the carcasses and caused the feather follicles to protrude	1 hour static ice slush with air injection	Dickens and Whittemore, 1995
Chicken carcasses	0.6%	Reduction of 1.16 $\log_{10}$, darkened the carcasses and caused the feather follicles to protrude	1 hour with paddle chiller	Dickens and Whittemore, 1995
Chicken breast skin	5%	2.5 $\log_{10}$ reduction on loosely *attached S* .Typhimurium populations	Chiller for 60 minutes, 0°C	Tamblyn and Conner, 1997
Chicken breast skin	5%	2.0 $\log_{10}$ reduction on firmly attached *S.* Typhimurium populations	Scalder for 2 minutes, 50°C	Tamblyn and Conner, 1997
CHEMICAL TREATMENT – Calcium hypochlorite				
Chicken carcasses	20 ppm available chlorine	No reduction	15 minutes, 25°C	Nassar et al., 1997
Chicken carcasses	50 ppm available chlorine	No reduction	15 minutes, 25°C	Nassar et al., 1997
Chicken carcasses	100 ppm available chlorine	3/10 inoculated (10^4CFU) carcasses negative after treatment, but yellow appearance and strong chlorine smell	15 minutes, 25°C	Nassar et al., 1997
Chicken carcasses	200 ppm available chlorine	7/10 inoculated (10^4CFU) carcasses negative after treatment, but yellow appearance and strong chlorine smell	15 minutes, 25°C	Nassar et al., 1997
CHEMICAL TREATMENT – Sodium hypochlorite				
Chicken carcasses	200 ppm available chlorine	99.9% reduction in *Salmonella* count; did not affect odour or flavour of the cooked meat	15 minutes, 25°C	Morrison and Fleet, 1985
Chicken breast skin	400 ppm	2.3 $\log_{10}$ reduction on loosely *attached* *S.* Typhimurium populations and 1.3 $\log_{10}$ reduction on firmly attached *S.* Typhimurium populations	Chiller for 60 minutes, 0°C	Tamblyn, Conner and Bilgili, 1997
Chicken breast skin	800 ppm	2.5 $\log_{10}$ reduction on loosely *attached* *S.* Typhimurium populations and 1.9 $\log_{10}$ reduction on firmly attached *S.* Typhimurium populations	Chiller for 60 minutes, 0°C	Tamblyn, Conner and Bilgili, 1997

SUBSTRATE	CONTROL LEVEL	REDUCTION REPORTED	CONDITIONS	REFERENCE
		CHEMICAL TREATMENT – Lactic acid		
Chicken carcasses	0.75%	4/10 inoculated (10^4CFU) carcasses negative after treatment, discoloration, slimy skin and tears in skin	pH 2.78, 15 minutes, 25°C	Nassar et al., 1997
Chicken carcasses	0.75%	5/10 inoculated (10^4CFU) carcasses negative after treatment, slimy skin and tears in skin	pH 2.68, 15 minutes, 25°C	Nassar et al., 1997
Chicken skin	1%	"Significant reduction"	Inoculated skin washed by agitating solution for 30 minutes at 25°C	Hwang and Beauchat, 1995
Chicken carcasses	1.0%	10/10 inoculated (10^4CFU) carcasses negative after treatment, slimy skin and tears in skin	pH 2.47, 15 minutes, 25°C	Nassar et al., 1997
Chicken skin	1%	2.2 $\log_{10}$ reduction	Pre-chill spray to inoculated chicken skin for 30 seconds at 206 kPa and 20°C.	Xiong et al., 1998
Chicken skin	2%	2.2 $\log_{10}$ reduction	Pre-chill spray to inoculated chicken skin for 30 sec at 206 kPa and 20°C.	Xiong et al., 1998
Chicken breast skins	o4%	2 log reduction of firmly attached cells, but bleaching and off odour	Scalder for 2 minutes at 50°C	Tamblyn and Conner, 1997
Chicken breast skins	6%	2 log reduction of loosely attached cells, but bleaching and off odour	Chiller for 60 minutes, 0°C; scalder for 2 minutes at 50°C	Tamblyn and Conner, 1997
		CHEMICAL TREATMENT – Mandelic acid		
Chicken breast skins	6%	2 log reduction of loosely attached cells, but bleaching and off odour	Chiller for 60 minutes, 0°C; scalder for 2 minutes at 50°C	Tamblyn and Conner, 1997
Chicken breast skins	4%	2 log reduction of firmly attached cells, but bleaching and off odour	Chiller for 60 minutes, 0°C	Tamblyn and Conner, 1997
		CHEMICAL TREATMENT – Malic acid		
Chicken breast skins	4%	2 log reduction of both loosely and firmly attached cells, but bleaching and off odour	Chiller for 60 minutes, 0°C	Tamblyn and Conner, 1997
		CHEMICAL TREATMENT – Propionic acid		
Chicken breast skins	4%	2 log reduction of firmly attached cells, but bleaching and off odour	Chiller for 60 minutes, 0°C	Tamblyn and Conner, 1997
		CHEMICAL TREATMENT – Tartaric acid		
Chicken breast skins	6%	2 log reduction of firmly attached cells, but bleaching and off odour	Scalder for 2 minutes at 50°C	Tamblyn and Conner, 1997

SUBSTRATE	CONTROL LEVEL	REDUCTION REPORTED	CONDITIONS	REFERENCE
CHEMICAL TREATMENT – Peroxidase catalysed chemical dip (PC)				
Broilers	0.1 M citric acid + 0.1 M sodium citrate, ratio 1:1.5	"Significant reduction"	pH 5.0, 30 minutes	Bianchi et al. 1994
CHEMICAL TREATMENT – Hydrogen peroxide				
Chicken carcasses	2%	3/10 (10^4CFU) carcasses negative after treatment, bleaching, bloating and brown spots on skin	pH 4.40, 15 minutes, 25°C	Nassar et al. 1997
Chicken carcasses	3%	7/10 (10^4CFU) carcasses negative after treatment, bleaching, bloating and brown spots on skin	pH 4.77, 15 minutes, 25°C	Nassar et al. 1997
CHEMICAL TREATMENT – Sodium metabisulphite				
Chicken breast skin	1%	No reduction.	Three application methods: chiller for 60 minutes, 0°C; scalder for 2 minutes, 50°C; dip for 15 seconds, 23°C	Tamblyn, Conner and Bilgili, 1997
CHEMICAL TREATMENT – NaOH				
Chicken skin	0.05%	"Significant reduction"	Inoculated skin washed by agitating solution for 30 minutes at 25°C	Hwang and Beauchat, 1995
CHEMICAL TREATMENT – AvGard® (TSP)				
Broiler carcasses	100 g/kg w/w	Greater than 2 log reduction	Immersion tank for 15 seconds	Coppen, Fenner and Salvat, 1998
CHEMICAL TREATMENT – Trisodium phosphate (TSP)				
Post-chill chicken carcasses	10%	Significant reduction (ca 1.6–1.8 logs) at both 1 and 6 days post-treatment. Although 50°C-TSP gave 0.4 log greater reduction than 10°C, the difference was not significant.	Dipped in solution at 10°C or 50°C for 15 seconds.	Kim et al., 1994
Chicken skin	1%	"Significant reduction"	Inoculated skin washed by agitating solution for 30 minutes at 25°C	Hwang and Beauchat, 1995
Chicken skin	1% plus 5% Tween 80	Reduction improved from the use of 1% TSP alone	Inoculated skin washed by agitating solution for 30 minutes at 25°C	Hwang and Beauchat, 1995
Chicken skin	5%	2.1 $\log_{10}$ reduction	Pre-chill spray to inoculated chicken skin for 30 seconds at 206 kPa + 20°C.	Xiong et al., 1998
Chicken breast skin	8%	1.6 $\log_{10}$ reduction on loosely *attached* S. Typhimurium populations and 1.8 $\log_{10}$ reduction on firmly attached S. Typhimurium populations	Chiller for 60 minutes, 0°C	Tamblyn, Conner and Bilgili, 1997

SUBSTRATE	CONTROL LEVEL	REDUCTION REPORTED	CONDITIONS	REFERENCE
Chicken breast skin	8%	1.8 $\log_{10}$ reduction on firmly attached *S.* Typhimurium populations	Dip for 15 seconds, 23°C	Tamblyn, Conner and Bilgili, 1997
Chicken breast skin	8%	1.5 $\log_{10}$ reduction on loosely *attached S.* Typhimurium populations	Scalder for 2 minutes, 50°C	Tamblyn, Conner and Bilgili, 1997
Chicken carcasses	10%	Salmonellae not detected on 25-g neck skin sample	pH 12, 15 seconds, 20°C	Whyte et al., 2001
Chicken skin	10%	2.2 $\log_{10}$ reduction	Pre-chill spray to inoculated chicken skin for 30 second at 206 kPa and 20°C.	Xiong et al., 1998

CHEMICAL TREATMENT – Cetylpyridinium chloride (CPC)

SUBSTRATE	CONTROL LEVEL	REDUCTION REPORTED	CONDITIONS	REFERENCE
Chicken skin	0.1%	CPC spraying reduced numbers by 0.9 to 1.7 log units (87 to 98%). Generally, 50°C spraying showed greater reduction than 15°C, but the difference was not always significant.	Solution sprayed against inoculated skin samples at 15°C or 50°C for 1 minute, at 138 kPa.	Kim and Slavik, 1996
Chicken skin	0.1%	Reduction ranged from 1.0 to 1.6 log units (90 to 97.5%). Longer immersion times were more effective. Based on amount of CPC used, immersion appears more cost effective than spraying CPC.	Immersion of inoculated skin surface at room temperature for either 1 minute, 1 minute + 2 minutes holding without CPC, or 3 minutes	Kim and Slavik, 1996
Chicken skin	0.1%	1.5 $\log_{10}$ reduction	Pre-chill spray to inoculated chicken skin for 30 seconds at 206 kPa and 20 C.	Xiong et al., 1998
Chicken skin	0.5%	1.9 $\log_{10}$ reduction	Pre-chill spray to inoculated chicken skin for 30 seconds at 206 kPa and 20°C.	Xiong et al., 1998

CHEMICAL TREATMENT – Grapefruit seed extract (DF-100)

SUBSTRATE	CONTROL LEVEL	REDUCTION REPORTED	CONDITIONS	REFERENCE
Chicken skin	0.1%	1.6 $\log_{10}$ reduction	Pre-chill spray to inoculated chicken skin for 30 seconds at 206 kPa and 20°C.	Xiong et al., 1998
Chicken skin	0.5%	1.8 $\log_{10}$ reduction	Pre-chill spray to inoculated chicken skin for 30 seconds at 206 kPa and 20°C.	Xiong et al., 1998

SCALD TREATMENTS

SUBSTRATE	CONTROL LEVEL	REDUCTION REPORTED	CONDITIONS	REFERENCE
Chicken carcasses	Scald temperatures of 52 C, 56°C and 60°C	Carcasses scalded at 52°C or 56°C showed ~0.3 to 0.5 log reduction greater than those at 60°C	52°C for 2.0 minutes, 56°C for 1.5 minutes and 60°C for 1.0 minute	Slavik, Kim and Walker, 1995

SUBSTRATE	CONTROL LEVEL	REDUCTION REPORTED	CONDITIONS	REFERENCE
Chicken carcasses	Counter-current scalding and post-scald spray	Changes contributed to substantial improvement in the bacterial quality of carcasses, but additional interventions in the chilling process (such as chlorination of chill water) are necessary	240 ml of 60°C water was sprayed on each carcass at a pressure of 40 lbs/sq inch (psi)	James et al., 1992b
CHILL WATER IMMERSION TANK TREATMENTS				
Chicken carcasses	Chill water without chlorination	Prevalence increased during immersion chilling	I hour in typical drag-through chiller	James et al., 1992a
Chicken carcasses	Chill water with 25 ppm chlorination	Prevalence remained constant with chlorination	I hour in typical drag-through chiller	James et al., 1992a
RADIATION – Cobalt 60				
Chicken carcasses	3 k gray	5/10 inoculated (10^4CFU) carcasses negative after treatment, no effect on colour, appearance or smell	57 minutes/kGy of radiation	Nassar et al. 1997
Chicken carcasses	4 k gray	8/10 inoculated (10^4CFU) carcasses negative after treatment, no effect on colour, appearance or smell	57 minutes/kGy of radiation	Nassar et al. 1997
Chicken carcasses	7 k gray	10/10 inoculated (10^4CFU) carcasses negative after treatment, no effect on colour, appearance or smell	57 minutes/kGy of radiation	Nassar et al. 1997
RADIATION – Gamma radiation				
Mechanically de-boned chicken meat (MDCM)	3.0 kGy	Reduction of 6.38 logs	Cesium-137 gamma radiation source, irradiated in air, at +20°C	Thayer and Boyd, 1991
RADIATION – Ultraviolet				
Halved broiler breast with skin on	2 000 λWs/cm^2	80.5% reduction	2 cm^{-2} skin pieces were inoculated with 50 λl of solution containing 7×10^5 CFU/ml and the UV intensity was kept constant at 81.7 λWs/cm^2 while the treatment times were 20, 40, 60, 90 and 120 seconds.	Summer et al., 1996
Halved broiler breast with skin on	82 560 to 86 400 λWs/cm^2	61% reduction compared with untreated halves. No negative effect on colour or increased rancidity of the meat	Halves were inoculated 5 minutes prior to exposure at a wavelength of 253.7 nm	Wallner-Pendleton et al., 1994
AIR SCRUBBING				
Broiler carcasses	Diffused air, 158.6 kPa in tap water	Water only: 32/40 positive. Air scrubbed: 9/40 positive.	30 minutes	Dickens and Cox, 1992

SUBSTRATE	CONTROL LEVEL	REDUCTION REPORTED	CONDITIONS	REFERENCE
		LINE SPEED		
Broiler carcasses	Processed at 70, 80, and 90 birds per minute	Prevalence did not change significantly with processing line speeds.		Brewer et al., 1995

7.4 REFERENCES CITED IN CHAPTER 7

Bianchi, A., Ricke, S.C., Cartwright, A.L., & Gardner, F.A. 1994. A peroxidase catalyzed chemical dip for the reduction of *Salmonella* on chicken breast skin. *Journal of Food Protection*, **57**: 301–304, 326.

Brewer, R.L., James, W.O., Prucha, J.C., Johnston, R.W., Alvarez, C.A., Kelly, W., & Bergeron, E.A. 1995. Poultry processing line speeds as related to bacteriologic profile of broiler carcasses. *Journal of Food Science*, **60**: 1022–1024.

Campbell, D.F., Johnston, R.W., Campbell, G.S., McClain, D., & Macaluso, J.F. 1983. The microbiology of raw, eviscerated chickens: a ten-year comparison. *Poultry Science*, **62**: 437–444.

Cason, J.A., Bailey, J.S., Stern, N.J., Whittemore, A.D., & Cox, N.A. 1997. Relationship between aerobic bacteria, salmonellae and Campylobacter on broiler carcasses. *Poultry Science*, **76**(7): 1037-1041.

Coppen, P., Fenner, S., & Salvat, G. 1998. Antimicrobial efficacy of AvGard® carcass wash under industrial processing conditions. *British Poultry Science*, **39**: 229–234.

Corry, J., James, C., James, S., & Hinton, M. 1995. *Salmonella, Campylobacter* and *Escherichia coli* 0157:H7 decontamination techniques for the future. *International Journal of Food Microbiology*, **28**: 187–196.

Dickens, J.A., & Cox, N.A. 1992. The effect of air scrubbing on moisture pickup, aerobic plate counts, Enterobacteriaceae, and the incidence of *Salmonellae* artificially inoculated broiler carcasses. *Poultry Science*, **71**: 560–564.

Dickens, J.A., & Whittemore, A.D. 1994. The effect of acetic acid and air injection on appearance, moisture pick-up, microbiological quality, and *Salmonella* incidence on processed poultry carcasses. *Poultry Science*, **73**: 582–586.

Dickens, J.A., & Whittemore, A.D. 1995. The effects of extended chilling times with acetic acid on the temperature and microbial quality of processed poultry carcasses. *Poultry Science*, **74**: 1044–1048.

Glynn, J.R., & Brandley, D.J. 1992. The relationship between infecting dose and severity of disease in reported outbreaks of *Salmonella* infections. *Epidemiology and Infection*, **109**: 371–388.

Health Canada. [2000]. Risk assessment model for *Salmonella* Entritidis. Unpublished Health Canada document.

Hwang, C.-A., & Beuchat, L.R. 1995. Efficacy of selected chemicals for killing pathogenic and spoilage microorganisms on chicken skin. *Journal of Food Protection*, **58**: 19–23.

Izat, A., Druggers, C., Colberg, M., Reiber, M., & Adams, M. 1989. Comparison of the DNA probe to culture methods for the detection of *Salmonella* on poultry carcasses and processing waters. *Journal of Food Protection*, **52**: 564–570.

James, W.O., Brewer, R.L., Prucha, J.C., Williams, W.O., & Parham, D.R. 1992a. Effects of chlorination of chill water on the bacteriologic profile of raw chicken carcasses and giblets. *Journal of the American Veterinary Medical Association*, **200**: 60–63.

James, W.O., Prucha, J.C., Brewer, R.L., Williams, W.O., Christensen, W.A., Thaler, A.M., & Hogue, A.T. 1992b. Effects of countercurrent scalding and postscald spray on the bacteriologic profile of raw chicken carcasses. *Journal of the American Veterinary Medical Association*, **201**: 705–708.

Kim, J.-W., & Slavik, M.F. 1996. Cetylpyridinium chloride (CPC) treatment on poultry skin to reduce attached *Salmonella*. *Journal of Food Protection*, **59**: 322–326.

Kim, J.-W., Slavik, M.F., Pharr, M.D., Raben, D.P., Lobsinger, C.M., & Tsai, S. 1994. Reduction of *Salmonella* on post-chill chicken carcasses by trisodium phosphate (NA_3PO_4) treatment. *Journal of Food Safety*, **14**: 9–17.

Lillard, H.S. 1980. Effect on broiler carcasses and water of treating chilled water with chlorine and chlorine dioxide. *Poultry Science*, **59**: 1761–1766.

Mintz, E.D., Cartter, M.L., Hadler, J.L., Wassell, J.T., Zingeser, J.A., & Tauxe, R.V. 1994. Dose-response effects in an outbreak of *Salmonella enteritidis*. *Epidemiology and Infection*, **112**: 13–23.

Morrison, G.J., & Fleet, G.H. 1985. Reduction of *Salmonella* on chicken carcasses by immersion treatments. *Journal of Food Protection*, **48**: 939–943.

Nassar, T.J., Al-Mashhadi, A.S., Fawal, A.K., & Shalhat, A.F. 1997. Decontamination of chicken carcasses artificially contaminated with *Salmonella*. *Revue scientifique et technique de l'Office international des epizooties*, **16**: 891–897.

Nauta, M.J. 2000. Separation of uncertainty and variability in quantitative microbial risk assessment models. *International Journal of Microbiology*, **57**: 9–18.

Parkin, R.T., & Balbus, J.M. 2000. Variations in concepts of "Susceptibility" in Risk Assessment. *Risk Analysis*, **20**: 603–611.

Rejnmark, L., Stoustrup, O., Christensen, I., & Hansen, A. 1997. Impact of infecting dose on severity of disease in an outbreak of foodborne *Salmonella enteritidis*. *Scandinavian Journal of Infectious Diseases*, **29**: 37–40.

Slavik, M.F., Kim, J.-W., & Walker, J.T. 1995. Reduction of *Salmonella* and *Campylobacter* on chicken carcasses by changing scalding temperature. *Journal of Food Protection*, **58**: 689–691.

Sumner, S.S., Wallner-Pendleton, E.A., Froning, G.W., & Stetson, L.E. 1996. Inhibition of *Salmonella typhimurium* on agar medium and poultry skin by ultraviolet energy. *Journal of Food Protection*, **59**: 319–321.

Tamblyn, K.C., & Conner, D.E. 1997. Bactericidal activity of organic acids against *Salmonella typhimurium* attached to broiler chicken skin. *Journal of Food Protection*, **60**: 629–633.

Tamblyn, K.C., Conner, D.E., & Bilgili, S.F. 1997. Utilization of the skin attachment model to determine the antibacterial efficacy of potential carcass treatments. *Poultry Science*, **76**: 1318–1323.

Thayer, D.W., & Boyd, G. 1991. Effect of ionizing dose, temperature, and atmosphere on the survival of *Salmonella typhimurium* in sterile, mechanically deboned chicken meat. *Poultry Science*, **70**: 381–388.

USDA-FSIS. 1998. *Salmonella* Enteritidis Risk Assessment. Shell Eggs and Egg Products. Final Report. Prepared for FSIS by the *Salmonella* Enteritidis Risk Assessment Team. 268 pp. Available on Internet as PDF document at: www.fsis.usda.gov/ophs/risk/contents.htm.

US SE RA. *See* USDA-FSIS, 1998.

Wallner-Pendleton, E.A., Sumner, S.S., Froning, G.W., & Stetson, L.E. 1994. The use of ultraviolet radiation to reduce Salmonella and psychrotrophic bacterial contamination on poultry carcasses. *Poultry Science*, **73**: 1327–1333.

Whyte, P., Collins, J.D., McGill, K., Monahan, C., & O'Mahony, H. 2001. Quantitative investigation of the effects of chemical decontamination procedures on the microbiological status of broiler carcasses during processing. *Journal of Food Protection*, **64**: 179–183.

Xiong, H., Li, Y., Slavik, M.F., & Walker, J.T. 1998. Spraying chicken skin with selected chemicals to reduce attached *Salmonella typhimurium*. *Journal of Food Protection*, **61**: 272–275.

8. DATA GAPS AND FUTURE RESEARCH NEEDS

One of the important outcomes of the risk assessment work was the compilation of a wealth of information on *Salmonella* in eggs and broiler chickens. The organization of these data in the structured risk assessment format has enabled the identification of the key gaps that exist in the data. This can provide guidance for future research work and help ensure that it is targeted towards generating and collecting the most useful and relevant data. These data and research needs are outlined below.

8.1 HAZARD CHARACTERIZATION

In order to improve the hazard characterization, additional outbreak and epidemiological data are needed. More specifically, these data should indicate cell number in the implicated food, amount of food consumed, accurate estimates of the size of ill and exposed populations, and accurate characterization of the population, including age profiles, medical status, sex and other potential susceptibility factors.

The impact of the food matrix was not incorporated into the hazard characterization due to the limitations of available data. So that these issues can be more completely addressed in future work, there is a need for characterization and quantification of the impact of the food matrix effects and also host-pathogen interactions and virulence factors and their effect on the probability of infection or illness, or both. Quantitative information to facilitate estimating the probability of developing sequelae following illness is also required.

As this is a developing science, the optimal models have not yet been developed. Therefore, new dose-response models that improve the ability to estimate the probability of illness would be useful.

8.2 EXPOSURE ASSESSMENT OF *S.* ENTERITIDIS IN EGGS

Data relating to the biology of *S.* Enteritidis in eggs is needed. This need is seemingly universal in its application to previous and future exposure assessments.

Additional studies on the numbers, and factors that influence the survival and growth of *S.* Enteritidis in naturally (yolk) contaminated intact shell eggs are needed, as information is currently available for only 63 intact shell eggs. Enumeration data of *S.* Enteritidis in raw liquid egg are also required. Additional data concerning the numbers of *S.* Enteritidis in raw liquid egg before pasteurization would assist in reliably predicting the effectiveness of such a regulatory standard concerning egg products.

More data on the prevalence of *S.* Enteritidis in breeder and pullet flocks and the environment, as well as in feedstuffs, is needed to adequately assess the benefit of pre-harvest interventions. In particular, associations between the occurrence of *S.* Enteritidis in these pre-harvest steps and its occurrence in commercial layers should be quantified.

Better data on time and temperature, specifically in relation to egg storage, and to preparation and cooking, would serve to build confidence in the modelling results. The importance of time and temperature distributions in predicting growth of *S.* Enteritidis in eggs, combined with the lack of reliable data to describe these distributions, highlights the need for these data. Furthermore, new studies are needed on the relationship between cooking time, cooking method and cooking temperature and the death of *S.* Enteritidis.

More studies are needed on the survival and growth of *S.* Enteritidis in eggs, particularly as a function of egg composition and the attributes of infecting strains of organism (e.g. heat sensitivity).

8.3 EXPOSURE ASSESSMENT OF *SALMONELLA* IN BROILER CHICKENS

As indicated earlier in the document, the lack of good quality data, prior to the end of processing in particular, limited the scope of this exposure assessment. In relation to primary production, the information available was mainly prevalence data, but for some regions of the world – including Africa, Asia and South America – even that was limited. In addition, information on study design, specificity or sensitivity of the analytical methodologies used was lacking. Very few quantitative data were available. A similar situation was observed for the processing stage. In addition, data tended to be old, and knowledge of processing practices was not readily available. In order to address these deficiencies, the areas where data collection and research efforts need to focus are identified below.

- Data on prevalence for many regions of the world regarding *Salmonella* in broilers during production and at slaughter, and on carcasses post-processing, together with information on study design.

- Microbial ecology studies to determine sources and numbers of the pathogen.

- Studies on the correlation between within-flock prevalence levels and the number of *Salmonella* cells shed in faeces or on birds.

- Precise estimates of the numbers of organisms per bird for all stages of the exposure pathway and improvements in the sensitivity and availability of cost-effective methods to enumerate small populations of *Salmonella.*

- Between-bird (bird-to-bird) cross-contamination data suitable for modelling this phenomenon at the pre-harvest, transport and processing stages.

- Data on the survival of *Salmonella* under chilling and freezing conditions. This information will improve the predictive microbiology component of exposure assessments relevant to international trade in poultry products.

- Specific consumption data and information about food preparation practices for most geographical locations, preferably presented as portion size and frequency of consumption rather than average consumption per day.

- Information on the distribution of time and temperature for storage and cooking in domestic kitchens in a variety of national environments.

- Data on the magnitude of cross-contamination in the domestic kitchen and the pathways for cross-contamination.

If an attempt were made to extend the risk assessment to assess more fully pre-slaughter interventions, then more data would be required on the prevalence of *Salmonella* in feed and replacement stock, and on fasting prior to slaughter. Data on the effect of scalding, de-feathering, evisceration, washing and chilling processes, as well as other decontamination treatments, are needed to effectively model the benefits of control interventions at the levels of processing.

9. THE APPLICATION OF MICROBIOLOGICAL RISK ASSESSMENT

Quantitative microbiological risk assessment is intended to answer specific questions of importance to public health. For microbiological risk assessment to deliver benefits it needs to be purposefully incorporated into the decision making process. This implies a change in the way nations approach food safety and public health decisions. The novelty of microbiological risk assessment is that it quantifies the hazard throughout the food production chain and directly links this to the probability of food-borne disease. The risk assessments of *Salmonella* in eggs and broiler chickens present an example of the potential of this approach.

The increased use of microbiological risk assessment will result in new capacity building needs. The exercise of producing this risk assessments has been a learning experience and since it is comprehensive, it can also provide a basis for future training efforts and applied research. These risk assessments are a resource that can be used by many parties including the Codex Alimentarius and national authorities. Ensuring their applicability and utility to all regions and countries is a priority for future work in FAO and WHO.

An important prerequisite for microbiological risk assessment is the need for an interdisciplinary approach. There is a dual need to develop the capacity for microbiological risk assessment skills and expertise within all the relevant disciplines (microbiology, modelling, epidemiology, etc.) and to ensure that these disciplines become effectively integrated into the risk assessment process. Transparency must be maintained throughout the risk assessment process from the initial stages of building the risk assessment team, to data collection and analysis.

This exercise in conducting risk assessment at the international level has underlined the need for data to be acquired from all regions and for the development of countries' capacities to conduct risk assessments. The development of these capacities requires an infrastructure for the surveillance of food-borne disease and the monitoring of microbial hazards in foods throughout the food-chain and the effect of processing and other factors on the micro-organism. It also requires human resources with the technical skills needed to conduct microbiological risk assessment.

There is a considerable amount of useful information made available through these risk assessments for both risk assessors and risk managers. The concepts presented are generic, and may be directly adaptable or considered as stand-alone modules. For those planning to undertake a quantitative microbiological risk assessment the models developed can be used as a template for undertaking risk assessment for these pathogen-commodity combinations at regional or national levels. The data used in the models, however, must reflect the food item, raw material, manufacture, retail conditions, and consumption habits as well as the characteristics of the population within the region under consideration.

These *Salmonella* risk assessments provide information that may be useful in determining the impact that intervention strategies have on reducing cases of salmonellosis from contaminated eggs and poultry.   This information is of particular interest to the Codex Alimentarius in their work on the elaboration of standards, guidelines and related texts. Furthermore, in undertaking this work a number of lessons were learned with regard to making optimal use of risk assessment as a decision support tool. In order to meet the needs of risk managers, the risk assessment must be clearly focussed. This can be achieved by adequate planning, good communication and a strong interface between the risk assessors and the risk managers. To ensure that risk assessment contributes to management decisions that can be successfully implemented, there needs to be communication from the outset with other relevant stakeholders such as the food industry and consumers.

In conclusion, the risk assessments provide an example of a format for organising the available information in a readable way, and connecting pathogen contamination problems in food with human health outcomes. They provide scientific advice and analysis that may be useful for establishing regulatory policies for control of foodborne disease in different countries. In addition, the risk assessment process has identified important data gaps, and includes recommendations for future research, which can be used to allocate resources to priority areas.

These are the first microbiological risk assessments to be undertaken at the international level. During the course of the work it was recognized that MRA is still a developing science, yet, every effort has been made to provide a valuable and unique resource for those undertaking risk assessments and addressing the problems associated with *Salmonella* in eggs and broiler chickens.